DETONATION NANODIAMONDS

DETONATION NANODIAMONDS

Science and Applications

edited by

Alexander Ya. Vul'

Olga A. Shenderova

PAN STANFORD PUBLISHING

Published by

Pan Stanford Publishing Pte. Ltd.
Penthouse Level, Suntec Tower 3
8 Temasek Boulevard
Singapore 038988

Email: editorial@panstanford.com
Web: www.panstanford.com

British Library Cataloguing-in-Publication Data
A catalogue record for this book is available from the British Library.

Detonation Nanodiamonds: Science and Applications

ISBN 978-981-4411-27-1 (Hardcover)
ISBN 978-981-4411-28-8 (eBook)

Printed in the USA

Contents

Preface

Diamond has been well known since time immemorial, and applications of this material, truly unique in its beauty, hardness, and chemical stability, keep expanding as years go by. For hundreds of years mankind has used natural diamonds, but it was only from the mid-1950s that industrial production of diamond on a commercial scale in reactors capable of maintaining the required high pressures, of tens of thousands of atmospheres, and temperatures of about 1,500 degrees Kelvin was announced to the world.

Annual production of such artificial diamonds ranging in size from a few hundred to a few microns, the so-called micropowders, has presently reached a level of hundreds of thousands of carats per year.

Presently, when most technologies are moving with confidence from the microscale into the nanometer-scale world, demand has arisen for diamonds of the corresponding size. Such nanodiamonds were first synthesized in the Soviet Union in the 1960s and their industrial production initiated in the late 1980s. The starting raw material for nanodiamond synthesis was carbon, present originally in explosives, and the high pressure and temperature needed for formation of the diamond structure from carbon atoms were reached as a result of the explosion itself. The short duration of the explosion accounted for the small size of the diamond crystallites, which measured a few billionths of a meter only.

As is now obvious, work conducted on the explosive method of synthesis was not made public for a period of time. As a consequence, this method of preparing nanodiamonds from the carbon of explosives was discovered over and over again independently by different research groups.

From the publication of the pioneering works on the so-called detonation nanodiamonds made in Russia and the United States in 1988, scientists have passed a long way indeed.

The first studies were naturally devoted to the investigation of the synthesis process and were aimed at increasing the fraction of the carbon atoms contained within the explosive eventually incorporated into the diamond lattice of the final product of the synthesis. Possibly, the most challenging problem in the process turned out to be the development of the technology for isolating the diamond crystallites from the final product, that is, the detonation carbon formed during the explosion. Detonation carbon is actually a mixture of nanodiamonds; particles of amorphous, graphite-like carbon; and impurities entering the material from the starting explosive and the reactor walls.

This work was paralleled by an investigation of the processes involved in the nanodiamond-graphite structural phase transition. The high temperature persisting during the unavoidable drop of pressure after the explosion brought about reverse transformation of a part of the formed nanodiamonds into graphite, thus lowering the efficiency of the synthesis process. It was, however, found that under certain conditions this structural phase transition passes through an intermediate stage in which onion-like carbon or multilayer fullerenes form. Moreover, if acted upon by an electron beam, the buckyonions themselves can reconvert back into nanodiamonds. This has revealed an intimate relationship between the fullerenes discovered in 1980s and nanodiamonds.

The emerging possibility of studying structural transformations on the nanoscale was naturally attractive to scientists. No less intriguing was finding nanodiamonds in meteorites. This finding alone could have been sufficient to account for the interest in nanodiamonds revealed in basic sciences, but it is the industrial application of nanodiamonds that is behind the attention focused presently on the investigation of their properties.

Studies conducted during recent years showed that nanodiamonds, new nanosized building blocks, can be used to advantage in devising nanocomposite materials, nanoelectronics components, selective adsorbents, and catalysts.

Application of nanodiamonds improves considerably the quality of microabrasive and polishing compositions, lubricating oils, abrasive instruments, polymer compositions, rubbers, and magnetic recording systems and offers the possibility of growing diamond films on various substrates.

One of the most attractive applications of nanodiamonds turned out to be their use in biology and medicine as biomarkers and for targeted drug delivery.

A significant factor that has apparently played a major role in stimulating interest in detonation nanodiamonds is that among the new carbon nanostructures discovered at the turn of the century—various types of fullerenes and nanotubes—detonation nanodiamonds were one of the first to be produced on an industrial scale and, thus, are commercially available for use in nanotechnologies.

In our opinion, progress in methods of synthesis of detonation nanodiamonds and clear insight into the broad scope of their applications have approached at the beginning of this century the stage where detonation nanodiamonds are considered one of the most attractive carbon-based materials for nanotechnologies. By this time, reports on nanodiamonds produced in detonation synthesis began to appear at increasing frequency at international "diamond devoted" conferences. In the period of 2003–2008, the first specialized conferences were held and reviews and monographs dealing with this material appeared.

Studies conducted in recent years in Russia, the United States, Japan, China, and a number of European countries have produced a wealth of comprehensive information on the specific features of the structure of nanodiamond particles. Truly unique possibilities have opened up as a result of chemical modification of their surface, leading to considerable promise for their use in numerous applications. The results of these studies are summarized in this monograph. The authors of the chapters of the monograph are world-renowned experts who have published a number of review papers in this area.

The monograph addresses successively the specific features of the production technology and the effect of the technological parameters on the structure and physicochemical properties of nanodiamonds, assesses the possibilities inherent in purposefully directed chemical modification of the surface and methods employed in structural modification of nanodiamonds to produce nanographite, discusses various approaches used in the investigation of nanodiamonds as a carbon nanostructure, and describes and critically analyzes the potential promised by the use of nanodiamonds in various areas of technology and medicine.

There are two significant features that distinguish the present monograph. First, rather than being just a collection of individual reviews, it is a book written with a common theme. Second, the monograph is published in both Russian and English, an approach that, in our opinion, will be a step forward in developing a common terminology and a common view of many problems, both already solved and those still remaining open, that bear on detonation nanodiamonds.

The monograph addresses a broad audience of readers interested in nanotechnologies, and we sincerely hope that it will be found useful both by specialists in the field and senior students who are still looking for challenging problems to direct their efforts to.

Alexander Ya. Vul'
Olga A. Shenderova

Chapter 1

Carbon at the Nanoscale

Alexander Ya. Vul'[a] and Olga A. Shenderova[b]
[a] *Ioffe Physical-Technical Institute, Solid State Electronics Department, 26 Polytechnicheskaya, St. Petersburg 194021, Russia*
[b] *International Technology Center, 8100 Brownleigh Dr., Raleigh, NC 27617, USA*
alexandervul@mail.ioffe.ru

1.1 Introduction

We believed it appropriate to define in the first chapter the place occupied by detonation nanodiamond (DND) in the large family of carbon nanostructures, which were discovered at the turn of the 21st century, and to offer the main definitions that will be used below.

We would like to start with a citation of the outstanding Russian chemist, D. I. Mendeleev, who wrote ". . . none of the elements reveals such a capability to complication to the extent we see it in carbon. There are still no grounds which would permit us to suggest the limit to polymerization of a carbon, graphite or diamond molecule; the only thing we know for certain is that they contain C_n, where n is a large figure" [1].

Detonation Nanodiamonds: Science and Applications
Edited by Alexander Ya. Vul' and Olga A. Shenderova

ISBN 978-981-4411-27-1 (Hardcover), 978-981-4411-28-8 (eBook)
www.panstanford.com

This rich variety of forms rests on the electronic structure of a single carbon atom with a $1s^2 2s^2 2p^2$ electron shell configuration in the ground state and four valence electrons in the outer $2s$ and $2p$ shells, which can take part in the formation of chemical bonds. The difference in energy between the $2p$ and $2s$ energy levels in the carbon atom is small (4 eV) to be compared with the gain in energy involved in formation of chemical bonds. Therefore the electron wave functions of these four electrons can mix (hybridize), thus varying the occupancy of the one $2s$ and three $2p$ orbitals and, in this way, increasing the binding energy among neighboring carbon atoms.

It is the mixing of the single $2s$ orbital with one, two, or three $2p$ orbitals that defines the sp^n ($n = 1, 2, 3$) hybridization of atoms and the existence of three main allotropic modifications of carbon "macro" materials—diamond, graphite, and carbyne—that can be traced to three types of hybridization of electronic orbitals, sp^3, sp^2, and sp, respectively.

Figure 1.1a displays schematically the corresponding hybrid orbitals, the tetrahedral, the planar triangular, and the linear one, with the valence angles of $109°28'$, $120°$, and $180°$, respectively,

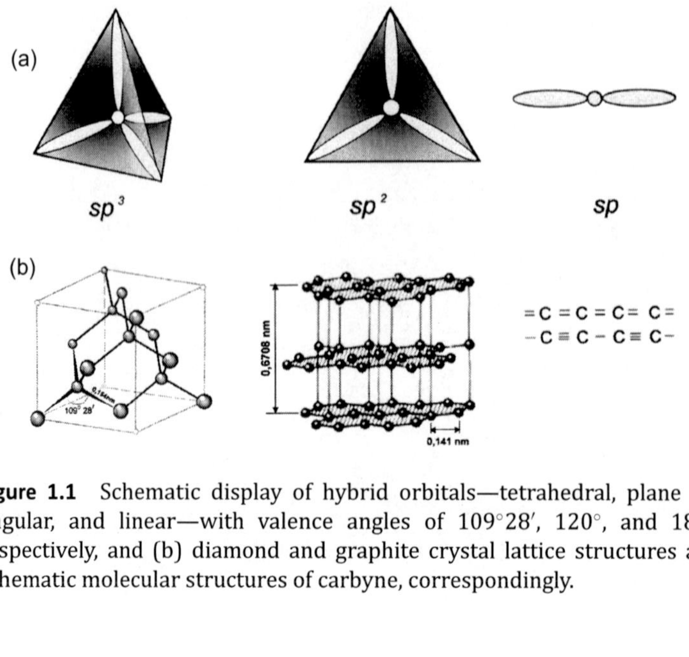

Figure 1.1 Schematic display of hybrid orbitals—tetrahedral, plane triangular, and linear—with valence angles of $109°28'$, $120°$, and $180°$, respectively, and (b) diamond and graphite crystal lattice structures and schematic molecular structures of carbyne, correspondingly.

Table 1.1 Parameters of bulk crystalline diamond (cubic lattice symmetry)

Property	Value
Lattice constant, nm	0.3567
Interatomic separation (bond length), nm	0.154
Angle between valence bonds	$109^\circ 29'$
Coordination number	4
Number of atoms in unit cell	8
Density, g/cm^3	3.5174
Binding energy, kJ/mole	244.9
Debye temperature, K	1,860
Thermal conductivity for natural diamond, type II *a*, W/mK	2,000
Thermal expansion, linear, at $T = 300$ K, K^{-1}	$1 \cdot 10^{-6}$
Bulk modulus, GPa	450
Young modulus, GPa	1,100
Energy gap, eV	5.5
Index of refraction	2.4
Permittivity at $(10^2–10^4)$ Hz	5.7

Abbreviation: GPa, gigapascal.

and Fig. 1.1b the structures of the cubic crystal lattice of diamond and graphite, the principal macroscopic allotropic forms of carbon. Table 1.1 lists relevant information on the main properties of "macro" diamond.

Other elements of the fourth group of the periodic system of elements, for instance, Si and Ge, manifest sp^3 electronic shell hybridization only. Because these chemical elements contain only the occupied core *p* atomic orbitals, repulsion between the valence electrons and the *p* electrons of the core shell specifies the direction of the hybridized orbitals and, thus, makes hybridization less labile. For instance, if we assume the core *p* orbitals to be oriented along the *x*-, *y*-, and *z*-axes, the hybridized orbitals in Si and Ge should be aligned with <111> type directions. No such directions of "hampered" hybridization exist in the carbon atom [4].

It is the numerous possibilities of different hybridizations of electronic shells that culminate in the immense variety of isomeric forms of carbon—about 95% of known organic compounds (about 10 million of them) contain carbon atoms in their structure.

In the recent decades, a great number of new carbon nanostructures, including the already well known ones (fullerenes, nanotubes, graphene, onions, and nanodiamond (ND)), and more exotic structures, such as nanohorns, nanobells, nanopeapods, and nanofoams [2], have been discovered. Consider now in more detail the methods employed in the classification of these nanostructures.

1.2 Classification of Carbon Nanostructures

There are several methods used in the classification of carbon nanostructures. One could, for instance, proceed from the topological dimension [3, 4], with the fullerenes and ND particles being assigned to zero-dimensional structures, nanotubes and diamond nanorods to one-dimensional, graphenes and diamond nanoplates to two-dimensional, and ultrananocrystalline diamond films to three-dimensional ones.

Another approach assumes as a basis for classification the characteristic size of the nanostructure. This approach covers not only nano- but also microscopic carbon materials, such as carbon fibers. Taking into account the differences in shape and spatial arrangement of structural elements (pentagons and hexagons in the case of fullerenes) or chirality (for nanotubes) may also serve as a basis for the classification of carbon nanostructures [5, 6].

We believe the approach to classification, in which the immense diversity of carbon nanostructures is assumed to stem from the capability of carbon atoms to form chemical bonds with differently hybridized *sp* atomic orbitals and, accordingly, different spatial localization of the electron clouds defining the topology of the carbon nanostructure, the aspect illustrated convincingly above, to be the most informative and convenient to use [7, 8]. This approach employed in [4, 7, 9] combines in a common and schematically simple way all, including the very recently discovered carbon nanostructures, and, in addition, suggests a transition from the nano- to the well-known macrostructures—diamond and graphite (Fig. 1.2).

As seen from Fig. 1.2, this classification is based essentially on three types of potentially realizable hybridizations of the *sp* atomic orbitals of carbon and, accordingly, three allotropic modifications of the "macro" carbon materials. These are the sp^3 hybridization in diamond, sp^2 in graphite and graphene, and sp in carbyne. All the other, both nano- and macroforms of allotropic carbon modifications, can be assigned to an intermediate type of atomic orbital hybridization, sp^n. For instance, for the fullerenes, carbon onions, and nanotubes with a partially curved surface, $2 < n < 3$. For the C_{60} fullerene such a noninteger value of $n = 2.28$ was first used in the well-known monograph devoted to fullerenes [10]. The same values $2 < n < 3$ can be employed also to describe the type of bonds dominating amorphous carbon and the so-called diamond-like films prepared by chemical vapor deposition (CVD). To the relatively small group of allotropic modifications featuring hybridization of an intermediate type with $1 < n < 2$ can be assigned various monocyclic carbon structures, mostly belonging to organic compounds.

It appears pertinent to note here that because carbyne discovered in 1960 [11] was accessible only in minute amounts precluding comprehensive investigation of its physical properties, its introduction into our scheme (Fig. 1.2) reflects our desire to construct an integral picture that would combine the various types

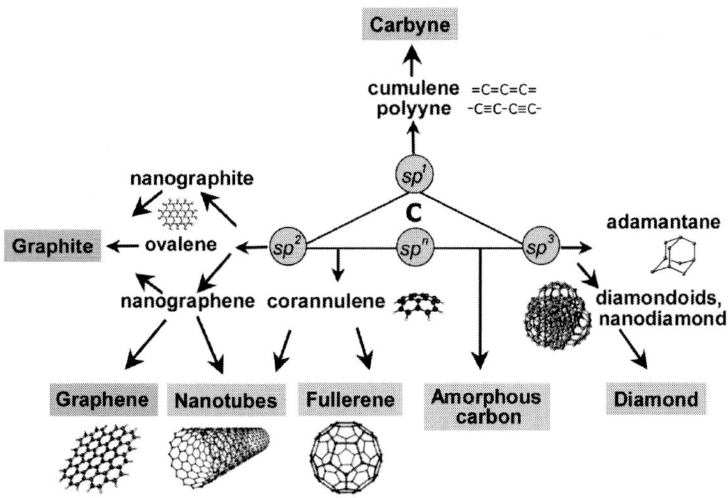

Figure 1.2 Schematic of the family of carbon nanostructures.

of carbon nanostructures. Carbyne is believed to consist of an array of chains of linearly bonded *sp*-hybridized carbon atoms, in which the electrons involved in π bond formation can be brought into resonance to form alternating single and triple (double) bonds (of the type of polyyne) or repeating double bonds (of the type of cumulene). Such chains are compacted by van der Waals forces to produce crystals. Learning a detailed structure of carbyne requires further studies [12], but it is believed presently that parallel chains of atoms also contain zigzags.

Thus, the family of nanocarbon structures includes nanographite and nanographene with sp^2 hybridization, sp^3-hybridized ND, and a numerous group of fullerenes and nanotubes with atomic orbitals coupled by hybridization of an intermediate type sp^n (Fig. 1.2).

Interestingly, carbon forms also structures representing combinations of nanostructures of various types, for instance, the nanopeapod made up of fullerenes encapsulated in nanotubes or the so-called bucky-nanodiamond, in which the diamond core is surrounded by fullerene-like shells.

If we combine carbon structures in a separate group based on their characteristic size and specific hybridization, it appears only logical to include in this scheme the corresponding molecular structures as well (Fig. 1.2). Such structures may be the molecules of adamantane with sp^3 hybridization, ovalene constructed from carbon hexagons with the corresponding sp^2 hybridization, and corannulene made up of a pentagon surrounded by five hexagons, a characteristic topological substructure of fullerenes that is responsible for mixed hybridization. Cumulene may serve as a typical example of sp^1-hybridized molecules.

Incidentally, molecular structures with a diamond-like shape are not restricted to adamantane made up of three cyclohexane rings forming a unit diamond cage.

The subclass of the so-called lower diamondoids includes also diamantane and triamantane, whose structure contains two and three diamond cages, respectively. The so-called higher diamondoids, consisting of more than three diamond cages, are intermediate in size between adamantane, the smallest molecule with the structure of cubic diamond containing 10 carbon atoms only, and ND particles a few nanometers in size.

The higher diamondoids were extracted from oil in the form of nanosized rods, helixes, discs, pyramids, etc., only a few years ago [12, 13]. Attempts at synthesizing higher diamondoids have thus far been unsuccessful, with the only exception of antitetramantane. By contrast, lower diamondoids, extracted from oil much earlier than the high-molecular members of the family of diamondoids, can be synthesized.

1.3 Main Types of Carbon Nanostructures

Because finding monographs and reviews dealing with carbon nanostructures would not present any difficulties for a person interested in the subject, we are going to limit ourselves in this section only to a brief account of the history of the discovery, specific features of the structure, and the areas where the most well-known members of the "nanocarbon family" find application.

1.3.1 *Fullerenes*

Fullerenes were discovered, using the well-known graphic expression, "at the pen's point," with publication in 1973 of a theoretical paper [14]; in this paper, calculations performed by the Hückel method predicted stability of an unusual molecule that consisted of 60 carbon atoms making up a highly symmetric structure in the form of a truncated icosahedron. This geometry was proposed for calculation by Ivan Stankevich, a member of the same research group. Remarkably, already this first study announced determination of the energies of the empty and occupied orbitals, including the triple degeneracy of the lowest empty orbital, which did not radically change in later calculations making use of more sophisticated methods. Interestingly, in 1970 E. Ōsawa expressed the idea of stability of C_{60}, but it was not supported by any calculations. Because both pioneering studies were published in their native languages before science became globally internationalized, they remained practically unknown outside the countries where they evolved, did not produce any effect on progress in science, and were revealed by the world's scientific community only after the experimental discovery of fullerenes in 1985 [15]. For the full story of the discovery of fullerenes, with all its fascinating details, the reader can be referred to Ref. [16].

The C_{60} fullerene, the best studied thus far, is a closed carbon cluster in the form of a truncated icosahedron. The carbon atoms are arranged on the spherical surface at the corners of 20 regular hexagons and 12 regular pentagons, with each hexagon sharing edges with 3 hexagons and 3 pentagons, while each pentagon shares edges with hexagons only. Thus, each carbon atom in a C_{60} molecule shares corners with two hexagons and one pentagon and, thus, is indistinguishable from other carbon atoms [10].

The easiest to synthesize and, hence, the best studied are the C_{60} and C_{70} fullerenes with the characteristic dimensions of 0.7–0.8 nm; their main properties originate from the peculiar electronic structure, whose main feature is delocalization of π electrons over the stressed three-dimensional ellipsoidal surface.

Two main methods were proposed for the synthesis of fullerenes, graphite evaporation in an electric arc (the arc method) [17] and hydrocarbon combustion [18].

The first method enjoyed broad recognition in the past and still remains apparently the most popular one for use in laboratory conditions. The fullerene soot produced contains 10–15 wt.% fullerenes with a typical ratio of C_{60}:C_{70} $\approx$ 80:20, and about 1–2 wt.% higher fullerenes C_n ($n > 70$). The high cost of the starting raw material was a major obstacle on the way to commercial production of fullerenes by this technique. Progress in development of the second method culminated in the beginning of industrial fullerene production in 2005 [19].

Even when touching very briefly on the subject of fullerenes, one cannot pass by their most exotic modification, the so-called endohedral fullerenes, whose noncarbon atoms are encapsulated in a hollow carbon shell [20].

The very possibility of the existence of such structures rests on the size of the inner cavity of 0.7 nm in the fullerene molecule being considerably in excess of the characteristic effective diameter of atoms and molecules, 0.1–0.4 nm. Of most interest among foreign atoms are those of metals. This special subclass of the fullerene derivatives, which is unique from the topological viewpoint and is designated in the general form by M@C_n, was discovered in the particular case of La@C_{82} [21]. While the metal atom cannot leave the carbon cage, it retains its inherent magnetic properties and influences the electronic state of the carbon structure, with part of the electron density of the metal atom being transferred to the carbon cage.

Endohedral metallofullerenes are produced with a small yield by electric arc synthesis with the electrodes filled by the corresponding metals. Detailed relevant information on the technology of their preparation and properties can be found in a number of review papers, primarily of the H. Shinohara group [22].

In the mid-1990s, promising areas of application of fullerenes were discussed in considerable detail (see, e.g., Ref. [23]). Nevertheless, despite the rich variety and uniqueness of their properties, the original hopes for large-scale application of fullerenes have not been realized. There are essentially three major directions of applied research in this area. The first, and fairly obvious one, is broadening of the scope of new chemical compounds, promising developments in the organic and inorganic chemistry of fullerenes (see, e.g., Refs. [24, 25]). The second direction lies along biomedical applications of fullerenes [26, 27], including the use of endometallofullerenes as contrast agents for magnetic resonance imaging (MRI) [28]. And, finally, the third one encompasses development of polymer-fullerene compositions for large-area flexible solar cells.

In the beginning of the 1990s an attractive idea was put forward of using separation of photogenerated carriers between a fullerene, a weak acceptor, and a polymer, a donor [29]; the efficiency η of these photoconvertors remained, however, $\eta \leq 4\%$ for a long time [30], and only in the recent years progress in the technology of polymer-fullerene compositions has permitted raising it to $\eta \approx 8\%$ [31].

1.3.2 *Carbon Onions*

There are grounds to suggest that carbon onions, in other words, multilayered nesting polyhedral or spherical carbon structures, which in the first publications were dubbed "Russian dolls," were first experimentally observed by S. Iijima by high-resolution transmission electron microscopy (HRTEM) as far back as in 1980

[32]. A widespread interest in the new carbon nanostructure has evolved, however, only after a series of publications by D. Ugarte (see Refs. [33–38]).

As reported in the first publication [33], arc discharge-produced particles of carbon soot exposed to electron irradiation in a Philips EM430 ST electron microscope to doses 10–20 times higher than those used in normal operation undergo continuous rearrangement into quasi-spherical particles made up of concentric graphite layers. Interestingly, while during the first 10 minutes of irradiation the samples reveal only a trend to curling, in 20 minutes they transform practically completely into spherical particles. These particles turn out to be stable and not influenced anymore by electron irradiation even with a small number (2–4) of layers curled up.[a]

A close examination of the structural details reveals that the spherical particles are made up of concentric carbon spherical cages. This motivated a hypothesis, generally adopted presently, that the planar graphite structure cannot be the most stable allotropic form of carbon at the nanoscale.

The mechanism driving generation of such symmetric low-entropy forms in an arc discharge under the conditions of random condensation of the carbon vapor rests on the need to restrict the number of dangling bonds and to lower the energy of the structure, which is reached through curling of the plane carbon sheets.

Turning now back to the fullerenes, we have to note that the structure of C_{60} may be considered as a hexagonal graphite sheet of monatomic thickness, with the pentagons inside restricting all dangling bonds.[b]

The remarkable stability of the fullerene C_{60} should be assigned to its having no neighboring pentagons, while its spherical form permits symmetrical distribution of the stress arising as a plane graphite sheet separating the atoms curls up.

The action of an electron beam can to a certain extent be compared to high-temperature heating because it increases the

---

[a] D. Ugarte calls the spherical particles bucky-onions to distinguish them from polyhedron-onions. Spherical onions have two specific features, namely, perfect sphericity and the size of the first inner shell being close to that of the C_{60} molecule (0.7–1.0 nm).

[b] By definition, graphene is a graphite sheet of monatomic thickness (see Section 1.3.4 below).

mobility of carbon atoms. Indeed, the mobility of carbon atoms should increase both because of the absorption of energy and as a result of the rupture of bonds under electron excitation. Moreover, high-energy particles are able to transfer momentum to the nuclei of carbon atoms by driving them out of the lattice sites in knock-on collisions. Therefore high-temperature annealing should initiate transformation of carbon nanostructures into the onions, similar to what is observed to occur in electron irradiation.

Such structural phase transitions occurring in carbon nanostructures under high-temperature annealing were observed experimentally and will be discussed in Section 1.4.

As follows from calculations [39], onions should indeed remain stable under intense electron irradiation so that for a number of atoms larger than 20 the spherical carbon cluster is the energetically preferable form, while for larger particles combining several hundred atoms the preferable form is a multilayer structure of nesting spheres coupled by weak van der Waals interaction.

The assumption that onions are a stable form of carbon nanoparticles motivated a study of the possibility of their preparation by application of other treatments of carbon soot. These studies were crowned with success. In particular, transformation of carbon soot obtained by the arc discharge technique into polyhedral onions under high-temperature (up to 2,400°C) annealing in vacuum was successfully demonstrated [35].

Among the methods of preparation of carbon onions, most popular, from the viewpoint of commercial availability of the starting material and the simplicity of the technology employed, have recently been high-temperature (with $T > 1,000^{\circ}\mathrm{C}$) DND annealing [40][a] and various techniques used in the heat treatment of carbon soot (see, e.g., Ref. [41]). It appears pertinent to refer in this connection the recently published paper [42], which reports that by combustion of naphthalene one can reach an output of onions of 0.4 g/h and produce in this way giant onions with the number of layers of up to 54.

[a]The issues associated with the DND–onion structural phase transitions will be considered in Section 1.4 of the present chapter.

Without going into details of possible promising applications of carbon onions, we shall discuss here only the ones that have been attracting most of the interest in the recent years. Among them is their use as antifriction material, both directly and in the form of additives to lubricating oils [43]; application in composite materials intended for screening against electromagnetic radiation [44]; and development of hydrophilic coatings [41].

1.3.3 *Carbon Nanotubes*

The discovery of carbon nanotubes is dated to 1991, when S. Iijima demonstrated by high-resolution electron microscopy the presence of multilayered nanotubes in the deposit forming in arc discharge between graphite electrodes [45]. Interestingly, similar structures of curled-up graphene sheets—carbon nanofibers obtained by pyrolysis of a mixture of benzene with hydrogen—were known to exist long before that [46]. These carbon nanofibers had, however, a substantially larger diameter ranging from 2 nm to 50 nm and did not feature, in contrast to nanotubes, a perfect cylindrical structure [47].

Nanotubes were originally fabricated by the arc discharge technique, to be subsequently complemented by other methods, among them laser ablation, pyrolysis, and chemical deposition from the gas phase in the presence of iron-group catalyst metals.

The main problems that had to be solved in the technology of carbon nanotubes dealt with development of methods to control their diameter, length, and chirality, as well as the number of layers and degree of purification.

Carbon nanotube technology has reached presently the level satisfying industrial production requirements, with parallel development of approaches to synthesis of single-, double-, and multilayer nanotubes, as well as of ordered stacks of nanotubes [48–50].

Significantly, catalytic chemical vapor deposition (CCVD) is constantly gaining in popularity. It is already used on a semi-industrial scale in fabrication of single- and multilayer nanotubes with controllable parameters [51].

Carbon nanotubes are unique in the dependence of their conductivity on the diameter and chirality (the angle of graphene plane orientation with respect to the nanotube axis), that is, the degree of the nanotube "twist." This feature is closely related to variation of the nanotube bandgap width within the 0–1 eV interval and, accordingly, of conductivity—from metallic to semiconducting—only with a change in the nanotube topology [52].

This fundamental property of changing the conductivity by orders of magnitude without the use of any dopant, as is the case with classical semiconductors, is an ideal feature for a new generation in electronics [53].

Demonstration in 1998 of a laboratory prototype of a field-effect transistor using one monolayer nanotube for the channel may be considered as the first symbolic event in this direction [54].

Such studies are becoming a common practice in many laboratories of the world [55, 56].

Among other nanotube applications, already realized and tested in laboratory conditions, are field emitters [57, 58] and materials for fabrication of supercapacitors [59], electrochemical power supplies, and fuel elements [51]. Due to its unique mechanical properties (the Young modulus of a nanotube, about 1 TPa, is larger than that of diamond [60]), nanotubes can and are already successfully employed as a component for improvement of the mechanical characteristics of polymer materials [61].

1.3.4 *Graphene*

By definition, graphene is a planar, single-layer sp^2-hybridized carbon atom. Interest in the investigation of this new carbon nanostructure emerged in 2004, when one experimentally demonstrated for the first time the stability of graphene in free state [62], which was in marked contrast with theoretical concepts accepted at the time [63] and assuming a monolayer to be curled as is the case with carbon nanotubes.

Significantly, already this pioneering publication revealed high room-temperature carrier mobility, which was as high as 10^4 $cm^2/V{\cdot}sec$ and exceeded by far that of silicon, the main material of present-day electronics. Subsequent experimental studies confirmed the validity of treating carriers in graphene as massless Dirac fermions [64] and culminated in observation of the quantum

Hall effect at room temperature [65]. It is presently believed that due to its unique electron dispersion law graphene will permit development of ultrafast electronics devices, primarily transistors capable of operating in the THz frequency range [66].

Studies of the properties of graphene and of its application potential are restrained presently by the absence of an appropriate technology of obtaining this material in the form of layers or films of a large enough area. Interestingly, even the authors of the pioneering publication [62] do not display too much of an optimism as concerns development of a suitable technology for fabrication of graphene of a large enough area and believe composition materials to be more likely to appear in the near future [67].

Most of the experimental studies reported by now relate to graphene samples obtained by mechanical exfoliation of highly oriented pyrolythic graphite (HOPG) [62] or by thermal destruction of silicon carbide (SiC) in high vacuum or a neutral atmosphere (see, e.g., Ref. [67]). The first method has not thus far been successful in producing graphene films with linear dimensions above a few microns because of the mechanical stresses developing in the film exceeding its ultimate strength, while the second approach meets with the limitations imposed on the quality of the silicon carbide used and, besides, does not permit separation of the graphene thus obtained from the substrate.

The reviews that have appeared in the recent years [68, 69] give a clear enough idea of the level reached presently in graphene technology and show convincingly that the problem of fabricating monolayers of perfect crystalline graphene of an area large enough (a few square cm) to make the design of an array of transistors on one substrate realistic remains unsolved. By contrast, the CVD technique permits one to grow on a metallic foil acting as a catalyst large-area films consisting of several graphene layers. These films can be transported from the metal onto another flexible insulating (polymer) substrate and used successfully in this form as optically transparent electrodes in solar power cells [70, 71].

In our opinion, CVD is at present the most promising technology to support "graphene-based electronics." This stems from the possibility it provides of monitoring many parameters of the process and of scaling appropriately the process. The CVD potential was demonstrated convincingly in the technology of fabrication of ordered carbon nanotube arrays (see above). At the same time, fabrication of graphene monolayers by reducing graphite oxide appears to be potentially promising [72]. This method offers an obvious advantage of controlling the bandgap width by varying appropriately the degree of reduction of the graphite oxide. In this connection, demonstration of controllable reduction of graphite oxide films to graphene with a DVD optical drive laser appears an impressive achievement [73].

A realistic analysis suggests, however, that practical application of graphene in the nearest future may be expected to become realizable only in the cases where large-area films of monolayer thickness are not required. In addition to the above-mentioned application in the form of optically transparent electrodes, the use of graphene in supercapacitors [74] is another promising area.

In discussing the properties and applications of graphene, one should not overlook nanosized graphite particles, because the characteristics of these materials are determined by the presence of the so-called edge π electronic states [75]. These issues will be considered in Chapter 8.

1.3.5 *Nanodiamonds*

Strictly speaking, the term "nanodiamond" (ND) is applied in modern publications to several subjects: crystalline grains of nanocrystalline CVD diamond films, nanocrystals prepared by fragmentation of large crystals grown at high temperatures and pressures from graphite, ND powders and suspensions obtained by detonation and laser synthesis, and ND crystals discovered in meteorites.

We are not going to dwell here on CVD-grown nanocrystalline diamond films, referring the reader instead to the corresponding chapters of the monograph [76], and will consider here mainly the fabrication of crystalline diamond powder, with the main emphasis placed on DNDs—the principal subject of this monograph.

1.3.5.1 Synthesis of diamond from graphite in the presence of metal catalysts

The publication announcing the first successful synthesis of diamond from graphite is dated 1955, when the equipment designed at General Electric permitted reaching the pressures and temperatures required for realization of the corresponding phase transition [77]. Depending on the metal catalyst used, the required pressure and temperature vary somewhat and lie, as a rule, in the ranges P = 5.0–7.0 GPa, $T = 1{,}300 - 1{,}800^{\circ}$C [78]. Incidentally, an increase of the temperature and pressure brings about an increase of the graphite–diamond transformation coefficient, while an increase of synthesis time results in a growth of the diamond crystals obtained in size. At present, synthesis of diamond from graphite at high temperatures and pressures, the so-called high-pressure–high-temperature (HPHT) synthesis is a well-developed procedure permitting one to reliably obtain submillimeter-sized crystals of gem quality [79, 80]. This provided a basis to use such crystals as a starting material for obtaining diamond nanocrystals less than 100 nm in size by mechanical milling. Such nanocrystals find application in medicine (see Chapters 5 and 10). Despite the complexities involved in the process of milling and in subsequent isolation of the nanosized fraction, fabrication of nanopowder from HPHT diamonds has been solved successfully in recent years [81, 82]. It should be added that production of NDs by this technique is a compound and costly process, which apparently cannot be scaled up to industrial production.

1.3.5.2 Detonation nanodiamonds

While the detonation-based synthesis of diamond particles was discovered more than 40 years ago, relevant publications started to appear in 1988. At that time, in *Reports of the Academy of Sciences* [83] and *Nature* [84] papers appeared that most of the authors tend to refer to as the pioneering ones, although there are a number of earlier publications as well. Details of the thrilling story describing repeated discovery of the detonation synthesis of diamond by various research groups can be found in the historical review of V. Danilenko, one of the pioneers in detonation diamond production (see Ref. [85]).

In the 1990s, studies of detonation diamonds were naturally devoted to development of an appropriate technology of their production and chemical purification, with investigation of the structure of this material and phase transitions occurring in it having been dealt with primarily at the Institute of Catalysis [86, 87] and the Ioffe Institute [88, 89]. An extensive bibliography covering NDs of all kinds can be found in Ref. [90].

In our opinion, progress in the technology of fabrication of DND, together with a clear understanding of the broad region of its possible application, besides the fairly obvious one as a means for grinding and polishing, became noticeable at the turn of the 21st century. At that time, reports dealing with DNDs started to appear ever more frequently at international diamond conferences. In the first decade of the new century, conferences dedicated to DNDs [91, 92] were held and reviews [93, 94] and a bibliographic index [90] were published.

The detonation-based method of diamond synthesis reduces essentially to detonation of powerful explosives, as a rule, of a mixture of trinitrotoluene with hexogen in a closed chamber; the pressure and temperature at the shock wave front reach the region in which diamond is thermodynamically stable, and this gives rise to the formation of diamond crystals directly from the carbon of the explosives. Because after the detonation both the pressure and the temperature drop in a matter of a few fractions of a microsecond, and the thermodynamic parameters leave the scope of the diamond stability region within a few fractions of a microsecond, the size of the diamond crystals thus formed turns out to be confined to the nanometer range.

Nano- and microscale diamonds can be synthesized by passing the detonation wave through graphite or a mixture of graphite with a catalyst metal confined to a closed chamber. It is this diamond synthesis, called dynamic, that was realized originally in the United States and the USSR. The exclusion of graphite from the process of ND synthesis is an obvious advantage of the detonation method. The technological process of the detonation synthesis of ND is discussed in considerable detail in the second chapter and the structure of DND in the third chapter.

1.3.5.3 Laser method of nanodiamond synthesis

Considered from a purely physical viewpoint, pulsed laser ablation (PLA) of a solid target is similar in many respects to dynamic synthesis of ND from graphite and, thus, has something in common with the detonation method.

The PLA method, which is directly associated with the appearance of powerful rubidium lasers in the early 1960s, is presently widely employed in the synthesis and modification of a broad range of nanomaterials.

Synthesis of nanoparticles by laser ablation of targets immersed in liquids is a comparatively new approach, which has been gaining in popularity during the recent two decades. The PLA method reveals a number of advantages over the multistep techniques, providing a basis for "wet" chemistry, which introduce contaminations originating from intermediate reagents and produce agglomerated structures whose surface is functionalized substantially less than that of DND particles. The method of direct laser ablation eliminates the need in chemical precursors and offers a possibility to produce "clean" nanoparticles. Laser ablation being a strongly nonequilibrium process, the structure and composition of the produced nanoparticles are likewise strongly nonequilibrium, which opens a way to synthesizing new nanostructures with unusual properties.

The PLA-based production of diamond particles was first realized in 1992 by Ogale et al. [95], who used a graphite target immersed in benzene and irradiated by nanosecond-range laser pulses. The laser beam is focused on the graphite surface. Later, in 1998, Polo et al. demonstrated the formation of diamond particles in laser ablation of graphite in a hydrogen atmosphere, that is, in a "dry" version [96]. Ablation of graphite immersed in various liquids is, however, the most efficient method of all, and it enjoys presently the widest popularity [97–104].

Yang et al. were the first to demonstrate synthesis of *nano*crystalline diamond from a graphite target in combination with water [97] and, somewhat later, with acetone [98]. The authors point out the presence of both cubic and hexagonal systems in the reaction products [97–99].

In a later publication [100], the authors detected atomic hydrogen in the ablation plume, which permitted a conclusion that atomic hydrogen must be present for the diamond phase to form. Piers et al. obtained similar diamond structures using water and cyclohexane as liquids [104]. Jiang et al. used a low-power laser with pulses 1.2 μs long [101, 102] and obtained single (not aggregated) diamond particles only 3–6 nm in size. Electron microscope studies of such ND particles revealed a very high degree of their twinning [102]. On the basis of thermodynamic calculations, the authors come to the conclusion that twinning is a process energetically favorable for an increase of stability of ND particles within a certain particle size range. Another interesting feature of the studies reported by this group of authors is observation of strong photoluminescence in the case where ND particles are passivated by polyethylene glycol [102].

A diamond phase can probably be obtained by PLA without a graphite target as well. There was a communication [103] about the use of an iron substrate (acting as a catalyst) immersed in a hydrocarbon liquid (a source of carbon) and irradiated by femtosecond laser pulses. The result was a mixture of diamond-like carbon with hexagonal diamond deposited on iron nanoparticles.

An analysis of the above publications indicates a wide range of experimental conditions (laser parameters, target, type of liquid) tested to determine the combination favorable for formation of diamond structures by PLA. It is believed that this process evolves in the following way.

Interaction of high-power laser pulses with the substrate produces the so-called laser plume or a cloud of reaction products consisting of the evaporated substrate material and, partially, the surrounding liquid [104]. These evaporated substances form bubbles inside the liquid. As the amount of the evaporated material increases, the bubbles expand and, as the pressure and temperature reach a certain critical combination, collapse. At the collapse of the cavitation bubbles, the temperature may become as high as a thousand degrees and the pressure several gigapascals [105]. It is believed that it is these extremal conditions that account for the formation of new nanomaterials in the PLA process [104].

While attempts at commercialization of ND produced by PLA have started recently [106], the laser ND is still two orders of magnitude more expensive than the DND of comparable quality.

1.3.5.4 Diamond synthesis by ultrasonic cavitation

The temperature and pressure required for diamond synthesis can be reached also in ultrasonic cavitation, which is accompanied by collapse of the cavitation bubbles formed in properly chosen liquids [107]. As specified above, depending on the actual cavitation conditions, the temperature and pressure inside a collapsing bubble may reach very high instantaneous values of 10^5-10^6 bar and 1,000 K, respectively [108].

Two research groups announced the possibility of formation of a diamond material of two types, more specifically, of nanocrystalline diamond polycrystals [109] and of single-crystal micron-sized particles [110]. The particles obtained by cavitation destruction of benzene were found to be made up of aggregates of nanoparticles 10–30 nm in size [107].

Micron-sized diamond crystals (5–10 μm) were produced by cavitation synthesis from a suspension of graphite particles (100–200 μm) in organic liquids [110]. The main factor determining the yield of the method is the composition of the cavitation liquid. The yield of diamond relative to the weight of the starting graphite was 10%. A series of aromatic C–H–O oligomers with a low saturated vapor pressure and a high boiling temperature were used as cavitation liquids.

1.3.5.5 Nanodiamonds in meteorites

In 1987, Lewis et al. [111] announced the discovery of NDs in meteorites. After a careful dissolution of the various minerals making up a meteorite by strong acids, a whitish or grayish powder was isolated from various so-called chondrites (primitive, the most ancient meteorites), which produced diamond peaks in X-ray diffraction patterns. Moreover, it turned out that these NDs date as far back as before the formation of the solar system, because they contain anomalous isotopes of krypton and xenon of extrasolar origin [112].

The solar system formed in the collapse of an interstellar gaseous cloud containing presolar dust differing in origin. In the course of evolution of the solar nebula a sizable fraction of this material transformed to produce new minerals, and its genesis was lost. A fraction of the presolar material remained, however, preserved in asteroids and meteorites. Besides NDs, grains of other compounds of presolar origin were identified, in particular, graphite, carbides of various metals, corundum, spinel, and other compounds. NDs dominate, however, among these presolar minerals in mass (up to 1,400 ppm of the chondrite mass).

The origin of these NDs still remains unclear. Most probably, different populations of NDs relate to different events in interstellar space. The anomalous xenon (Xe-HL) is usually assigned to supernovas-stars coming to anend in their evolution in a catastrophic explosive process. Thus, the origin of part of cosmic NDs relates directly to events that occurred close to supernovas.

Several hypotheses have been proposed to account for formation of NDs: low-pressure condensation of carbon in the expanding gas, by analogy with the method of chemical deposition of carbon from the gas phase used to grow diamond films; metamorphic transformation of graphite or amorphous carbon in shock waves produced in collisions of particles moving with a high velocity; conversion of graphite under intense ultraviolet (UV) irradiation; and transformation of a carbon-containing substance under bombardment of high-energy ions [112].

To reveal the mechanisms responsible for formation of NDs of interstellar origin, Doulton compared the microstructure (number and topology of twins) of NDs extracted from meteorites and those obtained in laboratory experiments by chemical gas-phase low-pressure deposition and metamorphic transformation of graphite under bombardment by shock waves and from the carbon of explosives [113]. The results obtained in the study permitted a conclusion that interstellar NDs formed most probably in metastable conditions in low-pressure gas-phase deposition. Note also that NDs of interstellar origin are smaller in size than DNDs; indeed, the average diameters of the NDs extracted from the Murchison and Allende meteorites are only 2.6 nm and 2.8 nm, respectively.

Note also that although diamonds and NDs were discovered in rocks of meteorite craters [114], they were produced in the shock wave–induced conversion of carbon-containing materials in collisions of meteorites-asteroids with the earth.

A recent infrared (IR) spectroscopic study of the emission spectrum of a nebula suggests that fullerenes are also part of interstellar matter [115]. The authors believe that 0.1–0.6% of interstellar carbon exists in the form of fullerenes.

1.4 Structural Phase Transitions between Different Forms of Nanocarbon

The carbon nanostructures considered above are intimately related. This finds convincing support in the structural phase transitions converting one form of carbon into another. We have already discussed the formation of carbon onions from graphite initiated by an electron beam [33], as well as conversion of onions into NDs in collisions with electrons or ions [116, 117] and transformation of NDs successively into quasi-spherical and polyhedral onions [40, 88] and, subsequently, into nanographite under high-temperature annealing [118, 119].

The experimental observation of successive transformation of a CVD polycrystalline diamond film into a uniform graphite layer 150 nm thick in high-temperature annealing in vacuum, followed by transformation of the same film into a diamond film by irradiation by 1.25 MeV electrons [120], appears an impressive achievement.

The influence of fullerenes on the HPHT synthesis of diamond from graphite can be related with the ease with which carbon nanostructures convert into one another. Adding as little as 0.1 wt.% C_{60} fullerenes to a mixture of graphite with catalyst metal increases 1.3 times the output of diamond from graphite at the same temperatures and pressures [121]. The authors assign the mechanism underlying this peculiar catalytic effect to the fullerenes triggering the cycloaddition reaction at the edges of graphene sheets, which initiates formation of the adamantane-like structure, a precursor of a diamond crystal forming from graphite [122]. In other words, the authors associate the influence of fullerene molecules with triggering of the mechanism accounting for the change of the carbon atom hybridization type from sp^2 to sp^3.

In analyzing the structural phase transitions coupling various carbon nanostructures one cannot overlook the issue of their thermodynamic stability.

In the classical phase diagram of carbon plotted on the pressure–temperature plane one will not find any of the nanocarbon forms discussed above. This is in no way strange; indeed, the classical phase diagram does not take into account the object size. The note that an analysis performed on the nanoscale would change the common notion of stability of different allotropic modifications of an element, and that the phase diagram of nanoparticles should be considered in three dimensions, that is, pressure, temperature, and size, could be met fairly long ago, and this statement, if applied to carbon nanostructures, appears now pretty obvious [7]. It is supported now by semiquantitative calculations, which have shown that it is not graphite but rather diamond that is a thermodynamically stable form of carbon for particles less than 4 nm in size [123, 124]. Computer simulation [125] likewise suggests that a stable form for a carbon particle with the number of atoms about 300 is a diamond cluster whose surface is coated by fullerene-like caps.

It is the gain in energy resulting from a decrease in the number of dangling bonds in a graphene sheet that can be identified with the force controlling the surface curling process and formation of curved fullerene-like structures and of the fullerenes themselves.

The remarkable stability of the concentric onions suggests that transformation into multilayered structures is the way by which a system reaches stabilization by triggering the interlayer van der Waals interaction energy [126] and that it is these onions that represent the most stable form of small carbon particles.

While this idea should certainly be checked, calculations cannot provide a definite answer to the question of onion stability. Calculational models based on empirical potentials bring us to the conclusion that the structure with a minimum energy is the plane-face icosahedron, with the curling concentrated at the corners of the polyhedra (the pentagons). First-principle calculations, however, indicate that the C_{240} structure (a closed sphere next after C_{60}) has two local energy minima, with one of them being a faceted icosahedron and the other representing a close to spherical and slightly more stable structure [127].

The issue of onion stability is directly related to the shape of a single ND particle. What is the shape of such a particle? Is it a perfect polyhedron? This point is discussed in considerable detail in Ref. [128], where the authors, on the basis of HRTEM measurements, come to the conclusion that NDs, at least those obtained by detonation synthesis, have a polyhedral shape. It should be pointed out, however, that the conclusion about the shape of a single particle [128] was drawn from a study not of single nanoparticles but rather of DND aggregates (see Chapters 2 and 3 for more details on DND agglomeration).

1.5 Closing Remarks

As we have mentioned before, in this chapter we wanted to point out the place occupied by DNDs in the large family of carbon nanostructures, where each of them has these or those attractive properties. The wish to combine in an optimum way the most remarkable features of each of the nanostructures has initiated in the recent years development of the technology of hybrid nanostructures, which can be illustrated by ND-decorated carbon nanotubes [129] and graphene deposited on a substrate with ND particles [130].

We believe that there are several features that facilitate distinction DNDs out from among other members of the "family." First, it is the stability of the diamond core of the DND particle, in clear contrast to the comparative ease with which the DND surface undergoes functionalization by various groups; this issue will be a subject of discussion in several chapters to follow.

Second, there is a possibility of comparatively easy transformation of DNDs into other nanocarbon forms, such as the onion structure (spherical and polyhedral); a comprehensive discussion of this transformation and of the features originating from the presence of edge π electronic states will be presented in Chapter 8. And, finally, DND is presently a commercially available material produced on an industrial scale, which forms a reliable basis for the use of a number of its unique properties in numerous applications to be addressed in Chapters 9 and 10.

References

1. Mendeleev D. I. (1927) *Basics of Chemistry*, Vol. 1 (State, Moscow, Leningrad), p. 159 (in Russian).
2. Shenderova O., Barnard A., and Gruen D. (2006) *Ultrananocrystalline Diamond: Synthesis, Properties and Applications*, eds. Shenderova O., and Gruen D., "Carbon family at nanoscale" (William-Andrew), pp. 3–23.
3. Dresselhaus M. S., Dresselhaus G., and Saito R. (1999) *Nanotechnology*, ed. Timp G., "Nanotechnology in carbon materials" (AIP Press), pp. 285–331.
4. Inagaki M. (2000) *New Carbons* (Elsevier).
5. Ōsawa E., Yoshida M., and Fujita M. (1994) Shape and fantasy fullerenes, *MRS Bull.*, **19**(11), p. 33.
6. Cataldo F. (2002) The impact of a fullerene-like concept in carbon black science, *Carbon*, **40**, pp. 157–162.
7. Shenderova O. A., Zhirnov V. V., and Brenner D. W. (2002) Carbon nanostructures, *Crit. Rev. Solid State Mater. Sci.*, **27**(3/4), pp. 227–356.
8. Vul' A. Ya., and Sokolov V. I. (2009) Nanocarbon studies in Russia: from fullerenes to nanotubes and nanodiamonds, *Nanotechnol. Russ.*, **4**(7–8), pp. 397–414.
9. Heimann R. B., Evsyukov S. E., and Koga Y. (1997) Carbon allotropes: a suggested classification scheme based on valence orbital hybridization, *Carbon*, **35**, pp. 1654–1658.
10. Dresselhaus M. S., Dresselhaus G., and Eklund P. C. (1996) *Science of Fullerenes and Carbon Nanotubes* (San Diego, California, USA).
11. Heimann R. B., Evsyukov S. E., and Kavan L. (eds.) (1999) *Carbyne and Carbynoid Structures* (Kluwer Academic).
12. Dahl J. E., Liu S. G., and Carlson R. M. K. (2003) Isolation and structure of higher diamondoids, nanometer-sized diamond molecules, *Science*, **299**, pp. 96–99.

13. Carlson R. M. K., Dahl J. E. P., and Liu S. G. (2005) *Synthesis, Properties, and Applications of Ultrananocrystallline Diamond*, eds. Gruen, D. M., Vul' A., and Shenderova, O. A., "Diamond molecules found in petroleum" (Springer, Dordrecht, the Netherlands).
14. Bochvar D. A., and Gal'pern E. G. (1973) On the hypothetical system carbododecahedron, s-icosahedrone and carbon-s-icosahedron, *Dokl. Akad. Nauk USSR*, **209**, pp. 610–612 (in Russian).
15. Kroto H. W., Heath J. R., O'Brien S. C., Curl R. F., and Smalley R. E. (1985) C_{60}: Buckminsterfullerene, *Nature*, **318**, pp. 162–163.
16. Hargittai I. (2000) *Candid Science. Conversations with Famous Chemists* (Imperial College Press, London), p. 322.
17. Krätschmer W., Lamb L. D., Fostiropoulos K., and Huffman D. R. (1990) Solid C_{60}: a new form of carbon, *Nature*, **347**, pp. 354–358.
18. Howard H. B., McKinnon J. T., Makarovsky Y., Lafleur A. L., and Johnson M. E. (1991) Fullerenes C_{60} and C_{70} in flames, *Nature*, **352**, pp. 139–141.
19. Takehara H., Fujiwara M., Arikawa M., Diener M. D., and Alford J. M. (2005) Experimental study of industrial scale fullerene production by combustion synthesis, *Carbon*, **43**, pp. 311–319.
20. Bethune D. S., Johnson R. D., Salem J. R., de Vries M. S., and Yannoni C. S. (1993) Atoms in carbon cages: the structure and properties of endohedral fullerenes, *Nature*, **366**, pp. 123–128.
21. Chai Y., Cuo T., Jin C., Haufler R. E., Chibante P. F., Fure J., Wang L., Alford M., and Smalley R. E. (1991) Fullerenes with metals inside La@C_{82}, *J. Phys. Chem.*, **95**, pp. 7564–7568.
22. Shinohara H. (2000) Endohedral metallofullerenes, *Rep. Prog. Phys.*, **63**, pp. 843–892.
23. Vul' A. Ya., and Huffman D. (1998) Fullerene applications: the first steps from dreams to reality, *Mol. Mater.*, **10**, pp. 37–46.
24. Troshin P. A., and Lyubovskaya R. N. (2008) Organic chemistry of fullerenes: the major reactions, types of fullerene derivatives and prospects for practical use, *Russ. Chem. Rev.*, **77**, p. 323.
25. Teylor R. (1999) Lecture Notes on Fullerene Chemistry (A Handbook for Chemists) (London, UK).
26. Bakry R., Vallant R. M., Najam-ul-Haq M., Rainer M., Szabo Z., Huck C. W., and Bonn G. K. (2007) Medicinal applications of fullerenes, *Int. J. Nanomed.*, **2**(4), pp. 639–649.

27. Anilkumar P., Lu F., Cao L., Luo P. G., Liu J. H., Sahu S., Tackett K. N., Wang Y., and Sun Y. P. (2011) Fullerenes for applications in biology and medicine, *Curr. Med. Chem.*, **18**(14), pp. 2045–2059.

28. Kato H., Kanazawa Y., Okumura M., Taninaka, A., Yokawa T., and Shinohara H. (2003) Lanthanoid endohedral metallofullerenols for MRI contrast agents, *J. Am. Chem. Soc.*, **125**, pp. 4391–4397.

29. Sariciftci N. S., Smilowitz L., Heeger A. J., and Wudl F. (1992) Photoinduced electron transfer from a conducting polymer to Buckminsterfullerene, *Science*, **258**, pp. 1474–1476.

30. Frchet J. M. J., and Thompson B. C. (2008) Polymer–fullerene composite solar cells, *Angew. Chem., Int. Ed.*, **47**, pp. 58–77.

31. Deibel C., and Dyakonov V. (2010) Polymer–fullerene bulk heterojunction solar cells, *Rep. Prog. Phys.*, **73**, 096401 (39 pp.).

32. Iijima S. (1980) Direct observation of the tetrahedral bonding in graphitized carbon black by high resolution electron microscopy, *J. Cryst. Growth*, **50**, pp. 675–683.

33. Ugarte D. (1992) Curling and closure of graphitic networks under electron-beam irradiation, *Nature*, **359**, pp. 707–709.

34. Ugarte D. (1993) Formation mechanism of quasi-spherical carbon particles induced by electron bombardment, *Chem. Phys. Lett.*, **207**(4–6), pp. 473–479.

35. De Heer W. A., and Ugarte D. (1993) Carbon onions produced by heat treatment of carbon soot and their relation to the 217.5 nm interstellar absorption feature, *Chem. Phys. Lett.*, **207**(4–6), pp. 480–486.

36. Bacsa W. S., de Heer W. A., Ugarte D., and Chatelain A. (1993) Raman spectroscopy of closed-shell carbon particles, *Chem. Phys. Lett.*, **207**(4–6), pp. 346–352.

37. Ugarte D. (1994) High-temperature behaviour of "fullerene black," *Carbon*, **32**(7), pp. 1245–1248.

38. Ugarte D. (1995) Onion-like graphitic particles, *Carbon*, **33**(7), pp. 989–993.

39. Tomanek D., Zhong W., and Krastev E. (1993) Stability of multishell fullerenes, *Phys. Rev.*, **48**(20), pp. 15461–15464.

40. Kuznetsov V. L., Chuvilin A. L., Butenko Yu. V., Mal'kov I. Yu., and Titov V. M. (1994) Onion-like carbon from ultra-disperse diamond, *Chem. Phys. Lett.*, **222**, pp. 343–348.

41. Hua S., Baia P., Tianb F., Caoa S., and Sunb J. (2009) Hydrophilic carbon onions synthesized by millisecond pulsed laser irradiation, *Carbon*, **47**, pp. 876–883.

42. Choucair M., and Stride J. A. (2012) The gram-scale synthesis of carbon onions, *Carbon*, **50**, pp. 1109–1115.

43. Joly-Pottuz L., Matsumoto N., Kinoshita H., Vacher B., Belin M., Montagnac G., Martin J. M., and Ohmae N. (2008) Diamond-derived carbon onions as lubricant additives, *Tribol. Int.*, **41**, pp. 69–78.

44. Shenderova O., Grishko V., Cunningham G., Mosseenkov S., McGuire G., and Kuznetsov V. (2008) Onion-like carbon for terahertz electromagnetic shielding, *Diam. Relat. Mater.*, **17**, pp. 462–466.

45. Iijima S. (1991) Helical microtubules of graphitic carbon, *Nature*, **354**, pp. 56–58.

46. Oberlin A., and Endo M. (1976) Filamentous growth of carbon through benzene decomposition, *J. Cryst. Growth*, **32**, pp. 335–349.

47. Endo M., Kim Y. A., and Hayashi T. (2002) *Perspective of Fullerene Nanotechnology*, ed. Ōsawa E., "Pyrolytic carbon nanofibers and nanotubes: structure and applications" (Kluwer Academic, Dodrecht, Boston, London), p. 83.

48. Ando Y., Zhao X., Inoue S., and Iijima S. (2002) Mass production of multiwalled carbon nanotubes by hydrogen arc discharge, *J. Cryst. Growth*, **237–239**, pp. 1926–1930.

49. Kobayashi K., Kitaura R., Nishimura F., Yoshikawa H., Awaga K., and Shinohara H. (2011) Growth of large-diameter (~4 nm) single-wall carbon nanotubes in the nanospace of mesoporous material SBA-15, *Carbon*, **49**, pp. 5173–5179.

50. Hata K., Futaba D. N., Mizuno K., Namai T., Yumura M., and Iijima S. (2004) Water-assisted highly efficient synthesis of impurity free single-wall carbon nanotubes, *Science*, **306**, pp. 1362–1364.

51. Endo M., Hayashi T., Kim Y. A., and Muramatsu H. (2006) Development and applications of carbon nanotubes, *Jap. J. Appl. Phys., Part 1*, **45**(6a), pp. 4883–4892.

52. Saito R., Dresselhaus G., and Dresselhaus M. S. (1998) *Physical Properties of Carbon Nanotubes* (Imperial College Press, London), p. 259.

53. Tsukagoshi K., Yoney N., Uryu S., Aoyagi Y., Kanda A., Ootuka Y., and Alphenaar B. W. (2002) Carbon nanotube devices for nanoelectronics, *Physica B*, **323**, pp. 107–114.

54. Tans S. J., Verschueren A. R. M., and Dekker C. (1998) Room-temperature transistor based on a single carbon nanotube, *Nature*, **393**, pp. 49–52.

55. Maehashi K., Iwasaki S., Ohno Y., Kishimoto T., Inoue K., and Matsumoto K. (2010) Improvement in performance of carbon nanotube field-effect transistors on patterned SiO_2/Si substrates, *J. Electron. Mater.*, **39**(4), pp. 376–380.

56. Aikawa S., Rong Xiang, Einarsson E., Chiashi S., Shiomi J., Nishikawa E., and Maruyama S. (2011) Facile fabrication of all-SWNT field-effect transistors, *Nano Res.*, **4**(6), pp. 580–588.

57. Nguyen Tuan Hong, Ken Ha Koh, Soonil Lee, Ngo Thi Thanh Tam, and Phan Ngoc Minh (2010) Flexible carbon nanotube-array cathodes: fabrication and bending effect on field-electron emission, *J. Vac. Sci. Technol. B*, **28**(2), C2–C5.

58. Pimenov S. M., Frolov V. D., Zavedeev E. V., Abanshin N. P., Du H.-Y., Chen W.-C., Chen L.-C., Wu J.-J., and Chen K.-H. (2011) Electron field emission properties of highly dense carbon nanotube arrays, *Appl. Phys. A*, **105**(1), pp. 11–16.

59. Pan H., Li J., and Feng Y. P. (2010) Carbon nanotubes for supercapacitor, *Nanoscale Res. Lett.*, **5**, pp. 654–668.

60. Treacy M. M., Ebbesen T. W., and Gibson J. M. (1996) Exceptionally high Young's modulus observed for individual carbon nanotubes, *Nature*, **381**, pp. 678–680.

61. Spitalsky Z., Tasis D., Papagelis K., and Galiotis C. (2010) Carbon nanotube–polymer composites: chemistry, processing, mechanical and electrical properties, *Prog. Polym. Sci.*, **35**, pp. 357–401.

62. Novoselov K. S., Geim A. K., Morozov S. V., Jiang D., Zhang Y., Dubonos S. V., Grigorieva I. V., and Firsov A. A. (2004) Electric field effect in atomically thin carbon films, *Science*, **306**, pp. 666–669.

63. Landau L. D., and Lifshitz E. M. (1980) *Statisical Physics, Part I* (Pergamon Press, Oxford, England).

64. Novoselov K. S., Geim A. K., Morozov S. V., Jiang D., Katsnelson M. I., Grigorieva I. V., Dubonos S. V., and Firsov A. A. (2005) Two-dimensional gas of massless Dirac fermions in graphene, *Nature*, **438**, pp. 197–200.

65. Novoselov K. S., Jiang Z., Zhang Y., Morozov S. V., Stormer H. L., Zeitler U., Maan J. C., Boebinger G. S., Kim P., and Geim A. K. (2007) Room-temperature quantum Hall effect in graphene, *Science*, **315**, p. 1379.

66. Geim A. K., and Novoselov K. S. (2007) The rise of graphene, *Nat. Mater.*, **6**, pp. 183–190.

67. Emtsev K. V., Bostwick A., Horn K., Jobst J., Kellogg G. L., Ley L., McChesney J. L., Ohta T., Reshanov S. A., Rohrl J., Rotenberg E., Schmid A. K., Waldmann D., Weber H. B., and Seyller T. (2009) Towards wafer-size graphene layers by atmospheric pressure graphitization of silicon carbide, *Nat. Mater.*, **8**, pp. 203–207.

68. Soldano C., Mahmood A., and Dujardin E. (2010) Production, properties and potential of graphene, *Carbon*, **48**, pp. 2127–2150.

69. Wu Y. H., Yu T., and Shen Z. X. (2010) Two-dimensional carbon nanostructures: fundamental properties, synthesis, characterization, and potential applications, *J. Appl. Phys.*, **108**, 071301.

70. Li X., Zhu Y., Cai W., Borysiak M., Han B., Chen D., Piner R. D., Colombo L., and Ruoff R. S. (2009) Transfer of large-area graphene films for high-performance transparent conductive electrodes, *Nano Lett.*, **9**(12), pp. 4359–4363.

71. Chen Y. P., and Qingkai Yu. (2010) Nanomaterials: graphene rolls off the press, *Nat. Nanotech.*, **5**, pp. 559–560.

72. Dideykin A., Aleksenskiy A. E., Kirilenko D., Brunkov P., Goncharov V., Baidakova M., Sakseev D., and Vul' A. Ya. (2011) Monolayer graphene from graphite oxide, *Diam. Relat. Mater.*, **20**, pp. 105–108.

73. El-Kady M. F., Strong V., Dubin S., and Kaner R. B. (2012) Laser scribing of high-performance and flexible graphene-based electrochemical capacitors, *Science*, **335**, pp. 1326–1330.

74. Compton O. C., and Nguyen S. B. T. (2010) Graphene oxide, highly reduced graphene oxide, and graphene: versatile building blocks for carbon-based materials, *Small*, **6**(6), pp. 711–723.

75. Enoki T. (2012) Role of edges in electronic and magnetic structures of nanographene, *Phys. Scr.*, **T146**, 014008.

76. Gruen D. M. (2006) *Ultrananocrystalline Diamond: Synthesis, Properties and Applications*, eds. Shenderova O., and Gruen D., "Electron transport and the potential of unltrananocrustallline diamond as a thermoelectric material" (William Andrew, Norwich, New York, USA), pp. 157–180.

77. Bundy F. P., Hall H. T., Strong H. M., and Wentorf R. H. (1955) Man-made diamonds, *Nature*, **176**, pp. 51–55.

78. Abbaschian R., Zhu H., and Clarke C. (2005) High pressure–high temperature growth of diamond crystals using split sphere apparatus, *Diam. Relat. Mater.*, **14**, pp. 1916–1919.

79. Liu X., Jia X., Zhang Z., Li Y., Hu M., Zhou Z., and Ma Hong-an (2011) Crystal growth and characterization of diamond from carbonyl iron

catalyst under high pressure and high temperature conditions, *Cryst. Growth Des.*, **11**, pp. 3844–3849.

80. Hu M. H., Ma H. A., Liu W. Q., Zhang Z. F., Zhao M., Li Y., Guo W., Qin J. M., and Jia X. (2010) Synthesis and characterization of strip-shape diamonds from Fe-based alloy and graphite system under high pressure and high temperature, *J. Cryst. Growth*, **312**, pp. 2989–2992.
81. Morita Y., Takimoto T., Yamanaka H., Kumekawa K., Morino S., Aonuma S., Kimura T., and Komatsu N. (2008) A facile and scalable process for size-controllable separation of nanodiamond particles as small as 4 nm, *Small*, **4**(12), pp. 2154–2157.
82. Boudou J.-P., Curmi P. A., Jelezko F., Wrachtrup J., Aubert P., Sennour M., Balasubramanian G., Reuter R., Thore A., and Gaffet E. (2009) High yield fabrication of fluorescent nanodiamonds. High yield fabrication of fluorescent nanodiamonds, *Nanotechnology*, **20**, 235602.
83. Lymkin A. I., Petrov E. A., Ershov A. P., Sakovitch G. V., Staver A. M., and Titov V. M. (1988) Production of diamonds from explosives, *Dokl. Akad. Nauk USSR*, **302**, pp. 611–613 (in Russian).
84. Greiner N. R., Philips D. S., Johnson J. D., and Volk F. (1988) Diamonds in detonation soot, *Nature*, **333**, pp. 440–442.
85. Danilenko V. V. (2004) On the history of the discovery of nanodiamond synthesis, *Phys. Sol. State*, **46**(4), pp. 595–599.
86. Kuznetsov V. L., Aleksandrov M. N., Zagoruiko I. V., Chuvilin A. L., Moroz E. M., Kolomiichuk V. N., Likholobov V. A., Brylyakov V. A., and Sakovitch G. V. (1991) Study of ultra disperse diamond obtained using explosion energy, *Carbon*, **29**, pp. 665–668.
87. Kuznetsov V. L., Chuvilin A. L., Moroz E. M., Kolomiichuk V. N., Shaikhnutdinov Sh. K., and Butenko Yu.V. (1994) Effect of explosion comditions on the structure of detonation soot: ultradisperse diamond and onion carbon, *Carbon*, **32**, pp. 873–882.
88. Alexenskii A. E., Baidakova M. V., Vul' A. Ya., Davydov V. Yu., and Pevtsova Yu. A. (1997) Diamond-graphite phase transition in ultradisperse-diamond clusters, *Phys. Solid. State*, **39**, pp. 1007–1015.
89. Baidakova M. V., Vul' A. Ya., Siklitski V. I., and Faleev N. N. (1998) Fractal structure of ultradisperse-diamond clusters, *Phys. Solid. State*, **40**, pp. 715–718.
90. Vul' A., Dolmatov V. Yu., Gruen D. M., and Shenderova O. (eds) (2006) *Detonation Nanodiamonds and Related Materials*, Bibliography index, 2nd issue (Ioffe Physico-Technical Institute, St. Petersburg).

91. Proceedings of the First International Symposium "Detonation Nanodiamond. Technology, Properties and Applications" (2004), July 7–9, St. Petersburg, Russia, *Phys. Sol. State*, **46**(4).

92. Gruen D., Shenderova O., and Vul' A. Ya. (eds.) (2005) *Synthesis, Properties and Applications of Ultrananocrystalline Diamond*, NATO Science Series II: Mathematics, Physics and Chemistry, Vol. 192 (Springer).

93. Dolmatov V. Y. (2007) Detonation synthesis nanodiamonds: synthesis, structure, properties and applications, *Russ. Chem. Rev.*, **76**, pp. 339–360.

94. Schrand A. M., Hens S. A. C., and Shenderova O. A. (2009) Nanodiamond particles: properties and perspectives for bioapplications, *Crit. Rev. Solid State Mater. Sci.*, **34**, pp. 18–74.

95. Ogale S. B., Malshe A. P., Kanetkar S.M and Kshirsagar S. T. (1992) Formation of diamond particulates by pulsed ruby-laser irradiation of graphite immersed in benzene, *Sol. State Commun.*, **84**, pp. 371–373.

96. Polo M. C., Cifre J., Sanchez G., Aguiar R., Varela M., and Esteve J. (1995) Pulsed laser deposition of diamond from graphite targets, *Appl. Phys. Lett.*, **67**, pp. 485–487.

97. Yang G.-W., Wang J.-B., and Liu Q.-X. (1998) Preparation of nanocrystalline diamonds using pulsed laser induced reactive quenching, *J. Phys.: Condens. Matter*, **10**, pp. 7923–7927.

98. Yang G. W., and Wang J. B. (2001) Pulsed-laser-induced transformation path of graphite to diamond via an intermediate rhombohedral graphite, *Appl. Phys. A*, **72**, pp. 475–479.

99. Wang J. B., Zhang C. Y., Zhong X. L., and Yang G. W. (2002) Cubic and hexagonal structures of diamond nanocrystals formed upon pulsed laser induced liquid–solid interfacial reaction, *Chem. Phys. Lett.*, **361**, pp. 86–90.

100. Yang L., May P. W., Yin L., Smith J. A., and Rosser K. N. (2007) Growth of diamond nanocrystals by pulsed laser ablation of graphite in liquid, *Diam. Relat. Mater.*, **16**, pp. 725–729.

101. Sun J., Hu S.-L., Du X.-W., Lei Y.-W., and Jiang L. (2006) Ultrafine diamond synthesized by long-pulse-width laser, *Appl. Phys. Lett.*, **89**, pp. 183115–183117.

102. Hu S., Sun J., Du X., Tian F., and Jiang L. (2008) The formation of multiply twinning structure and photoluminescence of well-dispersed nanodiamonds produced by pulsed-laser irradiation, *Diam. Relat. Mater.*, **17**, pp. 142–146.

103. Hu A., Sanderson J., Zhou Y., and Duley W. W. (2009) Formation of diamond-like carbon by fs laser irradiation of organic liquids, *Diam. Relat. Mater.*, **18**, pp. 999–1001.

104. Pearce S. R. J., Henleya S. J., Claeyssensa F., Maya P. W., Hallamb K. R., Smitha J. A., and Rossera K. N. (2004) Production of nanocrystalline diamond by laser ablation at the solidy liquid interface, *Diam. Relat. Mater.*, **13**, pp. 661–665.

105. Shaw S. J., Schiffers W. P., Gentry T. P., and Emmony D. C. (1999) A study of the interaction of a laser-generated cavity with a nearby solid boundary, *J. Phys. D: Appl. Phys.*, **32**, pp. 1612–1617.

106. Panich A., Shames A., Levinson O., and Zousman B. (2012) Magnetic resonance study of nanodiamonds prepared by laser-assisted technique, *Diam. Relat. Mater.*, **23**, pp. 150–153.

107. Flynn H. G. (1986) Method and means for converting graphite to diamond. United States Patent 4,563,341, January 7.

108. Brennen C. E. (1995) *Cavitation and Bubble Dynamics* (Oxford University Press, Oxford).

109. Galimov É. M., Kudin A. M., Skorobogatskii V. N., Plotnichenko V. G., Bondarev O. L., Zarubin B. G., Strazdovskii V. V., Aronin A. S., Fisenko A. V., Bykov I. V., and Barinov A. Yu. (2004). Experimental corroboration of the synthesis of diamond in the cavitation process, *Dokl. Phys.*, **49**, pp. 150–156.

110. Khachatryan A. Kh., Aloyan S., Sargsyan R., Khachatryan V. A., May P. W., and Baghdasaryan V. S. (2008) Graphite-to-diamond transformation induced by ultrasonic cavitation, *Diam. Relat. Mater.*, **17**, pp. 931–937.

111. Lewis R. S., Ming T., Wacker J. E., Anders E., and Steel E. (1987) Interstellar diamond in meteorites, *Nature*, **326**, pp. 160–162.

112. Daulton T. L. (2005) *Synthesis, Properties and Applications of Ultrananocrystalline Diamond*, eds. Gruen D., Shenderova O., and Vul' A. Ya. NATO Science Series II: Mathematics, Physics and Chemistry, Vol. 192, "Nanodiamonds in the cosmos microstructural and trapped elements isotopic data" (Springer), pp. 49–63.

113. Daulton T. L. (2006) *Ultrananocrystalline Diamond: Synthesis, Properties and Applications*, eds. Shenderova O., and Gruen D., "Extraterrestrial nanodiamonds in the cosmos" (William Andrew), pp. 23–79.

114. Lipschutz M. E. (1961) The record in meteorite-IV: origin of diamonds in iron meteorites, *Geochim. Cosmochim. Acta*, **24**, pp. 83–88

115. Sellgren K., Werner M. W., Ingalls J. G., Smith J. D. T., Carleton T. M., and Joblin C. (2010) C-60 in reflection nebulae, *Astrophys. J. Lett.*, **722**(1), pp. L54–L57.

116. Banhart F., and Ajayan P. M. (1996) Carbon onions as nanoscopic pressure cells for diamond formation, *Nature*, **382**, pp. 433–435.

117. Zaiser M., Lyutovich Y., and Banhart F. (2000) Irradiation-induced transformation of graphite to diamond: a quantitative study, *Phys. Rev. B.*, **62**(5), pp. 3058–3064.

118. Kuznetsov V. L., and Butenko Yu. V. (2005) *Synthesis, Properties and Applications of Ultrananocrystalline Diamond*, eds. Gruen D. M., Shenderova O. A., and Vul' A. Ya, "Nanodiamond graphitization and properties of onion-like carbon" (Springer, the Netherlands), p. 199.

119. Osipov V. Yu., Enoki T., Takai K., Takahara K., Endo M., Hayashi T., Hishiyama Y., Kaburagi Y., and Vul' A. Ya. (2006) Magnetic and high resolution TEM studies of nanographite derived from nanodiamond, *Carbon*, **44**, pp. 1225–1234.

120. Lyutovich Y., and Banhart F. (1999) Low-pressure transformation of graphite to diamond under irradiation, *Appl. Phys. Lett.*, **74**(5), pp. 659–660.

121. Davidenko V. M., Kidalov S. V., Shahov F. M., Yagovkina M. A., Yashin V. A., and Vul' A. Ya. (2004) Fulerenes as a co-catalyst for high pressure—high temperature synthesis of diamond, *Diam. Relat. Mater.*, **13**, pp. 2203–2206.

122. Kidalov S. V., Sokolov V. I., Shakhov F. M., and Vul' A. Ya. (2005) Mechanism of the catalytic effect of fullerenes on the graphite–diamond phase transition at high temperature and pressure, *Dokl. Phys. Chem.*, **404**(Pt 1), pp. 179–181.

123. Gamarnik M. Y. (1996) Energetical preference of diamond nanoparticles, *Phys. Rev. B.*, **54**, pp. 2150–2156.

124. Badziag P., Verwoerd W. S., Ellis W. P., and Greiner N. R. (1990) Nanometre-sized diamonds are more stable than graphite, *Nature*, **343**, pp. 244–245.

125. Raty J.-Y., Galli G., Bostedt C., van Buuren T. W., and Terminello L. J. (2003) Quantum confinement and fullerenelike surface reconstructions in nanodiamonds, *Phys. Rev. Lett.*, **90**, 037401.

126. Tomanek D., Zhong W., and Krastev E. (1993) Stability of multishell fullerenes, *Phys. Rev.*, **48**(20), pp. 15461–15464.

127. York D., Lu J. P., and Yang. W. (1994) Density-functional calculations of the structure and stability of C_{240}, *Phys. Rev. B*, **49**, pp. 8526–8528.

128. Kuznetsov V. L., and Butenko Yu. V. (2006) *Ultrananocrystalline Diamond. Synthesis, Properties and Applications*, eds. Gruen D. M., and Shenderova O. A., "Diamond phase transition at nanoscale" (William Andrew, Norwich, New York, USA), pp. 405–475.

129. Vul' A. Ya., Reich K., Eidelman E., Terranova M. L., Ciorba A., Orlanducci S., Sessa V., and Rossi M. (2010) A model of field emission from carbon nanotubes decorated by nanodiamonds, *Adv. Sci. Lett.*, **3**, pp. 1–8.

130. Wang Yu, Jaiswal M., Ming Lin, Saha S., Özyilmaz B., and Kian Ping Loh (2012) Electronic properties of nanodiamond decorated grapheme, *ACS Nano*, **6**(2), pp. 1018–1025.

Chapter 2

Technology of Preparation of Detonation Nanodiamond

Aleksandr E. Aleksenskii
Ioffe Physical-Technical Institute, Solid State Electronics Department,
26 Polytechnicheskaya, St. Petersburg 194021, Russia
blin@mail.ioffe.ru

This chapter addresses the key problems involved in the technology of preparation of nanodiamond by detonation of strong explosives. As pointed out in Chapter 1, detonation synthesis builds upon two basic ideas:

- Use of the shock wave forming in the detonation of an explosive as a source of high pressure and temperature for synthesis of crystalline diamond from carbon material
- Formation of diamond crystals by self-assembly from carbon atoms of the explosive itself

We are turning now to a more detailed consideration of the technology of the process.

Detonation Nanodiamonds: Science and Applications
Edited by Alexander Ya. Vul' and Olga A. Shenderova

ISBN 978-981-4411-27-1 (Hardcover), 978-981-4411-28-8 (eBook)
www.panstanford.com

2.1 Formation of Diamonds in a Detonation Wave

Explosion is a well-known phenomenon that may be of either natural or artificial origin. By explosion one understands usually processes involving a fast liberation of energy accompanied, as a rule, by a shock wave—a powerful oscillation of the density of air or of another medium in which this explosion occurs.

The shock wave is a supersonic process of propagation of a perturbation in a material, which culminates in the formation of discontinuities in physical parameters. The existence itself of a shock wave should be traced to the nonlinear properties of the medium in which it propagates. For instance, when a conventional shock wave passes through the air, at points where the pressure is at a maximum the gas has a higher temperature because it heats under compression. This entails also an increase of the velocity of molecular motion in the air, which accounts for the pressure profile in a conventional shock wave being not exactly sinusoidal; indeed, the front of a sonic wave is slightly contracted, while the dropping flank tails off.

While at moderate sonic intensities this effect may be disregarded, at high vibration amplitudes the front of the sonic wave becomes plane to transform it into a shock wave. Propagation of a shock wave through any medium is accompanied by intense dissipation of energy, which accounts for a shock wave damping out fairly fast to become a conventional sonic vibration.

For the propagation of a shock wave in a medium to be stable, it should be sustained by a constant influx of energy. The energy required for a shock wave in a conventional explosion comes from chemical transformation of the starting explosives into reaction products. The shock wave generated in the explosive by the release of the energy of chemical transformation is called the detonation wave.

The chemical reactions occurring in an explosion are similar to those supporting burning. Just as in the case of organic materials burning in air, an explosion involves the formation of carbon dioxide from carbon, of water from hydrogen, and of elemental nitrogen from complex chemical compounds.

Unlike conventional burning taking place in excess of the oxidizer, which is oxygen of the air, the starting ratio of the fuel to the oxidizer in the chemical reaction of explosive transformation is specified exactly by the starting composition of the explosive used. All of the energy expended in supporting the detonation wave in an explosion is liberated only in the reaction of transformation of the starting explosive. This is easy to understand—indeed, all the other processes, among them heat transport, diffusion, convection, etc., have orders-of-magnitude-longer characteristic times. This is why one usually refers to the detonation process as adiabatic. It is believed that it is the adiabaticity of the detonation process that is at the base of the usually observed linear dependence of the detonation rate on the density of the starting explosive (for the same composition).

The main concepts concerning the nature of detonation waves were originally discussed in Refs. [1–3]. The present state of the problem is described in Ref. [4]. For a popular treatment of its aspects, the reader can be referred to Ref. [5].

After the passage of the detonation wave, chemical reactions in the products of detonation do not stop. Detonation products, for instance, water and carbon oxides, interact with one another; such chemical reactions are called secondary. If detonation occurs in air, the process becomes complicated by additional oxidation of the hot detonation products and their dispersion in the atmosphere. This aspect plagues an experimental analysis of the composition of the explosion products. One runs also into difficulties in trying to theoretically describe the chemical reaction involved in the explosive transformation, with an exception of a few limiting cases. This interferes also with accurate calculation of the heat released in the chemical reaction providing energy for the detonation wave. Significantly, precise prediction of the parameters of a detonation wave from the composition and parameters of the starting explosive is of prime importance for applications.

There is a simplified approach to this problem based on a theoretical model easy to visualize [2]. In this model, the explosive is considered to be a mixture of the fuel, hydrogen and carbon; oxygen as the oxidizer; and, in rare instances, chlorine, as well as bound nitrogen. The first to form in an explosive transformation are compounds providing the major contribution to the energetics of the detonation process, namely, elemental nitrogen and water. The remaining oxygen is distributed approximately in accordance with the equilibrium distribution for the detonation parameters in the reaction $2CO \leftrightarrows C + CO_2$.

The carbon-to-oxygen (C/O) ratio in the starting explosive and the detonation products is usually called the oxygen balance.

Within this simple model, all explosive materials (EMs) can be divided into four main groups:

1. Explosives with a positive oxygen balance, in which the amount of oxygen is larger than that required for oxidation of combustible elements of the explosive (called oxidizers).
2. Explosives with zero oxygen balance, with the amount of oxygen exactly equal to that of carbon and hydrogen. In this case, explosion results in complete burning of the combustibles, with all of the carbon converting into carbon dioxide.
3. Explosives with a weakly negative oxygen balance. The C/O ratio in them is such that the amount of oxygen present is insufficient for complete conversion of carbon to CO_2 but larger than that needed for becoming bound in CO. In this case, the C/O ratio satisfies the condition $0{,}5 < C/O < 1$. Such an explosive is exemplified by hexogen (research department explosive [RDX]) employed in the preparation of nanodiamond (Fig. 2.1).
4. Explosives with a strongly negative oxygen balance. The C/O ratio here is such that the amount of oxygen present is insufficient for total conversion of carbon to CO, that is, $C/O > 1$. Actually,

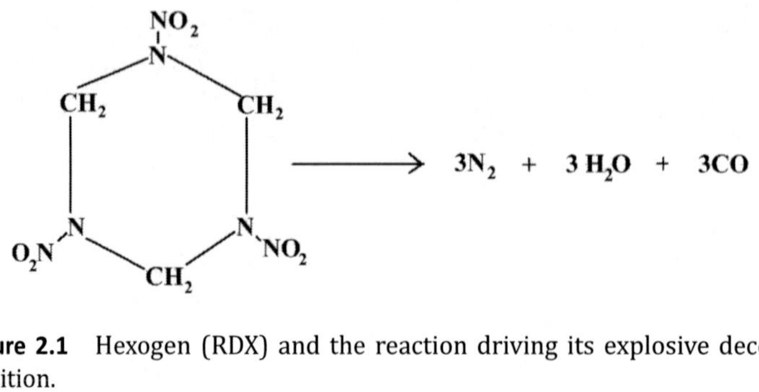

Figure 2.1 Hexogen (RDX) and the reaction driving its explosive decomposition.

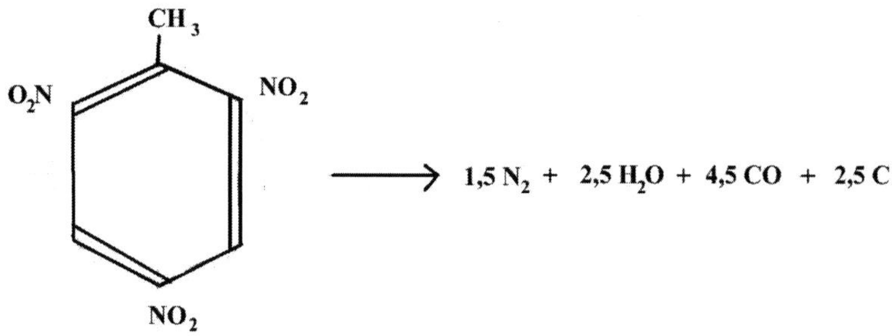

Figure 2.2 TNT and the reaction driving its explosive decomposition.

this means that part of the carbon will be released in free state. Trinitrotoluene (TNT) (trotyl) can serve as a specific example of such explosives (Fig. 2.2), and it is the mixture of TNT with hexogen that is employed in the synthesis of nanodiamond.

The elemental carbon released in the detonation of explosives with a strongly negative oxygen balance condenses naturally in the form that is most stable at the temperature and pressure reached in the detonation wave.

A comparison of the ranges of pressures and temperatures (PT parameters) with those corresponding to the region of diamond stability shown in the well-known diagram [6] (Fig. 2.3.) shows clearly that these ranges partially overlap. It is this overlap that underlies the conclusion of the possibility of segregating the condensed carbon phase in the form of diamond in detonation-based synthesis.

The thermodynamic functions of diamond and graphite differ noticeably. Accordingly, the heats of the reaction of chemical transformation in explosion will likewise differ depending on the form in which the condensed carbon will segregate. Therefore, calculation of the detonation rate for explosives with a strongly negative oxygen balance will yield two solutions, one of which corresponds to segregation of diamond and the other to that of graphite [3, 7].

Secondary reactions, primarily oxidation of the segregated carbon by oxygen of the air, make isolation of this carbon in the

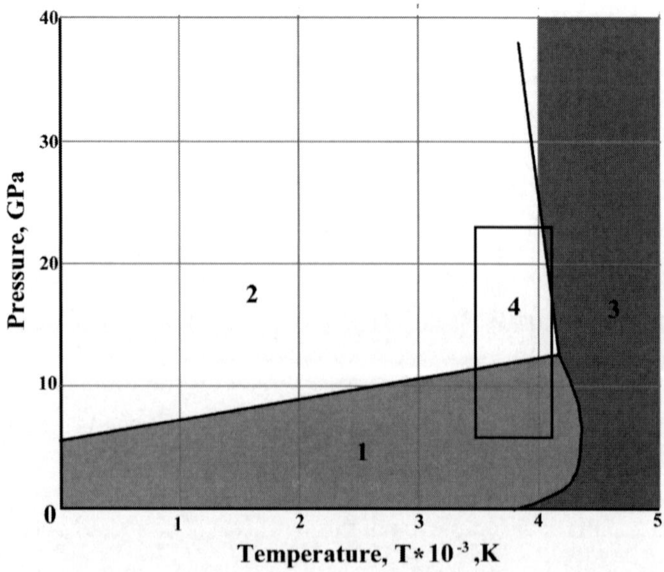

Figure 2.3 Phase diagram of carbon. The diagram specifies the regions of phase stability for 1, graphite; 2, diamond; 3, liquid carbon. Region 4 identifies the PT conditions reached in a shock wave.

form of diamond in an explosion conducted in standard conditions impossible.

If, however, the explosion is performed not in the open air but in special equipment providing adequate preservation of the products of the reaction [8], then one indeed will find in the products of the explosion an adequate amount of condensed carbon with a diamond structure. A characteristic feature of such detonation nanodiamonds (DNDs) is their small diameter, 4–6 nm, and their extremely narrow size distribution [9].

Thus, to obtain diamond in the detonation of explosives the following conditions should be met:

1. A strongly negative oxygen balance of the starting explosives, which would provide the presence of condensed carbon in the reaction products.
2. A high enough power of the starting explosives to ensure thermodynamic stability of diamond at the PT parameters of the

detonation wave. This condition is partially at variance with the first requirement.

3. Conduction of the explosion under special conditions that would provide preservation of the products of explosion and suppress secondary reactions.

2.2 Specific Features of Detonation Carbon Synthesis

As a starting material for the synthesis of nanodiamond one mostly uses a mixture of two explosives, TNT (60%) and hexogen (40%) [10]. The components of this mixture being commercial products, they contain small amounts (less than 1%) of impurities and of the various additives required by the specific technology employed. Selection of the specified composition is determined by several factors.

TNT and hexogen are economically affordable commercial explosives of a high enough power. All other explosives are much more expensive, the cost being partially governed by the substantially smaller production quantities.

The 60/40 mixture of TNT and hexogen is a standard product employed usually in explosives developed for a variety of purposes, and the fact that the component ratio of a standard explosive turned out to be approximately equal to that favorable for the maximum yield of nanodiamond compared to the starting explosives is purely fortuitous.

The charges of explosives intended for the production of DND are exploded in a special chamber. The carbon product obtained directly after the explosion and containing nanodiamond is called detonation carbon, with synonyms being detonation soot and diamond and detonation blends. Figure 2.4 illustrates the main components of the process of detonation diamond synthesis.

The explosion chamber (1) is a closed vessel with the walls made of high-strength steel; it is provided with a hermetical lid through which the charge of the explosive is introduced. The chamber can vary in volume from a few to 100 cubic meters [11]. The chamber is provided by an initiation system, a gas feed system, and a water supply and removal system. As a rule, the finished product is

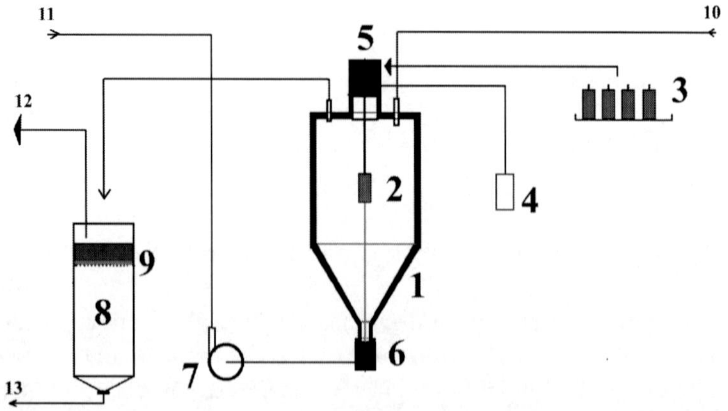

Figure 2.4 Schematic of the equipment employed in detonation synthesis. 1, detonation chamber; 2, the starting EM charge; 3, charge storage; 4, explosion initiation system; 5, loading device; 6, blocking device of the detonation chamber; 7, circulation pump; 8, vacuum filter for the produce; 9, finished detonation product on the filter; 10, gas supply; 11, supply of cooling water; 12, transfer of the finished product for subsequent treatment; 13, processed water.

removed from the explosion chamber by washing the walls down with water (hydro removal), but some constructions provide for removal of detonation soot by a gas flow with subsequent isolation of the finished product from the flow (pneumoremoval).

Besides collection of the reaction products, the explosion chamber has an additional function of suppressing the secondary reactions, the most serious of them being oxidation of the condensed carbon by oxygen of the air. To suppress this reaction, the chamber is filled by an inert gas, more specifically nitrogen, argon, or carbon dioxide.

The construction of the explosion chamber should also provide means for fast cooling of the reaction products. Because the detonation products released in the explosion expand (practically with the sound velocity), the pressure drops very fast. As a result, the synthesized nanodiamond may turn out in conditions of thermodynamic instability (see Fig. 2.5), which will inevitably initiate the reverse phase transition of diamond to graphite. To avoid this situation, one has to increase the cooling rate of the detonation

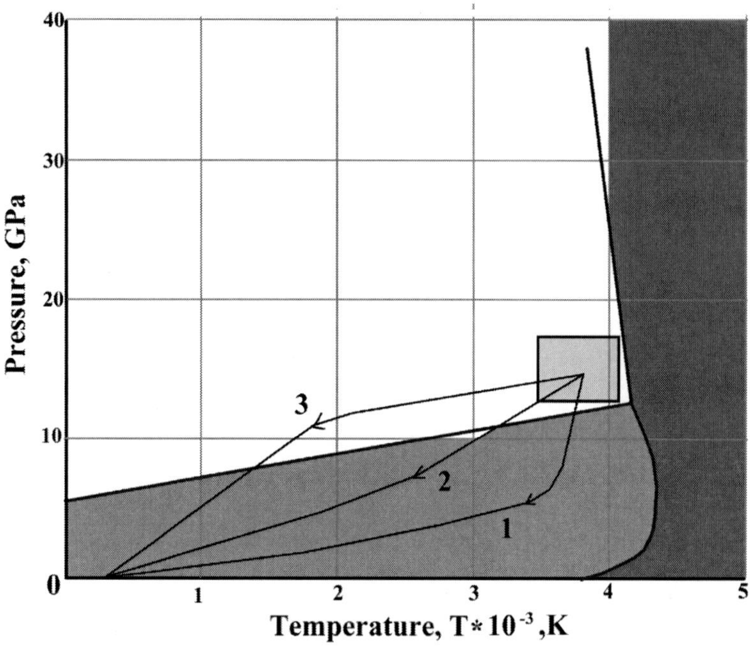

Figure 2.5 PT parameters of various modifications of detonation synthesis. 1, gas cooling; 2, wet (water) cooling; 3, ice cooling.

products as well. It is the ratio of the pressure-to-temperature drop rates that governs in the final count the efficiency of the detonation synthesis of nanodiamond, a point first stressed in Ref. [12].

One distinguishes among three major kinds of explosion-based synthesis, which are determined by the medium in which the reaction products are cooled.

Synthesis involving gas cooling (the "dry" process) is run in a gas-filled explosion chamber. The reaction product cooling rate is comparatively low, and the content of nanodiamond in the detonation carbon makes up 30–40 wt.%. This was the first technology developed for detonation synthesis of nanodiamond [11].

In the "wet" synthesis, cooling occurs with the explosion chamber filled partially with water [13]. The cooling rate of the products of detonation synthesis is here higher because this process is partially sustained by the heat released in the phase transition; the nanodiamond content in the detonation carbon is now 50–70 wt.%. This approach is presently the most widely employed technology of detonation synthesis. An essential technological feature of the wet detonation synthesis is placing the explosive charges into watertight shells. These shells are usually made of polyethylene; they collapse and crumble into fine powder under the action of the detonation wave. Therefore the content of organic impurities in the DND produced by wet synthesis is substantially higher than that demonstrated in the "dry" process.

Synthesis involving "ice" cooling takes place in the explosion chamber with EM charges encased in thick shells of ice [14]. The detonation products being cooled here in two phase transitions running in parallel, the nanodiamond makes up 80–95 wt.% of the detonation carbon produced by the ice synthesis. Actually, the additional shells for the charges are not used, which reduces the content of organic impurities in the detonation carbon of the ice synthesis to the level characteristic of the product of the dry technology. This technology was employed in the stage of experimental development of the method and was not employed subsequently on a commercial scale.

2.3 Isolation of Nanodiamond from Detonation Carbon

The product formed directly in detonation synthesis is detonation carbon, which, when dried, represents loose black powder with a packed density varying from 0.4 kg/L (wet synthesis) to 0.2 kg/L (for the dry process). The powder of light gray color produced by the "ice" technology has a density in excess of 1 kg/L.

As already mentioned, the content of nanodiamond in detonation carbon varies from 40% to 85%. The detonation carbon contains also oxygen and nitrogen (up to a few percent) and a number of metals as impurities. Figure 2.6 displays a typical X-ray diffraction pattern illustrating the phase composition of detonation carbon.

The figure demonstrates the strongly broadened lines from the (111) crystal planes of diamond and from the (0001) graphite plane.

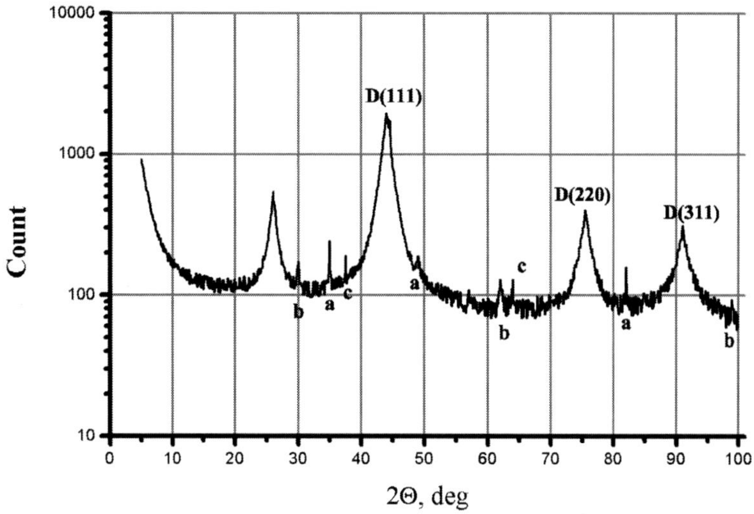

Figure 2.6 X-ray diffraction pattern. The peaks relate to a, SiO_2; b, TiO_2; c, Fe (metal); and D, diamond.

The nanodiamond particle size, as deduced from the width of the maxima, is 4–5 nm for both diamond and graphite. One observes also narrow weak lines originating from iron oxides and silicon dioxide. They can be related to the impurities entering the detonation carbon in the course of its formation as a result of the erosion of explosion chamber walls produced by the shock wave.

A significant part of the carbon is not combined in any ordered structure and, hence, does not manifest itself in X-ray diffraction patterns. At the same time, attempts at selective extraction by various solvents fail in isolating any organic compound in sizable amounts. All carbon components of detonation diamond that are other than diamond are frequently referred to as the nondiamond carbon forms (NDCFs).

Attempts at gravitational isolation of detonation carbon components and NDCFs in bromoform, a heavy liquid, did not meet with appreciable success.

Therefore to isolate nanodiamond from detonation carbon, one resorts to chemical methods based on the difference in the chemical properties of nanodiamond and other detonation carbon components. In this respect the impurities can be divided into the following groups:

1. Nondiamond forms of carbon that react with the oxidizer
2. Elemental metals and salts of metals that, while not interacting with the oxidizer, react with acids and, possibly, with alkalies
3. Oxides of nonmetals, for instance, SiO_2, which, while not being attacked by the oxidizer and acids, can react with alkalies

2.3.1 *Purification of Nanodiamond from Nondiamond Forms of Carbon*

Any method of selective oxidation of the nondiamond forms of carbon, that is, their transfer to the gas state by the reaction $C+2O_2 \rightarrow CO_2$, is based on the reactivity of the NDCFs being higher than that of diamond, with the difference increasing with increasing dispersion, defect content, and amorphism of the NDCFs. While theory does not preclude reduction of NDCFs to CH_4 in flowing hydrogen at 800°C, we have not thus far met any broadly cited publication on the subject.

Two regimes of NDCF oxidation, in the gas and liquid phases, are presently known, which differ in the moving phase used in the process as oxidizer.

The simplest and most widely used gas oxidizer is oxygen of the air. While ordinarily one employs the air of natural composition, use of gas-selective membranes for pre-enrichment of the air with oxygen is also a viable possibility. The temperature of the onset of the oxidation process can be lowered, as a rule, by employing a suitable activator or catalyst of oxidation. In most cases vanadium pentoxide (V_2O_5) is chosen for the purpose, although one can employ chromium, manganese, or iron compounds as well, the most active of which are acetylacetonates (Fig. 2.7) [15] or boron anhydride, B_2O_3 [16].

The process of gas-phase oxidation is usually run in standard drying or vacuum cabinets at temperatures of 200–350°C (depending on the catalyst used) provided by adequate thermostatting and air supply systems. The main attractive feature of this approach lies in its simplicity. On the other hand, the oxidation process here is

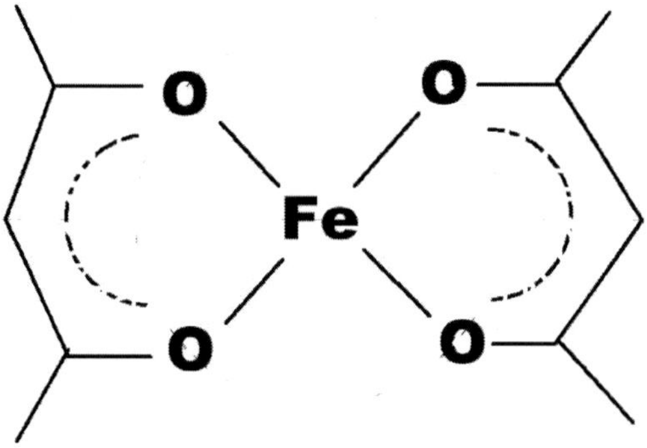

Figure 2.7 Iron acetyl acetonate.

difficult to control, its conditions are not the same over the mass of the detonation carbon involved, and the selectivity is not always satisfactory, because if the heating temperature rises above 400°C, part of the diamond can become oxidized as well.

Oxidation does not remove noncarbon impurities; indeed, their content even increases because of the presence of the catalyst. The catalyst or products of its destruction build up in the final product because they do not possess a noticeable volatility under the conditions employed in the oxidation. Therefore air (or any other gas-phase) oxidation requires inclusion of additional stages of purification for removal of inorganic impurities.

Ozone-based purification is a variety of gas-phase oxidation. As the oxidizer one uses in this method ozone (O_3) produced in coronal discharge [17]. The oxidation takes place here directly as the gas flows with a low velocity through the detonation carbon layer. The selectivity of this oxidation process is very high. Its parameters are easily controllable and practically the same for the whole batch of the product.

The main virtue of this method lies in the very high degree of purification from nondiamond carbon; it is apparently the highest among all methods known thus far. Ozone is a pure oxidizer, which does not introduce additional impurities into the final product. Its major shortcoming consists of the complexity of the equipment involved, high energy consumption, and, as a consequence, high cost.

One may need to provide also another stage of removal of inorganic impurities if their content is found to be unacceptably high.

The most widely used liquid-phase methods are those in which oxidation takes place in a medium possessing oxidizing properties. In this case, the oxidizing purification proper is paralleled by a partial dissolution of inorganic impurities. The conditions governing the process are the same for the whole batch, reliably controllable, and reproducible.

Several purification technologies based on liquid-phase oxidation are presently in use [18].

Purification by chromium anhydride or chromates dissolved in concentrated sulfuric acid is technologically the simplest method of all. The reaction is conducted in general-purpose glass equipment at atmospheric pressure and a temperature of 120–150°C. The reaction time is determined by the rate of dosing of the starting detonation carbon, which is chosen on the basis of the allowed level of foaming of the reacting mass. On completion of the dosing, the liquid phase is separated from the sediment of the isolated nanodiamond thus obtained, which is subsequently washed with water to remove the residue of the oxidizer.

The main drawback of this process is the large amounts of wastes containing chromium compounds. The final product obtained after the oxidation becomes also contaminated by Cr_2O_3 and CrO_2.

A similar process of purification run at about the same parameters can be realized with another inorganic oxidizer, potassium permanganate ($KMnO_4$). As in the preceding case, the principal drawbacks of the process are the large amount of wastes and contamination of the final product by MnO_2 impurity.

Purification of nanodiamond with 50–75% perchloric acid ($HClO_4$) is also conducted in glass equipment at atmospheric pressure. The reaction temperature of 80–100°C is lower than that used in oxidation with chromium compounds. The obvious advantages of this method are the absence of additional contamination of the product thus obtained by inorganic impurities and a low expenditure of the oxidizer.

The hydrochloric acid present in the reaction products reacts with elemental metal and metal oxide impurities by converting them to a soluble form. Regrettably, perchloric acid is an explosive compound, thus making other than small-scale application of this method impossible.

Oxidation of nondiamond carbon by hydrogen peroxide (H_2O_2) or a mixture of hydrogen peroxide with nitric acid is another well-known process [18]. This method has an attractive feature in that the reaction product is not contaminated additionally in the course of purification; quite the reverse, it becomes still better purified through dissolution of metal oxides in nitric acid. It does not, however, enjoy popularity because of the instability and explosion hazard associated with the use of concentrated hydrogen peroxide.

Purification of nanodiamond on an industrial scale is based on oxidation with nitric acid dissolved in concentrated sulfuric acid. Actually, nitric acid is a less active oxidizer than chromium salts or hydrochloric acid are, and therefore a satisfactory reaction rate can be reached only by providing a higher temperature, from 200°C upward.

The reaction proceeds in the liquid phase, and therefore for the reaction medium one employs high-boiling concentrated sulfuric acid or oleum (20% solution of SO_3 in H_2SO_4). Laboratory research requires that the reaction be conducted in high-quality glass or quartz capacity equipment, while in industrial conditions special brands of stainless steel are required for this purpose.

In this process, acids also react with metal and metal oxide impurities and reduce the associated level of contamination. To extract the finished nanodiamond, one resorts usually to dilution of the reaction mass because at conventional temperatures the time required to separate the nanodiamond deposit from the reaction mass is extremely long (concentrated sulfuric acid and oleum feature high viscosities).

This method is attractive in its industrial realization being comparatively simple, because the industry producing explosives is based on processes conducted in similar conditions and with similar equipment. The main drawback lies in the large amounts of wastes containing sulfuric acid in concentrations of up to 30%, which require for their recycling equipment that even surpasses in complexity the one employed in the purification process proper.

The best industrial method of nanodiamond isolation developed now employs oxidation of the nondiamond fraction of detonation carbon with nitric acid (diluted by water to 50–67% concentration). To ensure the presence of the liquid phase, the process is conducted under elevated pressure (80–100 atm) in equipment fabricated from special-purpose titanium alloys [11].

After the reaction has come to an end, there is no need to dilute the reaction mass in order to separate the sediment of the finished product, thus permitting one to realize the fairly simple recycling of the acid liquor. The nitrogen oxides released in the reaction can also be processed practically to the level of the starting acid, which provides close-to-complete, economically attractive oxidizer regeneration. The product obtained after proper washing is characterized by low levels of contamination by inorganic compounds because nitric acid forms well-soluble salts practically with all known metals.

The equipment designed for purification by nitric acid under pressure is technologically fairly complicated. The main drawback of this technology is the formation of an additional contamination, titanium dioxide (TiO_2), which is produced in corrosion of the reaction equipment. The content of the residual TiO_2 in nanodiamond may be as high as 0.3–0.5 wt.%, particularly if recycling is poorly organized and the recirculating oxidizer is not filtered. Since, however, TiO_2 is an inert and nonmagnetic compound, its presence in reasonable amounts can be tolerated in most applications.

2.3.2 *Purification of Nanodiamond from Metal Oxides*

Following chemical purification, the conventional DND contains nondiamond forms of carbon in insignificant amounts and residual impurities of metal salts and inert impurities (SiO_2, TiO_2) in concentrations of up to 1 wt.% [8].

While such a contamination level does not interfere with traditional industrial applications, in some cases it may be found intolerable. This particularly relates to impurities of transition metals, which should be assigned to their toxicity, para- and ferromagnetism, and catalytic activity. To cite an example, even straight heating above 700°C can reduce metal salts with the formation of carbides or metals possessing magnetic properties. It is this phenomenon that is assumed to underlie announcements of the discovery of magnetic carbon. This is why the process of purification of nanodiamond from impurities of transition metals is frequently referred to as magnetic purification.

To prepare magnetically pure samples, one can use additional treatment of nanodiamond by strong acids. Interaction of strong acids with metal oxides initiates formation of metal salts, which, as a rule, are water soluble. Hydrochloric acid (HCl) and hydrobromic acid (HBr) are the most appropriate candidates for this purpose.[a]

Straight washing of a nanodiamond suspension by acids, even under intense stirring, is insufficient for complete removal of metal impurities. The reason for this lies in agglomeration of nanodiamonds (formation of porous diamond microparticles 500–1,000 nm in size; see below on deagglomeration). Inside these agglomerates, the diffusion rate of the etchant is low, and to increase it, intense ultrasonic irradiation is necessary [19].

The washing finished, one has to remove the acid from the nanodiamond suspension. This can be done by multiple washings with water. The water itself used for the washing has naturally to be purified to not worse than American Society for Testing and Materials (ASTM) level 2, and its resistivity should be not less than 5 MOhm/cm.

2.3.3 *Purification of Nanodiamond from Inert Oxides*

Inert impurities of the type of SiO_2 or TiO_2 are contaminations most difficult to get rid of. Unlike the metal oxides, particles of this composition do not dissolve in acids and are not resolved by them. On the other hand, such oxides, particularly SiO_2, should react with alkalies.

A particular problem is meeting the requirement of high purity of the alkaline reagents involved in the purification process. For instance, standard sodium hydroxide (NaOH) may contain up to 10% NaCl or microscale impurities of heavy metals. Besides, NaOH is contaminated by carbonates and silicates in concentrations of

[a]These phenomena can be observed even for metal contaminations of about 10^{-5}–10^{-6} wt.%, and therefore monitoring the degree of nanodiamond purification from magnetic impurities by electron paramagnetic resonance is necessary (see below).

up to a few percent, a level higher than that of contamination in the nanodiamond proper to be purified. The same applies to potassium and lithium hydroxides (KOH and LiOH). Significantly, concentrated solutions of alkalies are capable of dissolving partially glass, particularly under heating, so that such treatment even increases the SiO_2 content in nanodiamond. For this reason, one has not thus far succeeded in developing a successfully operating technology of nanodiamond purification based on the use of alkaline reagents.

Another approach to purification from inert impurities is based on processing nanodiamonds with hydrofluoric acid (HF). This reagent interacts with inert impurities in the following reactions:

$$SiO_2 + 4HF \rightarrow 2H_2O + SiF_4$$

$$TiO_2 + 4HF \rightarrow 2H_2O + TiF_4$$

$$TiF_4 + 2HF \rightarrow H_2TiF_6$$

Obviously enough, all operations with hydrofluoric acid should be conducted in vessels made of polypropylene or fluoroplastic because this reagent attacks glass.

The technology of purification employed is close to that described in the preceding section. This method was successfully employed to obtain nanodiamond samples with ash contents below the detection threshold (ash free) but naturally after preliminary purification from magnetic impurities [20].

2.3.4 *Washing and Isolation of Nanodiamond*

Each cycle of water-washing of nanodiamond brings about a gradual decrease of the suspension acidity (increase of pH). Also, after purification the content of metal ions is low. We are thus witnessing elimination of both factors favoring secondary aggregation of the nanodiamond. After the threshold of pH-dependent coagulation ($pH = 2$) has been surpassed, nanodiamond particles reveal a sharp decrease in size in water, from a few microns to 100–500 nm. This manifests itself in formation of a highly stable suspension, with sedimentation of the particles taking up a time ranging from a few months to years. Therefore after the suspension has reached this stage, for further purification from acid residues one resorts to the use of a centrifuge.

The best purification of a nanodiamond suspension from acid residues can apparently be attained by dialysis or electrodialysis through a semitransparent, inert (celluloid) or ion exchange membrane.

In the case where a solid product has to be isolated from the suspension, the latter is dried. The process of nanodiamond isolation can be accelerated by preliminary concentration of the suspension with a centrifuge.

It goes without saying that the conditions chosen for drying and subsequent storage of the product should be chosen such as to exclude repeated contamination of the purified nanodiamond. One has, in particular, to envisage thorough purification of the drying cabinets and their blow-down only with well filtered air, as well as special measures in preparing the appropriate premises and containers for storage. One should also, wherever possible, stick to the corresponding requirements imposed by the electronics industry.

Interestingly, in development of the nanodiamond deagglomeration technology a totally unexpected side effect was revealed. Indeed, the product isolated from the solution containing single 4 nm particles was also found to be practically ash free. This issue will be treated in more detail in the next section.

2.3.5 *Methods of Deaggregation of Nanodiamond Particles and Preparation of Nanodiamond Suspensions*

As already mentioned (see Chapter 1 and the first section of this chapter), experimental data amassed in X-ray structural analysis, Raman spectroscopy, and transmission electron spectroscopy suggests convincingly that crystalline DND particles are 4–5 nm in size.

By contrast, studies of DND powder morphology by microscopic techniques, deriving the size of nanodiamond particles in water suspension from their dynamic characteristics (rate of precipitation or Brownian motion), or by light diffraction measurements yield totally different results; indeed, the particle sizes turn out to be, as a rule, two orders of magnitude larger. Repeated measurements support the absence in conventional industrial nanodiamond suspensions of particles a few nanometers in size. Moreover, the size of particles suspended in water media turned out to be noticeably dependent on the characteristics of the medium, primarily on the magnitude of pH [22]. Viewed on the macroscopic scale, this phenomenon is revealed as a strong change of the coagulation stability of nanodiamond suspensions under variation of the composition of the water medium. To cite an example, the precipitation time for a nanodiamond suspension in an acidic medium with pH = 2 is about 20 minutes, with the average diameter of the particles being 1–2 microns. At the same time, at pH = 5 the same suspension does not settle out for as long as a year, and the average size of particles in it is about 100 nm.

This phenomenon, namely, formation of complex DND from the 4 nm primary ones, was termed aggregation. Because the size of nanodiamond particles depends on the composition of the medium, this aggregation was referred to also as pH dependent. It is believed that it originates from interaction of the functional groups attached to the diamond surface with the surrounding liquid phase/water medium, which form a electric double layer (EDL) around each particle.[a]

Let us look more closely at the formation of a EDL around a DND particle in a water suspension (Fig. 2.8).

We have to recall that the EDL represents actually a disruption of the electric neutrality of a medium related closely to the phase boundary, a situation in which the particle surface acquires an excess charge, while in the thin layer of the liquid surrounding the particle an excess of ions of opposite sign appears, thus maintaining neutrality of the system as a whole. At the particle, the electric charge is localized in a monomolecular layer on its surface, while the oppositely charged ions form a relatively extended diffuse shell around the particle.

Because particles with a EDL interact primarily by their diffuse layers, they experience the conventional Coulomb repulsion

[a]The concept of a EDL, one of the key ideas underpinning colloidal chemistry, is described in considerable detail in [23, 24].

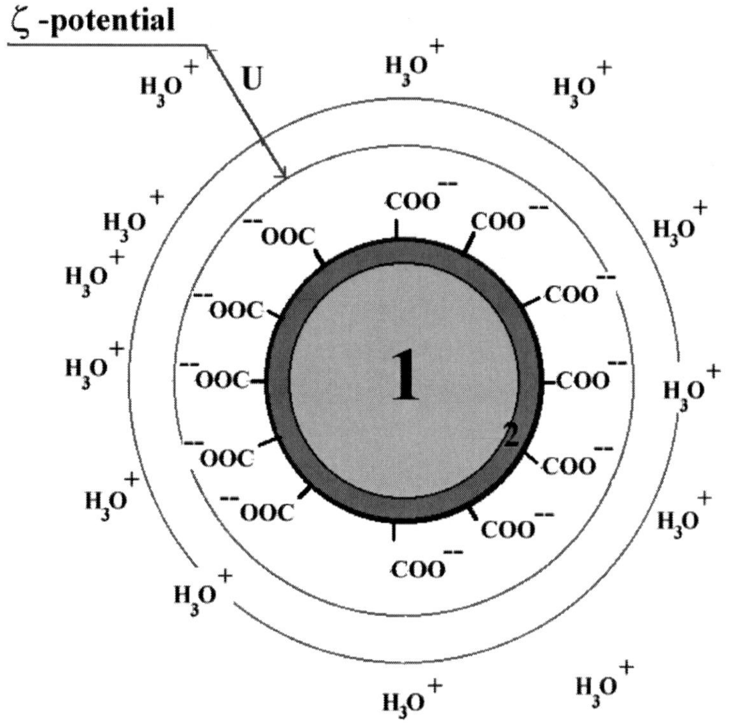

Figure 2.8 Schematic image of a EDL around a DND particle. 1, crystalline diamond core; 2, amorphized carbon layer coating the diamond core.

characteristic of charges of the same sign, which is inversely proportional to the squared distance separating the particles. If the potential barrier arising in the process is large enough compared with the energy of thermal motion, the colloid system is stable; indeed, particles cannot approach one another to distances at which the forces that are driven by induced dipole–dipole interactions and are inversely proportional to particle separation raised to the fourth power become significant (van der Waals interaction)—see Fig. 2.9.

The potential barrier hindering adhesion of particles is quantitatively referenced to the ζ potential, which can be defined as the potential difference between the surface of a particle and the boundary of the diffuse layer of oppositely charged ions (Fig. 2.8).

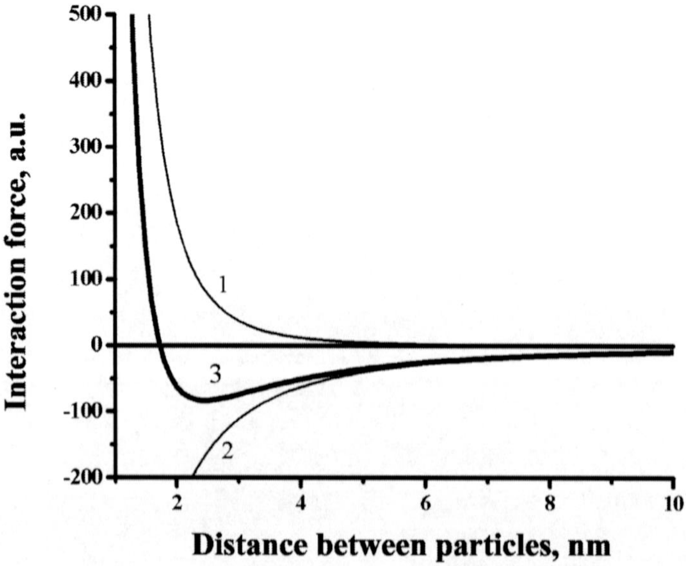

Figure 2.9 Forces involved in interaction between two particles with a EDL. 1, van der Waals forces of attraction; 2, electrostatic repulsion; 3, total force of interaction.

The sign of the ζ potential is determined by that of the particle surface charge.

The ζ potential can be measured experimentally by studying the electrokinetic phenomena, such as particle drift in an electric field (the electrophoretic potential) or formation of a potential difference when a liquid flows through the porous layer of the material under study (the streaming potential).

For reasons of a purely technical nature, one measures presently almost exclusively the gliding potential, and it is this potential that is usually referred to as the ζ potential. A more detailed popular description of methods employed presently in measurements of the ζ potential can be found in Ref. [25] and of the electroacoustic method used for the same purpose in Ref. [26].

Studies of the stability of DND water suspensions showed convincingly [27, 28] that at room temperature the ζ potential $|\zeta| > 30$ mV is large enough to ensure resistance to coagulation. Incidentally, the sign of the surface charge itself was found to be insignificant. We may note here that coagulation of DND particles is not always an instantaneous process, which allows formation of metastable states.

Formation of an excess charge on the surface of colloidal particles, as well as the concomitant appearance of the EDL, may be caused by different physical factors, such as selective adsorption of ions from the solution, interaction of surface chemical groups with water, or selective disruption along an appropriate lattice plane initiated by formation of small particles.

In the particular case of nanodiamond, formation of a EDL is believed to be caused by interaction with the water of functional groups coupled chemically with the nanodiamond surface [22].

The state of the nanodiamond surface subjected to interaction with water is probed by dzeta-metric titration, that is, by studying the dependence of the magnitude of the ζ potential on the pH of the medium [29].

In Fig. 2.10, curve 1 visualizes the dependence of the ζ potential on the pH of the suspension of dry-synthesized DND isolated from detonation soot by oxidation with a chromium-based mixture.

The curve is seen to follow a complex pattern that includes even reversal of the sign of the ζ potential. For this nanodiamond suspension sample, stability regions are observed to exist for pH $<$ 2–7 and pH $>$ 10.

Such a titration curve is characteristic of all DND suspensions isolated from detonation soot by the wet technique; this pattern can be explained as due to the presence of several functional groups on the surface of nanodiamond particles.

This explanation finds support from IR spectroscopy measurements [30].

Diamond is known not to have intrinsic absorption lines in the IR region because of the very high symmetry of its crystal lattice (deformation of such a lattice does not entail any changes in the dipole moment), and therefore, the absorption bands observed on curve 1 originate either from adsorbed compounds or from surface-attached functional groups (see Fig. 2.11).

The intense bands centered close to 3,400 cm^{-1}, 1,600 cm^{-1}, and 1,100 cm^{-1} should be assigned to vibrations of adsorbed water molecules. These bands coincide also with the OH group vibrations,

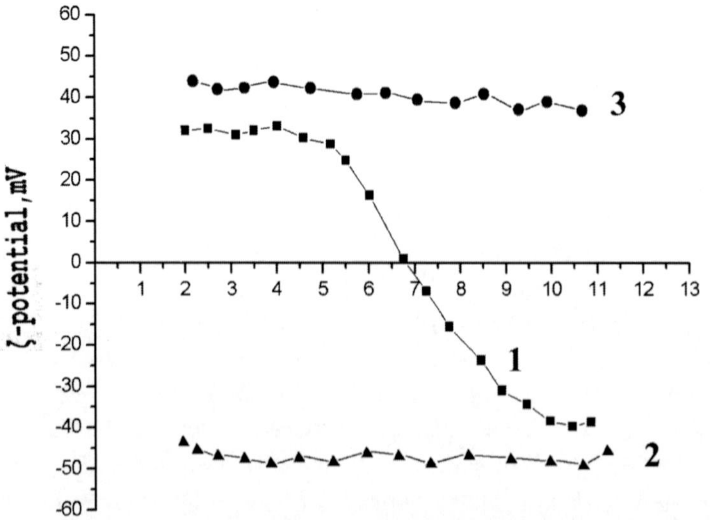

Figure 2.10 The ζ potential of DND particles plotted vs. pH. 1, industrial DND produced by dry synthesis; 2, deagglomerated DND prepared by oxidation; 3, deagglomerated nanodiamond obtained by reduction.

which makes unambiguous identification of the group, adsorbate or OH, they belong to a difficult problem. The band at 1,750 cm^{-1} can be attributed to the C=O stretching vibration of the carboxyl group. The bands at 3,020 cm^{-1} and 1,250 cm^{-1} belong to stretching and bending vibrations of the C–H bonds. In the 1,400–1,000 cm^{-1} region one can see a broad band, which is usually assigned to vibrations of interacting oxygen-containing functional groups that have not been reliably identified. There are also several additional bands (e.g., the ones at 1,050 cm^{-1} and 1,350 cm^{-1}), which can be associated with vibrations of complex ether or anhydride groups.

To sum up, the data obtained by IR spectroscopy evidences the existence of electrically active groups, at the very least, of two types, more specifically, of the carboxyl COOH and hydrocarbon CH, on the surface of a DND particle. We tend to associate the appearance of the negative ζ potential with the presence of the carboxyl, and of the positive ζ potential, with CH groups bonded to the surface of DND particles.

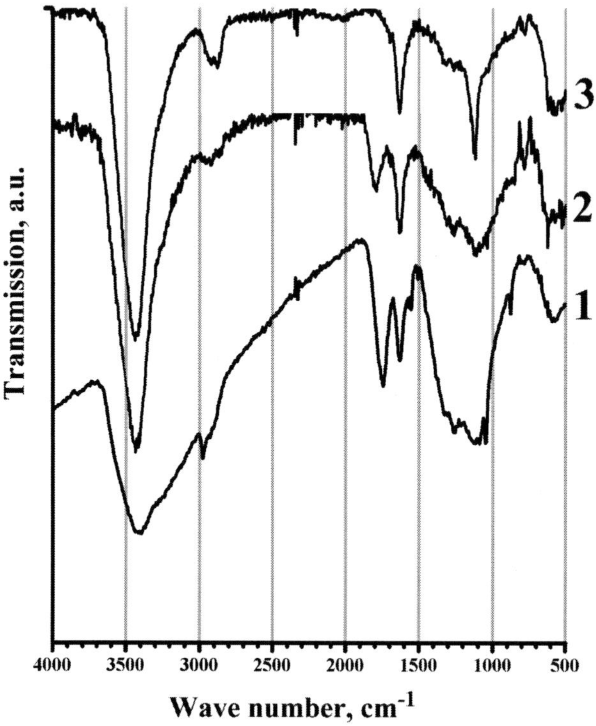

Figure 2.11 IR spectrum of nanodiamond. 1, industrial nanodiamond obtained by dry synthesis; 2, deagglomerated nanodiamond prepared by oxidation; 3, deagglomerated nanodiamond obtained by reduction.

As established already a long time ago, the surface of any diamond containing hydrogen in the form of CH, CH_2, or CH_3 acquires a positive potential in interaction with water [31]. This phenomenon is widely employed in diamond electrodes and chemical field-effect transistors/ion-sensitive field-effect transistors (ChemFETs/ISFETs). Although it is of considerable practical significance, the mechanism underpinning the formation of a positive potential on a diamond surface and, accordingly, of negative ions in the EDL remains unclear. The fact itself of the existence on the surface of such groups is remarkable because the hydrogenized nanodiamond surface is not very stable even in the air environment. But when on the surface of nanodiamond, they

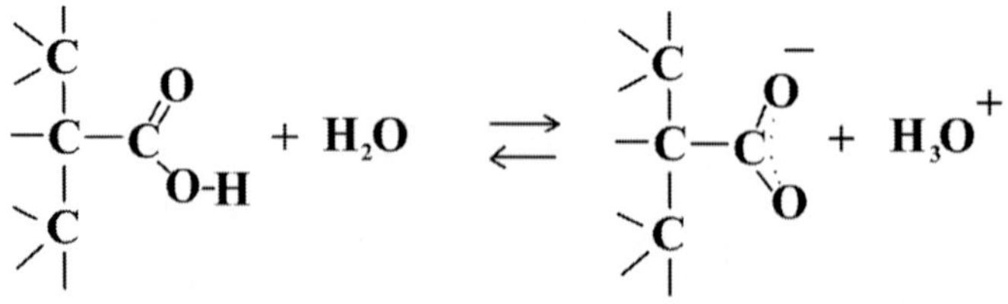

Figure 2.12 Electrolytic dissociation of a bonded carboxyl group.

resist complete removal even in such an aggressive oxidizer as the chromium mixture, where one would expect to reach complete oxidation.

The appearance of a negative potential on the surface of a DND particle, by contrast, does not seem strange and is assigned to the presence on the surface of immobilized carboxyl groups, which produce on the surface of nanodiamond in the course of electrolytic dissociation in water (Fig. 2.12) a negative potential, with a concurrent accumulation in the EDL of hydroxonium ions $(H_3O)^+$.[a]

This structure of the nanodiamond surface gives one grounds to class it among amphoteric ion exchangers. This approach furnishes an insight into the other properties of nanodiamond associated with aggregation.

The size of nanodiamond particles in a liquid is affected noticeably, among other factors, by ions of heavy metals, in particular, by iron, which are present in industrial nanodiamond. This phenomenon is attributed to ions of polyvalent metals being bonded to carboxyl groups in a bridging mode (Fig. 2.13) [22].

Another significant factor that defines the size of nanodiamond particles in a suspension is the total concentration of ions in the solution, whose electric fields affect the structure and characteristics of the solvent (dielectric constant, polarizability, viscosity). At low concentrations, this effect may be neglected, in which case we are dealing with ideal or infinitely diluted solutions. By contrast, if the total ion concentration is high enough (>0.5 mole/L), one has to invoke the Debye–Hückel theory of strong electrolytes, which takes into account changes in the solvent properties. This factor

[a]The hydroxonium ion is sometimes denoted by H^+, which is certainly wrong.

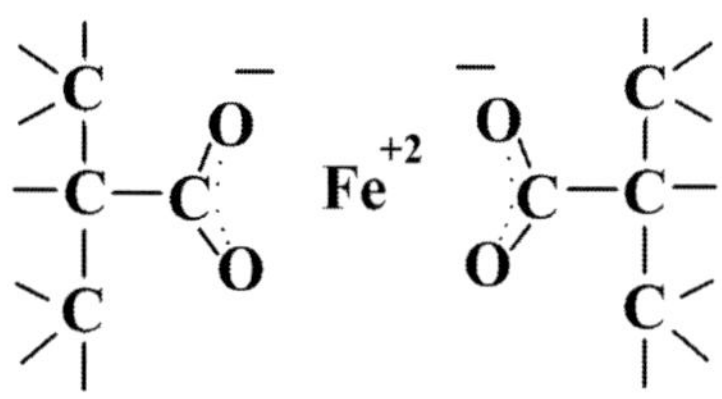

Figure 2.13 Nanodiamond particles coupled by a bridging iron ion.

can be quantitatively estimated in terms of the ionic strength of the solution, which is the half-sum of the products of the concentrations of all ions in the solution by their squared charge. For solutions of salts of simple single charged ions (NaCl) this factor is numerically equal to concentration. For more details bearing on this concept, the reader is referred to textbooks on physical chemistry [23–24].

Washing or any other process of removal of impurities from nanodiamond suspensions initiates all the three factors, improving the hydrosol stability, namely, dissociation of surface functional groups, reduction of the ionic strength of the solution, and removal of multicharged ion impurities. Indeed, in high-quality purification of industrial nanodiamond with micron-size particles one succeeds in obtaining a hydrosol with an average size of about 100 nm, from which a 20–50 nm fraction can be isolated by centrifuging, although in a small amount [32, 33].

Such a behavior is anything but ordinary for inorganic colloidal systems, in which the EDL forms, as a rule, by adsorption of ions from the solution and exists in equilibrium with the latter. Such systems are stable only for high enough electrolyte concentrations in water; in the case of dilution, conventional ashes with a EDL produced by adsorption coagulate.

Attempts at reducing the size of DND particles in suspensions below 100 nm by traditional methods, such as application of high-power ultrasonics or addition of surface-active substances, have not thus far met with success. Numerous experiments aimed at detection of 4–5 nm particles in DND hydrosols likewise have failed; this has stimulated the development of the concept of primary particles, 4–5 nm in size, of primary agglomerates with sizes below 100 nm and large secondary aggregates.

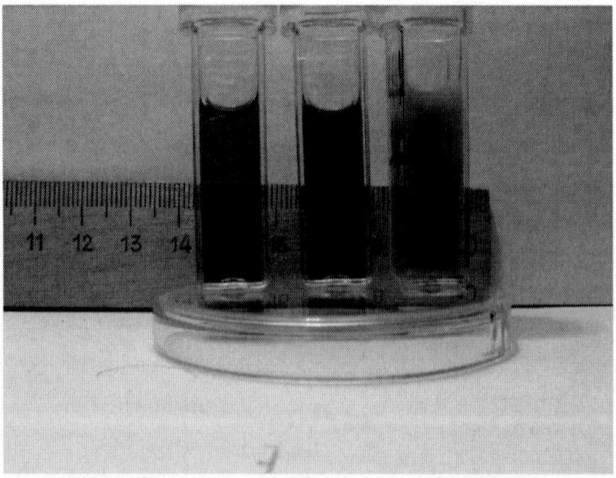

Figure 2.14 Deagglomerated nanodiamond: left to right–samples of $\zeta+$ type, of ζ type and of the starting nanodiamond of approximately the same concentration (0.2–0.3 wt.%).

There are reports of successfully breaking up the secondary and, partially, the primary aggregates by additional purification of industrially produced DND powder, followed by its ultrasonic treatment. One has succeeded also in isolating small fractions 20–50 nm or 60–80 nm in size by centrifuging [33]. But the need for preparing suspensions of 4–5 nm primary DND particles forced scientists to develop new methods.

This initiated a search for appropriate ways of breaking up agglomerates of DND aimed at obtaining free 4 nm particles. This problem remained unsolved for a long time until Ōsawa has succeeded in 2004 in realizing deagglomeration by milling the starting nanodiamond in mills with micron-size balls of zirconium oxide of cubic modification [34]. In 2010, two other methods of deagglomeration were demonstrated practically simultaneously [27, 28].

The method based on oxidation deagglomeration makes use of annealing commercial DND powder, purified preliminarily from magnetic impurities, in air at a temperature above 400°C, followed by ultrasonic treatment in purified water and subsequent isolation of the 4 nm fraction by centrifuging [27].

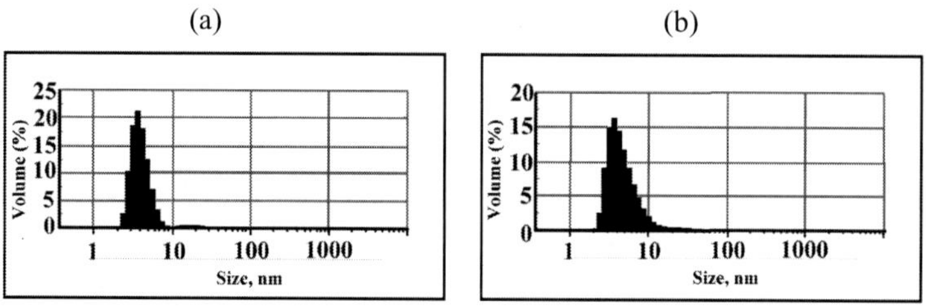

Figure 2.15 Volume distribution of particles of deagglomerated nanodiamond in water suspension. (a) obtained in the oxidation process (ζ - type); (b) obtained in the reduction process ($\zeta +$ type).

Reduction deagglomeration involves annealing of nanodiamond in hydrogen at a temperature of 500°C, with subsequent ultrasonic treatment of the annealed nanodiamond in purified water and isolation of the 4 nm fraction by centrifuging [28].

Hydrosols of deagglomerated nanodiamond are dark-brown liquids with a very weak opalescence, which can be traced primarily to residual impurities of agglomerates. The viscosity of the hydrosol grows nonlinearly with its concentration to convert it into a gel of black color at concentrations above 5 wt.%.

Significantly, the processes involved in DND deagglomeration contribute to purification as well. The residual ash content of the nanodiamond isolated from deagglomerated hydrosol is, as a rule, below the detection threshold (0.1%). As demonstrated by our studies, hydrosols of deagglomerated nanodiamond remain stable under storage and do not change their properties for a year [35].

Figure 2.15a,b displays size distributions of particles of deagglomerated nanodiamond (obtained by the dynamic light scattering technique) relating to the oxidation and reduction processes. In both cases, the maximum of the distribution falls on the primary particle size of ~4–5 nm.

Curve 2 in Fig. 2.10 relates the dependence of the ζ potential to the value of pH for deagglomerated nanodiamond obtained by the oxidation process.

We see immediately that the ζ potential remains negative over a broad range of pH values, which strongly suggests that there is only one functional group on the surface of the particles. This conclusion is borne out by the pattern of the IR spectrum obtained for an isolated deagglomerated nanodiamond (Fig. 2.11).

A ζ potential of ~40–50 mV is strong enough to ensure proper hydrosol stabilization.

Just as in the case of agglomerated nanodiamond, one observes here bands attributable to adsorbed water. The band at 1,750 cm^{-1} associated with the C=O stretching vibration of the carboxyl group is well developed. The intensity of the 1,400–1,100 cm^{-1} bands is markedly below the starting level. The 3,020 cm^{-1} and 1,250 cm^{-1} bands deriving from C–H bond vibrations are extremely weak in intensity.

Curve 3 in Fig. 2.10 displays the dependence of the ζ potential on pH for deagglomerated nanodiamond produced by the reduction process. While the ζ potential is seen to maintain a constant magnitude over a broad range of pH variation, as in the preceding case, it reverses sign. The pattern of the IR spectrum obtained for isolated deagglomerated nanodiamond (curve 3 in Fig. 2.11) suggests convincingly the relation of the ζ potential to C–H groups. Just as in the case of agglomerated nanodiamond, the spectrum reveals bands of adsorbed water and the strong bands at 3,020 cm^{-1} and 1,250 cm^{-1} associated with C–H bond vibrations. The broad band at 1,400–1,100 cm^{-1} is here fairly weak. There are no additional lines attributable to oxygen-containing groups like those seen in curve 2.

A ζ potential of +35–40 mV is strong enough to make the hydrosol stable.

With the above two cases of a ζ potential of a large magnitude and with opposite signs, we have exhausted possible methods of electrostatic stabilization of hydrosols. This poses an intriguing question of the factor that provides a high value of the ζ potential for deagglomerated nanodiamond obtained by milling [36].

The ζ-metric curve presented in the above-cited paper reveals a complex pattern of the ζ (pH) dependence with a high value of the ζ potential in the pH=3–6 region, where the deagglomerated suspension remains stable. For pH values outside of this region, particles are observed to coagulate. Thus the pattern observed here is similar to that of curve 1 in Fig. 2.10, although with a different maximum magnitude of the ζ potential.

The IR spectrum of the DND powder isolated from the suspension [36] resembles qualitatively that shown in Fig. 2.11 (curve 1). One observes also strong vibration bands of adsorbed water. The fairly strong band at 1,750 cm^{-1} relates to the stretching vibration of the C=H carboxyl group. The bands at 3,020 cm^{-1} and 1,250 cm^{-1} attributable to the stretching and bending vibrations of C–H bonds have a high intensity. As revealed by IR spectroscopy, the surface of particles of deagglomerated DND produced by milling likewise has, as a minimum, electroactive groups of two types, of the carboxyl COOH and the hydrocarbon C–H, but with a different relative content. Such an IR spectrum is capable already of accounting for the maximum value of the ζ potential as due to the relative content of the CH groups in the sample studied being higher.

The region of stability of the milled sample is, because of the complex composition of its surface, naturally more narrow compared with samples having a chemically modified surface. It is as obvious that far from any sample of industrial nanodiamond can provide a value of the ζ potential high enough to ensure stability of a DND suspension.

To sum up, data accumulated by now suggests that the surface of conventional DND has a complex structure with several types of bonded functional groups. Strong agglomerates can be disrupted only if the chemical composition of the surface can provide a strong enough ζ potential of the EDL required to stabilize a water suspension of DND particles. By purposeful chemical modification of the surface it is possible to reach a stable and large value of this potential, thus opening the way to obtaining a deagglomerated nanodiamond suspension that would be stable over a broad range of operating conditions.

Regrettably, the experimental progress reached up to now in deagglomeration has not brought us closer to understanding the nature of the strong agglomeration of DND particles and of the mechanism involved in deagglomeration, which makes this area a challenging field of research.

2.3.6 *Methods Employed in Control of the Purity of Detonation Nanodiamond*

Control of the impurity content is naturally the primary issue in obtaining high-purity nanodiamond. The wide variety of possible impurities makes selection of one universal method of analysis, which would permit high-precision determination of the nano-diamond composition in one measurement, an impossible task. Industry and laboratory practices resort to a variety of both chemical and physical methods, which are specific for a particular kind of contamination.

An overall estimate of the composition and degree of purity of nanodiamond can be obtained by X-ray diffraction analysis, which permits one to reveal any kind of impurity with a crystal lattice and evaluate the content of amorphous carbon.

X-ray diffraction is capable, however, of establishing the presence of impurities only on a semiquantitative level, with an accuracy of not better than 20%. The actual threshold of detection by this technique depends on the impurity and is in the range of a few percent.

To estimate the content of nondiamond forms of carbon in detonation soot and nanodiamond one can employ also a chemical method involving trial oxidation of the product by chromium oxide. This analysis involves treating the specimen product with a certain amount of standardized oxidizer mixture, more specifically a solution of chromium oxide in sulfuric acid. It is assumed that this procedure guarantees complete oxidation of NDCFs, with the component to be determined not left in the reaction products. On completion of oxidation, the remnant of the oxidizer present in the solution is found by titration with Mohr salt, $FeSO_4 \cdot (NH_4)_2SO_4 \cdot 6H_2O$, an easily oxidizable reagent.

Knowing the amount of the oxidizer left, one can readily derive that of the chromium oxide spent and, accordingly, the amount of the oxidized carbon by the reaction $3C + 4CrO_3 \rightarrow 3CO_2 + 2Cr_2O_3$. The accuracy of the method claimed to be on the order of 0.25%, it is not able to yield reliable results at lower contents of nondiamond carbon.

The total content of noncarbon impurities is estimated in most cases from the ash content of the nanodiamond. This technique is based on burning a nanodiamond sample in a muffle furnace in standard conditions (1,000°C). In this procedure, all forms of carbon and the carbides become oxidized. The ash obtained contains the nonvolatile oxides only. It is their mass referred to as the total mass of the sample that is the result of the analysis. The accuracy and detection threshold characterizing this technique are usually 0.1%; they can, however, be reduced to 0.02% if one employs crucibles of nonhygroscopic materials (thorium dioxide) and high-precision scales.

Significantly, burning is accompanied by concentration of the impurities, so after determination of the ash composition one can conduct qualitative identification of the noncarbon contaminants.

X-ray fluorescence analysis[a] is used to determine the elemental composition of complex samples for elements heavier than Na. X-ray fluorescence measurements offer a possibility of qualitative identification of almost any impurity in nanodiamond, irrespective of their crystal structure. Quantitative analysis is possible only if calibration against standard samples can be carried out, which for nanodiamond have not thus far been developed. To increase the sensitivity, determination of the elemental impurity composition can naturally be conducted in the ash residue.

The content of the light elements H, N, and C can be derived by the classical methods of elemental analysis, more specifically by burning a sample in an oxygen flow in standard conditions, with subsequent chemical or gas chromatographic identification of the reaction products—H_2O, N_2, and CO_2.

The most sensitive techniques employed in determination of the impurities of transition metal ions are based on the electron paramagnetic resonance (EPR) and nuclear magnetic resonance (NMR) methods. These methods, as applied to nanodiamond, are treated in considerable detail in Ref. [21] and Chapters 6 and 7.

By properly combining all the above methods, one can reach a good enough estimate of the degree of nanodiamond purity.

[a]This method being amply described in the literature, we are not going to treat it here in more detail.

References

1. Fickett W., and Davis W. (2000) *Detonation: Theory and Experiment*, 2nd Ed. (Dover Publications, Mineola, New York, USA).
2. Dremin A. N. (1999) Toward Detonation Theory (Shock Wave and High Pressure Phenomena) (Springer, Heidelberg).
3. Mader Ch. (2007) *Numerical Modeling of Explosives and Propellants*, 3rd Ed. (CRC Press, Boca Raton, Florida, USA).
4. Committee on Advanced Energetic Materials and Manufacturing Technology/National Research Council of the National Academies (2004) *Advanced Energetic Materials* (The National Academic Press, Washington, DC, USA).
5. Sikler R. (1992) Explosives Principles: An Essential Guide to Understanding Explosives and Detonation (Paladin Press, Boulder, Colorado, USA).
6. Bundy F., Hall H., Strong H., and Wentorf R. (1955) Man-made diamonds, *Nature*, **176**, pp. 51–57.
7. Urizar J., and Smith L. (1961) Detonation velocity of pressed TNT, *Phys. Fluids*, **4**, pp. 262–264.
8. Vul' A. Ya., Aleksenskiy A. E., and Dideykin A. T. (2009) *Nanoscience and Nanotechnologies*, eds. Harkin V., Bai C., and Kim S.-C., "Detonation nanodiamonds: technology, properties and applications" (Eolss, UK; Oxford, UK).
9. Staver A., Gubareva N., Lyamkin A., and Petrov E. (1984) Ultrafine diamond powders made by the use of explosion energy, *Combust. Explo. Shock+*, **20**(5), pp. 567–570.
10. Danilenko V. (2006) *Ultra Nanocrystalline Diamond*, eds. Shenderova O., and Gruen D., "On the discovery of detonation nanodiamond" (William Andrew, Norwich, New York, USA).
11. Dolmatov V. (2006) *Ultra Nanocrystalline Diamond*, eds. Shenderova O., and Gruen D., "Synthesis and post-synthesis treatment of detonation nanodiamond" (William Andrew, Norwich, New York, USA).
12. Aleksenskii A., Baidakova M., Vul' A. Ya., Davydov V., and Pevtsova Yu. (1997) Diamond-graphite phase transition in ultradisperse-diamond clusters, *Phys. Sol. State*, **39**, pp. 1007–1015.
13. Danilenko V. V. (2004) On the history of the discovery of nanodiamond synthesis, *Phys. Sol. State*, **46**, pp. 595–601.

14. Titov V., Anisichkin V., and Mal'kov I. (1989) Synthesis of ultradispersed diamond in detonation waves, *Combust. Explo. Shock+*, **25**(3), pp. 372–379.
15. Isakova V., and Isakov V. P. (2004) Thermal oxidation initiated in explosive detonation products by exothermal pyrolysis of metal acetylacetonates, *Phys. Sol. State*, **46**, pp. 622–624.
16. Chiganov A. S. (2004) Selective inhibition of the oxidation of nanodiamonds for their cleaning, *Phys. Sol. State*, **46**, pp. 620–623.
17. Petrov I., Shenderova O., Grishko V., Grichko V., Tyler T., Cunningham G., and McGuire G. (2007) Detonation nanodiamonds simultaneously purified and modified by gas treatment, *Diam. Relat. Mater.*, **16**, pp. 2098–2102.
18. Petrov I., and Shenderova O. (2006) *Ultra Nanocrystalline Diamond*, eds. Shenderova O., and Gruen D., "History of Russian patents on detonation nanodiamonds" (William Andrew, Norwich, New York, USA).
19. Aleksenskiy A., Baidakova M., Osipov V., and Vul' A. Ya. (2009) *Nanodiamonds: Applications in Biology and Nanoscale Medicine*, ed. Ho D., "The fundamental properties and characteristics of nanodiamonds" (Spinger, Heidelberg), pp. 55–77.
20. Pichot V., Comet M., Fousson E., Baras C., Senger A., Le Normand F., and Spitzer D. (2007) An efficient purification method for detonation nanodiamonds, *Diam. Relat. Mater.*, **16**, pp. 2098–2107.
21. Vul' A. Ya. (2006) *Ultra Nanocrystalline Diamond*, eds. Shenderova O., and Gruen D., "Characterization and physical properties of UNCD particles" (William Andrew, Norwich, New York, USA).
22. Aleksenskii A., Vul' A. Ya., and Yagovkina M. (2004) Intercalation of ultrafine-dispersed diamond in aqueous suspensions, *Phys. Sol. State*, **46**(4), pp. 685–686.
23. Holmberg K. (ed.) (2002) *Handbook of Applied Surface and Colloid Chemistry* (John Wiley & Sons, Hoboken, New Jersey, USA).
24. Shaw D., ed. (1992) *Introduction to Colloid and Surface Chemistry*, 4th Ed. (Elsevier Science BV, Amsterdam).
25. Xu R. L. (2006) *Particle Characterization: Light Scattering Methods* (Particle Technology Series) (Kluwer Academic Publishers, Dordrecht, the Netherlands).
26. Dukhin A., and Goetz P. (2002) Ultrasound for Characterizing Colloids: Particle Sizing, Zeta Potential, Rheology (Elsevier Science BV, Amsterdam).

27. Aleksenskii A., Eydelman E., and Vul' A. Ya. (2011). Deagglomeration of detonation nanodiamonds, *Nanosci. Nanotechnol. Lett.*, **3**, pp. 68–74.
28. Williams O., Hees A., Dieker C., Jäger W., Kirste L., and Nebel C. (2010) Size-dependent reactivity of diamond nanoparticles, *ACS Nano*, **4**, pp. 4824–4830.
29. Xu X., Yu Z., Zhu Y., and Wang B. C. (2005) Influence of surface modification adopting thermal treatments on dispersion of detonation nanodiamond, *J. Solid State Chem.*, **178**, pp. 688–694.
30. Osipov V., Aleksenskiy A., Shames A., Panich A., Shestakov M., and Vul' A. Ya. (2011) Infrared absorption study of surface functional groups providing chemical modification of nanodiamonds by divalent copper ion complexes, *Diam. Relat. Mater.*, **20**(8), pp. 1234–1238.
31. Fujishima A., and Einaga Ya. (2003) *Diamond Electrochemistry* (Elsevier Science BV, Amsterdam; BNC, Tokyo).
32. Chukhaeva S. I. (2004). Synthesis, properties, and applications of fractionated nanodiamonds, *Phys. Solid State*, **46**, pp. 625–630.
33. Shenderova O., Petrov I., Walsh J., Grichko V., Grishko V., Tyler T., and Cunningham G. (2006) Modification of detonation nanodiamonds by heat treatment in air, *Diam. Relat. Mater.*, **15**, pp. 1799–1804.
34. Krueger A., Kataoka F., Ozawa M., Fujino T., Suzuki Y., Aleksenskii A. E., Vul' A. Ya., and Ōsawa E. (2005) Unusually tight aggregation in detonation nanodiamond: identification and disintegration, *Carbon*, **43**, pp. 1722–1726.
35. Hees J., Kriele A., and Williams O. (2011) Electrostatic self-assembly of diamond nanoparticles, *Chem. Phys. Lett.*, **509**, pp. 12–15.
36. Sakurai H., Ebihara N., Ōsawa E., Takahashi M., Fujinami M., and Oguma K. (2006) Adsorption characteristics of a nanodiamond for oxoacid anions and their application to the selective preconcentration of tungstate in water samples, *Anal. Sci.*, **22**, pp. 357–362.

Chapter 3

Methods of Characterization and Models of Nanodiamond Particles

Marina V. Baidakova

Center of Nanoheterostructure Physics, Ioffe Physical-Technical Institute, 26 Polytechnicheskaya, St. Petersburg 194021, Russia

baidakova@mail.ioffe.ru

This chapter will trace the variation of our ideas bearing on the structure of nanodiamond particles, starting with the publication of the first papers on this subject, and how the results obtained by structure-sensitive methods and in theoretical calculations have contributed to the formation of the present-day concepts.

3.1 Structure of Detonation Carbon Particles

As already mentioned in the first chapter, detonation-based synthesis of detonation nanodiamond (DND) particles was discovered more than 40 years ago, but its significance was broadly recognized from publications that appeared in the late 1980s [1, 2]. It appears only natural that the original studies of DNDs dealt with development of the technology of their fabrication and chemical purification [1].

Detonation Nanodiamonds: Science and Applications
Edited by Alexander Ya. Vul' and Olga A. Shenderova

ISBN 978-981-4411-27-1 (Hardcover), 978-981-4411-28-8 (eBook)
www.panstanford.com

The structure of single particles in detonation carbon (DC) powder (detonation soot) and of the DNDs obtained after chemical purification of DC and investigated by high-resolution transmission electron microscopy (HRTEM) has been dealt with in numerous publications [3–5]. A noticeable part of these studies addressing the change the structure of these particles undergoes in the DND–graphite phase transition were conducted by the research group headed by V. Kuznetsov [6, 7].

The HRTEM measurements provided convincing support to the viewpoint that DND powder particles are actually diamond crystallites featuring a narrow distribution in size for an average dimension of 4–5 nm and that these crystallites can be coated, to a greater or lesser extent, by defective nanocarbon layers [4]. As evident from the microphotograph in Fig. 3.1, DC is actually a polycrystalline nanodiamond material; indeed, one clearly sees randomly arranged crystallites. The absence of any specific directions in their orientation can be readily verified by tracing the crystal planes making up the lattice, with a characteristic interplanar separation of 0.206 nm for (111)-type planes. The images of the crystallites are superposed on one another, and the thickness of a sample is not uniform over its area, which makes unambiguous conclusions concerning the shape of the crystallites proper and, all the more, the structure of their surface fairly difficult to derive. HRTEM studies provided convincing evidence for the crystal perfection and uniformity in size of nanodiamond crystallites in DND powder, although observations of crystallites with defects (see Fig. 5.3 in Chapter 5) and twinned crystals were also reported [4, 5].

The small size, 10–100 nm, of the image provided by HRTEM imposes certain limitations on powder studies. One meets, in particular, with difficulties in arriving at an unambiguous conclusion on the structure, namely, whether the structure revealed is characteristic of the specific region or of the object as a whole. This is an important point because quite frequently samples, and particularly DC powders, were found to be nonuniform in composition and structure. An additional difficulty encountered in studies by HRTEM, as well as, by the way, by Raman spectroscopy, arises in the possibility of the DND–carbon onion phase transition

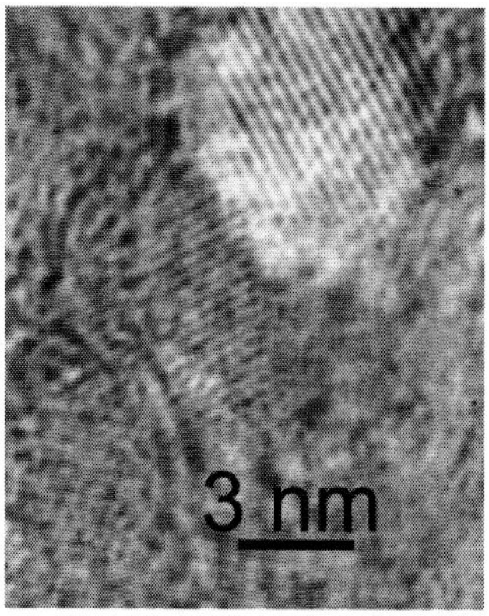

Figure 3.1 HRTEM image of DND powder. One clearly sees (111) crystallographic planes of DND particles. The DND particles are 4–5 nm in size. The author is grateful to A. A. Sitnikova for providing the image.

(see Chapters 1 and 8). Indeed, the formation of structurally perfect onion particles under high-energy electron irradiation was recorded unambiguously [8, 9] (see Chapter 1). This is why HRTEM was usually combined with other experimental methods, for instance, X-ray diffraction and Raman scattering. The researchers were also sure that one could draw from the experience accumulated in structural studies of microcrystalline carbon materials and nanodiamond films and that, in particular, inclusion of the phonon localization effect would permit quantitative description of Raman spectra (see Chapter 5 and references therein).

We are going to demonstrate subsequently the difficulties encountered by authors of the pioneering studies in treatment of both small-angle X-ray scattering (SAXS) curves and of Raman spectra of DND [10, 11] and how the results amassed in recent investigations have permitted one to come to an unambiguous interpretation of the experimental data obtained earlier.

Figure 3.2a displays X-ray diffraction patterns of DND samples obtained by oxidation of DC at various temperatures with such an active oxidizer as 50% nitric acid under pressure. The oxidizer activity could be increased by raising the acid temperature from 180°C to 260°C.

(a)

(b)

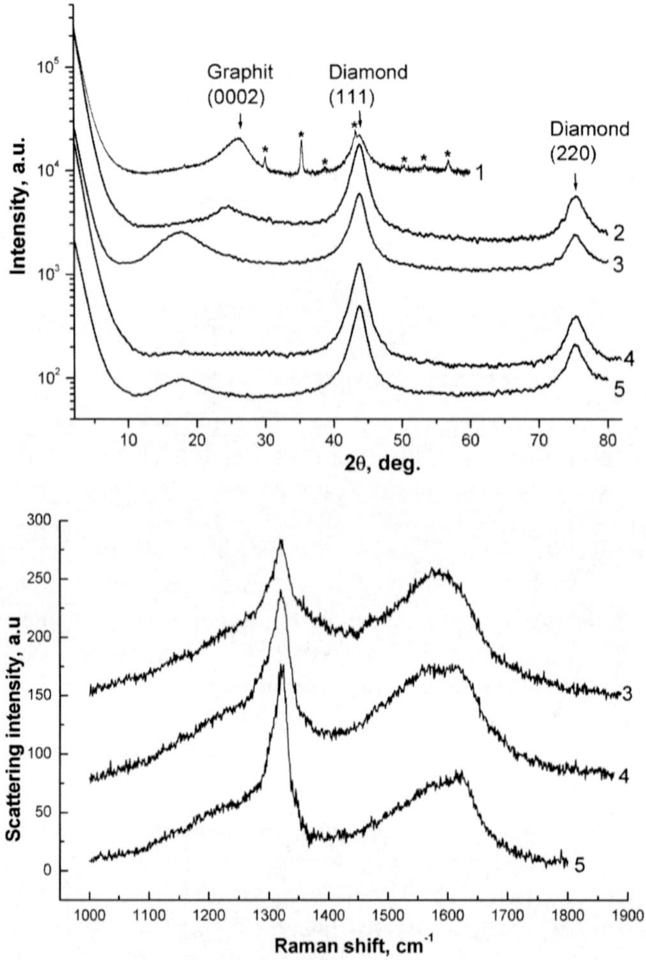

Figure 3.2 Diffraction patterns (a) and Raman spectra (b) of samples of DC obtained in successive oxidations by 50% nitric acid at different temperatures. 1, original DC; 2, 180°C; 3, 200°C; 4, 230°C; 5, 240–260°C. Asterisks identify diffraction maxima corresponding to Fe_3O_4. The X-ray radiation was CuK_α ($\lambda = 1.54$ Å). The curves are offset with respect to one another along the vertical axis for clarity. The author is grateful to V. Yu. Davydov for providing the Raman spectra. *Abbreviation*: Fe_3O_4, iron oxide

An X-ray diffraction pattern is conventionally divided into a SAXS curve (0.5° to 10°) and a diffraction trace obtained at large angles (wide-angle X-ray diffraction [WAXRD], 20° to 100°, CuK_α radiation). The angular interval between 10° and 20° contains information on the so-called average order and is usually disregarded because of the difficulties involved in interpretation of the features it contains.

Note the following two issues of significance encountered in WAXRD and SAXS interpretation.

Analysis of WAXRD curves can yield information, and, quite frequently, quantitative as well, on the phase composition of a sample (e.g., the ratio of the sp^2 to sp^3 phases in DND). But for this information to be representative, the crystallites in the sample should have a perfect enough crystal structure.

An analysis of the shape of SAXS curves may suggest conclusions concerning electron density distribution in a sample and, indirectly, the structure of both a single particle and the agglomerates they may combine to form. Significantly, the particles in the sample should not necessarily have a crystal structure. To illustrate this point, these can be pores in a uniform matrix or dense particles coated by a less dense (e.g., amorphous) shell.

As seen from Fig. 3.2a, DC samples represent a mixture of a diamond (sp^3) with a nondiamond phase, primarily in the form of graphite (sp^2) particles, with metals present as impurities. This becomes manifest in the presence of maxima at 43 grad (diamond), 26 grad (graphite), and narrow peaks originating from metals (asterisked in the figure). By proper treatment with 50% nitric acid at a temperature of 180°C, the graphite and the metal impurities can be removed, which becomes evident in the disappearance of the impurity peaks in the X-ray patterns. Accordingly, the maximum at 26 grad assumes an asymmetric shape and shifts toward smaller angles. This indicates that the treatment does not remove noncarbon two-dimensional structures made up of equidistant carbon nanoplates [10], which are arranged directly on the surface of the diamond particles. The possibility of formation of such structures, which consist of diamond particles coated by a shell of nanographene sheets, was predicted theoretically [12] and demonstrated experimentally [4].

Treatments conducted at temperatures of 200°C and higher leave in the X-ray images (curves 3 and 5 in Fig. 3.2) only the maxima attributable to the diamond phase and some additional maxima (haloes) in the small-angle region at 17 grad. That these maxima resemble haloes permitted their assignment to the presence of a nonuniformity in the electron density distribution over the sample with a characteristic size of 0.3 nm [13]. Because no particles of either the diamond or nondiamond carbon forms could survive the acid treatment, this halo was assumed [13] to indicate the existence of a shell with a complex structure that coats DND crystallites.

Samples oxidized at temperatures above 200°C revealed changes in the small-angle domain as well. Assuming the particles to be of spherical shape and of equal size, these changes were interpreted as resulting from a change in the morphology of the shell coating the DND crystallite; indeed, as the oxidation temperature was increased, the shell would become ever thinner and more uniform in structure [14, 15].

It appeared only natural to assume that the shell is made up of carbon atoms and has in the final stage of purification a structure resembling that of the carbon onion [10, 11, 16].

Additional information on the structure of these samples and its changes with purification was derived from Raman scattering studies (Fig. 3.2b).

There are three main features that should be taken into account in analyzing DND Raman spectra [3, 10, 17]. There is, first, the strong background luminescence generated by laser irradiation; second, the photon energy of the laser employed conventionally in Raman spectroscopy ($\lambda = 0.488$ nm; $h\nu = 2.54$ eV) is close to the energy of the $\pi - \pi^*$ resonance of sp^2-hybridized electrons, and therefore, the intensity of the Raman signal generated by the sp^2 phase is substantially higher than that of the sp^3 phase for equal amounts of the two; and, finally, third, the small size of the crystallite removes the veto on participation in scattering of phonons with wave vectors $q \approx 0$ imposed by the uncertainty relation.

These issues are considered in detail in Chapter 5.

Let us dwell in more detail on the two latest points in order to get a better understanding of how the concept of the DC particle structure evolved.

The Raman scattering spectrum of crystalline graphite was found to contain a single narrow band at 1,575 cm^{-1} [3], which is assigned to the planar E_{2g} mode of the Brillouin zone center [18, 19]. On the other hand, sp^3-hybridized carbon, that is, single-crystal diamond, demonstrates in the Raman spectrum a narrow symmetric line at a frequency of 1,332.5 cm^{-1} (with a width of about 2.0 cm^{-1}) [20], which derives from the transverse TO phonon of F_{2g} symmetry. At the same time, changes in the structure and size of the sp^2-hybridized phase become reflected in those of the spectrum near 1,600 cm^{-1} [21]. Indeed, microcrystalline graphite demonstrates broadening and displacement of the so-called *G* band at 1,600 cm^{-1} toward higher frequencies, accompanied by the appearance of a new band close to 1,620 cm^{-1}. In addition, an additional feature at 1,350 cm^{-1} appears, the *D* band. The broadening and frequency shift of the "diamond" mode band (1,332.5 cm^{-1}) toward lower frequencies are observed to occur in Raman spectra at the transition from bulk to nanocrystalline diamond [22, 23].

The above analysis permits a suggestion that the Raman spectra of DC samples measured after purification at 200°C with 50% nitric acid contain two bands of equal intensity, more specifically one at 1,325 cm^{-1}, which is actually a superposition of the diamond and the *D* band, and another one associated with microcrystalline graphite (*G* band, near 1,600 cm^{-1}). After an increase of the temperature of the treatment to 230°C, the diamond band grows markedly in intensity compared to the *G* band, with the diamond line becoming narrower and a shoulder emerging on its low-energy side.

Because for equal amounts of the phases the Raman intensity of the sp^2 phase signal is substantially higher than that of the sp^3 one, there were grounds to suggest that purification of DC in the above conditions brings about removal of the sp^2-hybridized graphite shell coating the surface of the diamond crystallite. This interpretation was at variance with the observation that the *G*-band intensity still remained noticeable as the acid treatment temperature was increased. Also, nuclear magnetic resonance (NMR) (see Chapter 7) and X-ray photoelectron and Auger spectroscopy [24] provided convincing evidence that DND subjected to oxidative treatment may contain trace amounts of the sp^2-hybridized phase.

Only recently an explanation of the marked intensity of the band at 1,600 cm^{-1} observed after a high-temperature DC treatment, which should have removed the sp^2 phase, has been found [25, 26]. As shown in Section 5.1 (Chapter 5), the Raman band at 1,600 cm^{-1} can be produced by bending vibrations of the OH groups, in other words by water adsorbed on the surface of a DND particle and in pores among them.

The existence of a layer of nanophase water adsorbed on the surface of DND particles in water solutions was demonstrated convincingly in Ref. [27]. The same study revealed the appearance on the surface of a DND particle of a highly ordered structure of adsorbed water and reconstruction of the structure of nanophase water when water was added after drying at a temperature above 630°C. The thickness of such a layer of nanophase water was estimated to be not more than 1 nm.

This suggests that the maximum at 17 grad observed in the diffraction curves (Fig. 3.2) derives from scattering primarily from the shell of nanophase water. This conclusion finds convincing support in the halo intensity being dependent on the sample humidity rather than on the degree of oxidation at which they were obtained.

It goes without saying that such a shell will not contribute noticeably to the small-angle part of the diffraction curve (less than 10 grad) because of the low water density.

Significantly, nanophase water can be contained in pores separating the DND particles as well (see further on). As follows from small-angle neutron-scattering measurements, such nanopores are closed, and therefore, they may contain water even under heating at moderate temperatures [16].

Summing up the information presented in this section, Fig. 3.3 visualizes schematically the structure of the DC particle and traces its variation in the course of the oxidation process.

DC represents a mixture of the nanodiamond and nanographite phases containing metal impurities, as shown in Fig. 3.3a. The diamond core in DC is coated completely by a shell of sp^2-hybridized carbon. The nanographite, metal impurities, and, partially, the sp^2-hybridized carbon can be removed by DC oxidation at 180–200°C, with particles of the oxidized DC representing, as before,

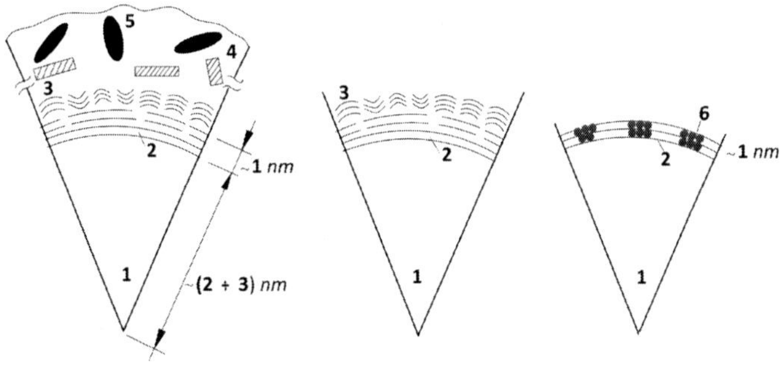

Figure 3.3 Model of a DC structure obtained in successive oxidations by 50% nitric acid. (a) Original DC, (b) DC after removal of graphite particles and impurities of metals and their oxides, and (c) DC after partial removal of the graphite-like shell. The figure identifies 1, diamond core; 2, onion-like carbon layers; 3, nanosized graphite-like platelets; 4, graphite particles; 5, inclusions of impurities and metal oxides; 6, a layer of adsorbed water. Stages (b) and (c) correspond to treatment at temperatures of 180°C and 240–260°C (curves 2 and 5 in Fig. 3.2), respectively. The graphite and metal impurity particles have a large dispersion in size for average sizes of 2 nm and 10 nm, respectively.

a diamond core in a graphite shell (as shown in Fig. 3.3b). Oxidation conducted under more stringent conditions terminates in a practically complete removal of the graphite shell from the surface of the diamond core; as a result, the larger part of its surface becomes hydrophilic, which manifests itself in the formation of a nanophase water shell, as shown in Fig. 3.3c.

It appears appropriate to note here that most of the studies of DC structure and of its variation induced by acidic purification were conducted on powder samples. It was essential that because of the high surface energy the powder of a substance in ultradisperse state consists not of single particles with a characteristic size of a few nanometers but rather of their conglomerates of up to a few microns in size. The DC and DND powders are certainly also no exceptions to the rule (see Chapter 2).

In particular, DND particles immersed in a water suspension form, depending on the pH of the solution, conglomerates measuring from a few hundred to even a few thousand nanometers. Dried

DND powder also combines to produce particles a few hundred nanometers in size [28, 29]. We shall enlarge on the problem of aggregation in more detail below (in the next section).

3.2 Agglomerates of Detonation Carbon Particles

If details in the surface structure of a nanodiamond cluster, and the thickness itself of the surface layer, are determined both by the kinetics of cooling of the detonation product and by the technique employed in chemical purification of DC from the nondiamond phase [14], the nanosized core proper will invariably be detected in DND, primarily by the universally accepted methods of X-ray diffraction [10] and transmission electron microscopy [4, 30]. One should not, however, disregard the fact that both the DC powder and the DND suspension consist of particles of a substantially larger size, as a rule, of about 200 nm. This means essentially that researchers deal not with nanosized but rather with submicron-size diamond particles, or, to be more precise, with aggregates of particles (see Ref. [31] and Chapter 2).

Conventional DND powder is actually a typical fractal structure exhibiting self-similarity under variation of scale [15], a point illustrated graphically in Fig. 3.4. The fractal structure is assembled

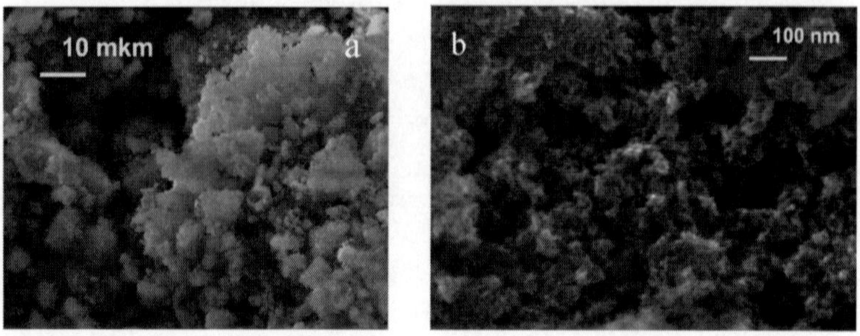

Figure 3.4 SEM patterns of DND powder. The photographs were obtained in different scales. The similarity between the patterns demonstrates the fractal nature of DND aggregates. The author is grateful to D. A. Sakseev for providing the photographs. *Abbreviation*: SEM, scanning electron microscopy.

of different-level aggregates. Earlier studies revealed primary aggregates about 100–200 nm in size [28, 29]. These aggregates combine, in their turn, to form larger but not as strong aggregates, because the density in DND powders, as in any fractal objects, falls off with increasing radius. Incidentally, secondary aggregates are assembled from the primary ones, and this regularity persists for all higher-level aggregates. It is structures of this type that exhibit self-similarity under variation of scale.

Because of the high surface energy involved, in the aggregation itself of nanosized particles into aggregates of submicron and micron size, there is nothing remarkable for a researcher altogether; what is remarkable, indeed, is the strength of the primary aggregates with dimensions of about 100–200 nm [3]. It turned out virtually impossible to break up these aggregates into single 4 nm diamond particles by traditional ultrasonic treatment of a suspension.

One succeeded for the first time in attaining nearly complete (up to 90%) breakup of primary aggregates and obtaining a suspension consisting of single nanodiamond particles (single-particle DND) [32, 33] with the use of a specially designed ball mill, in which DND powder immersed in a water medium was stirred intensely together with micron-size zirconium dioxide balls. This method of grinding (method of Ōsawa) has, however, at least two drawbacks, namely, the formation on each 4–6 nm particle of an sp^2 shell produced in a diamond–graphite phase transition on the surface of a particle as a result of local overheating in the course of grinding and contamination of the surface by zirconium dioxide [34, 35].

Significantly, removal of the zirconium dioxide impurity and of the sp^2 phase brings about reaggregation of particles of the suspension thus obtained [35].

This makes particularly interesting the announcement [36] about the use for grinding of such milling media as water-soluble substances that do not introduce contaminants (various salts and sugar), which would require subsequent purification by chemical methods. In this case, reaggregation does not take place even after drying of the suspension followed by its dispersion in water. The characteristic size of particles in the suspension is 5–20 nm (see Fig. 4 in Ref. [37]).

Speaking about methods employed in partial deaggregation of DND, one should not overlook the method of reducing aggregate size down to 50–60 nm, which involves the "graphitization-oxidation" process [38–40].

Incidentally, different DND applications impose different requirements upon the particle size [31]. Separation of DND into suspensions with particles of different size by centrifugation is described in many publications (see Ref. [37] and references therein). It was shown, in particular [41, 42], that ultrasonic treatment with a high-power sonicator followed by multistep ultracentrifugation permits isolation of a fraction of particles less than 30 nm in size with a yield of about 20 wt.%. It was demonstrated that the stability of a suspension is directly related to the ζ potential of the hydrosol, that is, it depends on surface functionalization [42], with annealing in air at temperatures of 350–450°C facilitating the process of DND fractionalization [41].

These observations received support in a demonstration [43] that treatment of purified DND in oxygen at temperatures of about 450°C, followed by centrifugation, terminates in total DND deaggregation; indeed, it was found possible to isolate a single-particle DND suspension with a maximum in size distribution at about 4 nm (see Chapter 2).

Not less remarkable is the finding that annealing in hydrogen at 500°C likewise facilitates the problem of isolation of a single-particle suspension from dispersed DND solutions in the subsequent centrifugation of the product [44]. Note that this method has obvious advantages because, in contrast to annealing in air, in the case of annealing in hydrogen DND does not burn even at temperatures of up to 800°C. The details of the functionalization responsible for the stability of the suspension in this case remain, however, unclear.

An analysis of the above results suggests that all methods bringing about partial deaggregation permit one to obtain a DND fraction with particle sizes of a few tens of nanometers. Significantly, this size is noticeably smaller than that of the primary aggregates (below 100 nm). This strongly suggests that there should exist structures made up of about 10 DND particles that are stronger and denser than the primary aggregates combining more than thousands of 4 nm particles. The existence of such structures was demonstrated by Ōsawa [45], and they were called core agglutinates.

The difference in DND particle size (from 20 nm to 60 nm) obtained by partial deaggregation can be attributed both to the products used being of different nature ("dry" and "wet" synthesis, method of chemical purification employed) and to different techniques of size determination.

Indeed, atomic force microscopy yielded reliable results for particle sizes of 40–60 nm [28]. In Ref. [42], for instance, instruments of two types were used for the purpose. In more recent studies, where particle sizes less than 30 nm have to be estimated, one presents usually data obtained by the dynamic light scattering (DLS) method. The problems associated with size determination of DND particles in hydrosols by this method are analyzed in Ref. [46]. We note that the characteristic size of the reference used widely in DLS calibration is 60 nm.

Summing up the results obtained in studies of the deaggregation processes, one can, on the basis of Ref. [47], suggest the following model of the structure of the aggregate.

This study analyzed the structure of aggregates in DND water suspensions with concentrations not over 8 wt.% by SAXS. This permitted one to exclude the effect of particle interaction and analyze the small-angle scattering curves in terms of the theory of scattering from single noninteracting spatial structures. Such spatial structures can obviously be either the 4 nm DND particles themselves or fractal aggregates of DND particles of a different order.[a]

Among the most interesting results presented in Ref. [47] is the reconstruction of the shape of the scattering DND particles in a water suspension. Two types of scattering structures were identified, those with a size of about 100 nm and fractal dimensionality $d = 2.2$ (mass fractal) and those with a size of about 20 nm and dimensionality $d = 2.0$. Significantly, the coincidence in size permits identification of two types of structures (in the terminology of Ref. [45]) as primary aggregates and core agglutinates.

---

[a]For the radiation chosen (CuK_α, $\lambda = 1.54$ Å), this method could provide information on scattering structures not over 200 nm in size.

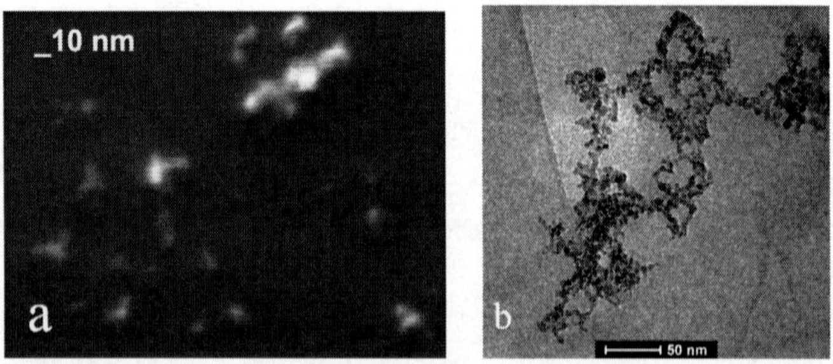

Figure 3.5 SEM and TEM images of DND aggregates that formed in secondary aggregation when suspensions of single 4 nm DND particles were dried on substrates of silicon (a) and graphite oxide (b), respectively. The shape and sizes of the core agglutinates in image (a) correlate with their models [47]. In photograph (b) one clearly sees the chains and loops predicted in Ref. [30]. The author is grateful to D. A. Sakseev and D. A. Kirilenko for providing the images. *Abbreviation*: TEM, transmission electron microscopy.

It was shown that primary aggregates represent actually extended spatial structures identical in shape to those displayed in the SEM microphotographs (see Fig. 3.5a). They represent agglomerates combined of 9–10 core agglutinates, which, in their turn, are dense formations of 4–5 DND crystallites. Also, core agglutinates retain the shape of faceted polyhedra characteristic of DND nanocrystallites.

It should be pointed out that DND fractions consisting of core agglutinates do not differ in structural characteristics from conventional DND powders; indeed, as follows from SEM images, they represent small DND aggregates [41]. It was not stressed either that their coherent-scattering region (CSR) is larger than 4 nm [30]. This suggests that core agglutinates approach most closely in nature polycrystalline assemblies of 4 nm DND crystallites.

Such core agglutinates should in no way be identified with accreted DND crystallites of the same orientation, or large-size crystals, which were observed more than once in TEM images [4] and whose fraction is a few percent. In the latter case, the CSR size determined by X-ray diffraction would have been substantially larger than 4 nm.

Let us see how the ideas concerning the primary aggregates varied with time and what are the forces that "glue" the core agglutinates. The model proposed originally considered the aggregate as a polycrystal combining 4 nm diamond crystallites. The strength of the aggregates was attributed here to the same factor as that holding together conventional polycrystals, that is, to chemical bonds coupling its component grains, which in this particular case are apparently covalent [48].

The recent attempt at relating the strength of the aggregates with the strong Coulomb interaction among spatially oriented DND particles was based on the assumption that these particles have the shape of polyhedra with distinct faceting. The corresponding calculations showed that such an approach is, in principle, valid [30, 49].

Experimental evidence for the significant effect exerted by group III metal impurities on DND aggregation provided a basis for proposing [31] and substantiating [43] a model attributing the strength of the primary aggregates to bridging bonds (see Chapter 2).

While all these models are shown to be well substantiated, they describe the properties of different structural units of the DND aggregate, namely, the core agglutinate, the primary aggregates, and higher-level aggregates (see Table 3.1).

Covalent bonding that couples DND aggregates combined in core agglutinates accounts for their stability with respect to the various attacks employed in deaggregation. And it is this factor that explains the difficulties encountered in earlier attempts to produce a suspension consisting of 4 nm single (primary) particles, whereas fractionation and production of particles 150 nm and even 60 nm in size were successfully realized.

Table 3.1 Structural units of a DND aggregate

Name	Size (nm)	Type of bond	References
Primary particle	4–5	—	
Core agglutinate	20–60	Covalent, electrostatic	[47, 48]
Primary aggregate	ca 150	Chemical, involving metal ions, electrostatic	[31]
Secondary and higher-level aggregates	above 1,000	van der Waals	[3, 31]

Indeed, the looseness and large extent of the structure of primary aggregates are essential prerequisites to their breakup in parts under application of a strong mechanical force, the essence of ultrasonic treatment and centrifugation. Fractionation and partial deaggregation must be preceded by chemical purification and washing with deionized water to remove the various impurities and acid traces that are conducive to aggregation [31, 46].

The polyhedral shape of both the core agglutinates and the DND particles themselves favors creation of Coulomb interaction among them, and this is why additional surface functionalization is required to prevent secondary aggregation in the course of drying of the product. Moreover, particles in the powders obtained after the drying of suspensions of deaggregated DND form extended structures of a regular shape whose existence was predicted by [30, 49].

Figures 3.5b displays clearly primary aggregates in the powders that were produced from suspensions of chemically purified DND fabricated in dry synthesis and from deaggregated DND, respectively. In samples obtained with the detonation product cooling mediated by kinetics different from that governing dry synthesis, the shape of the primary aggregates and, as a consequence, the porosity of large aggregates may be different. Incidentally, the shape of the core agglutinates and of primary aggregates can be studied not only by X-ray diffraction but also by transmission, scanning, and atomic force microscopy; therefore, conducting such studies on a large set of DND samples differing in both the kinetics of cooling of the detonation product and the methods of chemical purification and deaggregation employed would be certainly expedient for construction of a more comprehensive model of the DND aggregate structure.

3.3 On the Structure of the DND Particle

An analysis of the present-day concepts concerning the structure of a single DND particle has become possible only recently, after the successful solution of the deaggregation problems by milling [32, 48] and annealing, with subsequent ultrasonic treatment and centrifugation [43, 44]. While deaggregation has opened a way to producing TEM microimages of single particles, the literature still abounds with heated discussions of what the shape of a single 4 nm DND particle is. One considers versions of both spherical and faceted shapes and some combinations thereof [37]. The interest aroused in this problem is easy to understand; indeed, theoretical calculations suggest that the physicochemical properties of particles and of their hydrosols depend directly on their shape [5, 34, 50]. Indeed, the stability of a particle of spherical shape can be assigned to the presence on its surface of various functional groups, which stabilize the particles by terminating the dangling bonds. By contrast, in the case of a faceted particle one should invoke more complex mechanisms of surface stabilization (relaxation).

As already mentioned, the first microimages of DND particles were obtained by studying aggregated powders. While the particles observed in these images had a spherical shape, this effect appeared through overlap of the images of a large number of particles. That the particle core is actually a perfect diamond crystallite could not be subject to doubt. The only unsolved issue was the structure of the surface of the facets, more specifically to what extent their structure reproduces that of the corresponding surfaces of bulk diamond and whether it is mediated by the relaxation mechanisms typical of these surfaces.

HRTEM data that kept building up with time provided compelling evidence for DND particles being indeed of polyhedral (primarily octahedral or cuboctahedral) shape or as twinned crystals [5, 30, 37]. In this case, the truncated octahedral structure, which is most realistically tractable for mathematical description, is limited to the $\{111\}$ faces by 76% and to the $\{100\}$ faces by 24%. A particle graphitized to less than 76% will be stable in a diamond structure with a thin (one to two monolayers) shell [48]. Following the approach outlined in Ref. [12], this shell was viewed as fullerene-like skullcaps coating the $\{111\}$ faces of a crystallite. Another interesting concept of the idealized DND surface structure was proposed in Ref. [45]. The structure is visualized in the form of a holey giant fullerene with holes accommodating six $\{100\}$ facets. Interestingly, this representation was found to be suitable in describing the properties of single-particle suspensions produced by milling DND aggregates by the Ōsawa method, because this method resulted in additional amorphization (graphitization) of the particle surface.

Investigation of the optical properties of single-particle suspensions [50, 51] obtained by the method described in Ref. [43] resulted in a revision of the above concepts. It was proposed to disregard the effects associated with the curvature of nanodiamond particles because of the existence of clearly pronounced $\{111\}$ facets, in which case the surface of a nanodiamond particle should approach in properties that of a bulk diamond crystal, which was already studied by the time well enough.

It is well known that a decrease of surface energy is accompanied in crystals by surface reconstruction. In particular, neighboring carbon atoms on the $\{111\}$ diamond surface rearrange to form Pandey dimer chains [52] (see also Chapter 4). Thus, in real conditions the surface of DND particles is coated by various functional groups [53, 54] and Pandey chains.

One more effect, predicted theoretically and reinforced experimentally, suggests that DND particles are regularly faceted, thus indicating specific orientation of interfacing particles, which manifests itself in observation of linear superstructures, such as chains and loops formed in the drying of DND suspensions [30]. The mechanism responsible for the strong and long-range interaction among particles in a suspension and terminating in the formation of the above-mentioned linear structures was described in Refs. [30, 49]. It was shown that single 4 nm particles differ in their surface electrostatic potential, which is determined by the type of their faceting. The differences in potential energy characterizing different crystallographic orientations permit one to single out three types of preferential orientation among particles. It was shown that a negative electrostatic potential is thermodynamically preferable for the $\{111\}$ faces, particularly if they are completely graphitized, whereas the $\{100\}$ faces exhibit usually a positive potential. When immersed in a suspension, particles assume predominantly the orientation in which the charged faces would form interfaces with near-neutral $\{110\}$ and, under certain conditions, $\{111\}$ faces of the type $\{100\}|\{111\}$, $\{110\}|\{111\}$, and $\{111\}|\{111\}$.

It is well known that in a cubic crystal the angles between the (100) and (111) planes make up about 36 grad and between the two (111) planes 71 grad or 109 grad. The same angles are known to become reproduced in the shape of microcrystalline diamond particles fabricated, for instance, by chemical vapor deposition (CVD). It appeared only natural that $\{100\}|\{111\}$-type interfaces would have an angle of 109 grad between particles and the $\{111\}|\{111\}$ interfaces an angle of 36 grad, exactly what was observed in Ref. [30]. This results in the formation of linear structures resembling the one displayed in Fig. 3.5b.

Significantly, the $\{100\}|\{111\}$ interfaces were not observed in the HRTEM images presented in Ref. [30]. As an explanation, the authors suggest that in the course of a TEM study the sample becomes charged by the electron beam and that this charging is strong enough to destroy the electrostatic coupling among the $\{100\}|\{111\}$ interfaces, even in the case of the aberration-corrected HRTEM operated at 80 kV, which was employed by them.

The above models of the DND structure have a common underlying assumption of the diamond core crystal lattice consisting of sp^3-hybridized carbon atoms only. A core may be coated, partially or completely, by fragments of sp^2-hybridized carbon. These are the so-called core-shell models. Regrettably, not all experimental results can be described in the frame of this approximation. Indeed, the elegant experiment aimed at determination of the density of a scattering particle by properly selecting the contrast in a neutron-scattering study showed that the density of DND particles is noticeably lower than that of bulk diamond [16]. Significantly, the difference between the densities cannot be accounted for only by the presence of any one of the particle surface shells alone described above, and one has to introduce some transient layer separating the core from the lower-density shell [11, 16, 54–58]. The size of the diamond core is estimated [11, 54] to be 3 nm, with the transient region made up of two to four layers (0.8–1.2 nm) for a particle 3.5–4.5 nm in size.

There are several approaches to describing such a transient layer.

Interatomic separation in the transient region was shown [54, 57] to be smaller than that in the particle core, a characteristic feature for the mixed sp^3/sp^2 bonding of carbon atoms. Note that the ratio of the mixture of sp^3-/sp^2-ordered carbon in the transient region depends on the kinetics of cooling of the detonation product and of the method employed to isolate the DND from the latter [55, 58]. Both the fullerene-like shell and the holey part of the transient region with predominantly sp^2 hybridization can be removed by the corresponding oxidative technique [58]. But the fullerene-like shell forms immediately again because of the $\{111\}$ planes having become exposed [35]. Thus, the structure of the transient layer depends on the method and details of purification employed with the DC.

There is a somewhat different point of view concerning the absence of the sp^2 phase in the transient layer [56]. On comparing the calculations made by molecular dynamics simulation with experimental X-ray diffraction data obtained with synchronous radiation ($\lambda = 0.13$ Å), the authors came to the conclusion that tetrahedral packing transforms gradually into a fullerene-like shell to form a transient layer with distorted sp^3 hybridization. The surface of the $\{111\}$ faces will bear now graphite-like fragments, whereas the $\{100\}$ facets will preserve a diamond-like atomic arrangement.

Let us see how the concepts of the faceted shape of a particle and the existence of a transient layer correlate with the well-known models of formation of diamond nanoparticles in the course of the detonation process, for instance, with the model of DND formation from liquid carbon [4, 37, 59]. Viewed in terms of this model, the particles forming in synthesis from a liquid drop have originally a spherical shape. These particles represent actually a diamond core coated by a transient layer, as well as several agglomerated particles, that is, core agglutinates. Faceting appears as a result of graphitization, reconstruction of the diamond surfaces, and their interaction with the surrounding gases, which accompanies cooling of the detonation products [4]. The faceting of particles and agglutinates occurs with decreasing temperature and pressure in accordance with the different chemical activity and capacity of different diamond crystallographic planes to graphitization. Moreover, the oxidative treatment employed to isolate diamond particles from the DC is accompanied by partial oxidation of the DND

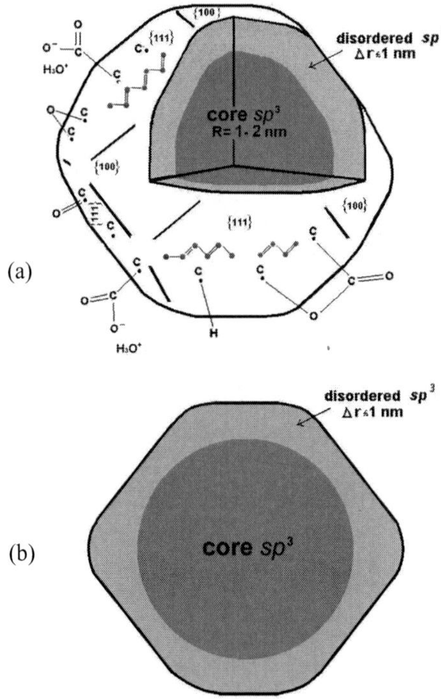

Figure 3.6 Schematic image of a single 4 nm DND particle in the form of a truncated octahedron. (a) Outer structure (with one segment removed), and (b) cross section. The {111} face shows different functional groups and Pandey chains. The cross section shows that the core, spherical in shape, is surrounded by a transient layer of sp^3-disordered carbon. Particle faceting makes the thickness of the transient layer different for the {111} and {100} faces.

surface and leads to additional faceting of particles with the most reactive faces in these conditions [4].

Because the most active in the processes of graphitization and oxidation are the (111)-type planes, it is these planes that form primarily the surface of the primary DND particles, with the thickness of the transient layer on {111} faces remaining the smallest up to emergence onto the core surface.

Significantly, the characteristic particle shape is determined mainly by the kinetics of cooling of the detonation product, and thickness of the transient layer, by the method employed in chemical purification of DC from the nondiamond phase. This is why some researchers report, in particular, on observation of particles of different shapes, up to spherical [38], and of different types of graphite-like fragments on the DND surface.

The above considerations suggest a conclusion that idealized particles have the shape of a polyhedron with distinct faceting, as shown in Fig. 3.6a. The particle is faceted both by partially graphitized $\{111\}$ faces and the $\{100\}$ faces retaining the diamond surface structure and, possibly, by $\{110\}$ faces.

One can single out three parts in the structure of a single DND particle, as shown in Fig. 3.6b. The central part is a diamond core built from sp^3 carbons. Directly adjoining the core is the transient layer formed primarily by disordered sp^3 carbon. This transient near-surface layer is coated partially by a fullerene-like shell or graphite-like fragments of sp^2-hybridized carbon. Carbon atoms arranged on areas not coated by sp^2 fragments bind atoms of carbon or oxygen to form various functional groups that saturate the dangling bonds.

It is the diamond core that accounts for the main "diamond" DND properties, more specifically its thermal and chemical stability, thermal conductivity, low electrical conductivity, quasi-abrasivity, and quasi-hardness; the existence of the transient layer offers an explanation for the unique electronic properties of DND. The outer functionalized coating is responsible for such, certainly not less important, characteristics of the substance as absorption, adsorption, chemisorption, and colloidal DND stability in liquid media [58].

3.4 Conclusion

Summing up, it should be stressed that one has succeeded in obtaining reliable structural data on such a complex material as DND only through a complex application of a variety of methods, including Raman scattering of light; small-angle X-ray and neutron scattering; and scanning, transmission, and atomic force microscopy.

It has been established that commercially produced DND is an aggregate of a complex hierarchical structure. A specific feature of

DND aggregates is their containing water, both adsorbed on the surface of particles and contained in closed pores. Interestingly, the presence of water is detected by structurally sensitive methods, which finds ready explanation in both DND particles themselves and the adsorbed water being nanometer scaled.

A single 4 nm DND particle has a complex structure made up of a spherical diamond core coated by a surface layer. The carbon atoms in this layer exist in a disordered sp^3-hybridized state. The surface of the particle reveals distinct faceting. Note that the $\{111\}$ surface bears graphite-like fragments, whereas the $\{100\}$ faces retain a diamond-like arrangement of atoms.

References

1. Lymkin A. I., Petrov E. A., Ershov A. P., Sakovitch G. V., Staver A. M., and Titov V. M. (1988) Production of diamonds from explosives, *Dokl. Akad. Nauk USSR*, **302**, pp. 611–613.
2. Greiner N. R., Philips D. S., Johnson J. D., and Volk F. (1988) Diamonds in detonation soot, *Nature*, **333**, pp. 440–442.
3. Vul' A. Ya. (2006) *Ultrananocrystalline Diamond. Synthesis, Properties and Applications*, eds. Gruen D. M., Shenderova O. A. "Characterization and Physical properties of UNCD particles" (William Andrew, Norwich, New York, USA), pp. 379–404.
4. Kuznetsov V. L., and Butenko Yu. V. (2006) *Ultrananocrystalline Diamond. Synthesis, Properties and Applications*, eds. Gruen D. M., Shenderova O. A. "Diamond phase transition at nanoscale" (William Andrew Publishing, Norwich, New York, USA), pp. 405–475.
5. Shenderova O. A., Zhirnov V. V., and Brenner D. W. (2002) Carbon nanostructures, *Crit. Rev. Sol. State*, **27**(3/4), pp. 227–356.
6. Kuznetsov V. L., Chuvilin A. L., Butenko Yu. V., Mal'kov I. Yu., and Titov V. M. (1994) Onion-like carbon from ultra-disperse diamond, *Chem. Phys. Lett.*, **222**, pp. 343–348.
7. Kuznetsov V. L., Chuvilin A. L., Moroz E. M., Kolomiichuk V. N., Shaikhnutdinov Sh. K., and Butenko Yu. V. (1994) Effect of explosion comditions on the structure of detonation soot: ultradisperse diamond and onion carbon, *Carbon*, **32**, pp. 873–882.
8. Banhart F., and Ajayan P. M. (1996) Carbon onion as nanoscopic pressure cell for diamond formation, *Nature*, **382**, pp. 433–435.

9. Banhart F. (2002) Structural transformations in carbon nanoparticles induced by electron irradiation. *Phys. Solid State*, **44**(3), pp. 399–404.
10. Alexenskii A. E., Baidakova M. V., Vul' A. Ya., Davydov V. Yu., and Pevtsova Yu. A. (1997) Diamond-graphite phase transition in ultradisperse-diamond clusters, *Phys. Solid. State*, **39**, pp. 1007–1015.
11. Mykhaylyk O. O. David Yu. M., Batchelder N., and Brydson R. (2005) Transformation of nanodiamond into carbon onions: a comparative study by high-resolution transmission electron microscopy, electron energy-loss spectroscopy, x-ray diffraction, small-angle x-rayscattering, and ultraviolet Raman spectroscopy, *J. Appl. Phys.*, **97**, 074302.
12. Raty J. Y., Galli G., Bostedt C., van Buuren T. W., and Terminello L. J. (2003) Quantum confinement and fullerenelike surface reconstructions in nanodiamonds, *Phys. Rev. Lett.*, **90**, pp. 037401–037404.
13. Iakoubovskii K., Baidakova M. V., Wouters B. H., Stesmans A., Adriaenssen G. J., Vul' A. Ya., and Grobet P. J. (2000) Structure and defects of detonation synthesis nanodiamond, *Diam. Relat.Mater.*, **9**, pp. 861–865.
14. Aleksenskii A. E., Baidakova M. V., Vul' A. Ya., and Siklitsky V. I. (1999) The structure of diamond nanoclusters, *Phys. Solid State*, **41**, pp. 668–671.
15. Baidakova M. V., Vul' A. Ya., and Siklitski V. I. (1999) Ultradisperse-diamond nanoclusters. Fractal structure and diamond-graphite phase transition, *Chaos Soliton. Fract.*, **10**, pp. 2153–2163.
16. Avdeev M. V., Rozhkova N. N., Aksenov V. L., Garamus V., Willumeit R., and Ōsawa E. (2009) Aggregate structure in concentrated liquid dispersions of ultrananocrystalline diamond by sall-angle neutron scattering, *J. Phys. Chem. C*, **113**(22), pp. 9473–9479.
17. Mikov S. N., Igo A. V., Gorelik V. S. (1995) Raman scattering of light on diamond quantum dots in a matrix of potassium bromide, *Phys. Solid State*, **37**, pp. 1671–1673.
18. Tuinstra F., and Konig J. L. (1970) Raman spectra of graphite. *J. Chem. Phys.*, **53**, pp. 1126–1131.
19. Nemanich R. J., and Solin S. A. (1979) First- and second order Raman scattering from finite size crystals of graphite, *Phys. Rev.*, **20**, pp. 392–401.
20. Solin S. A., and Ramdas A. K. (1970) Raman spectrum of diamond, *Phys. Rev. B*, **1**, pp. 1687–1698.
21. Yoshikawa M., Nagai M., Matsuki M., Fukuda H., Katagiri H., Isida H., and Ishitani A. (1992) Raman scattering from sp^2 carbon clusters, *Phys. Rev. B*, **46**, pp. 7169–7174.

22. Yoshikawa M., Mori Y., Maegawa M., Katagiri G., Ishida H., and Ishitani A. (1993) Raman scattering from nanometer-sized diamond, *Appl. Phys. Lett.*, **62**(24), pp. 3114–3116.

23. Obraztsov A. N., Timofeev M. A., Guseva M. B., Babaev V. G. Valullova Z. Kh., and Babina V. M. (1995) Comparative study of microcrystalline diamond, *Diam. Relat. Mater.*, **4**, pp. 968–971.

24. Dement'ev A. P., and Maslakov K. I. (2004) Chemical state of carbon atoms on the surface of nanodiamond particles, *Phys. Solid State*, **46**, pp. 678–680.

25. Osswald S., Mochalin V. N., Havel M., Yushin G., and Gogotsi Y. (2009) Phonon confinement effects in the Raman spectrum of nanodiamond, *Phys. Rev. B*, **80**, 075419.

26. Mochalin V., Osswald S., and Gogotsi Y. (2009) Contribution of functional groups to the Raman spectrum of nanodiamond powders, *Chem. Mater.*, **21**, pp. 273–279.

27. Korobov M. V., Avramenko N. V., Bogachev A. G., Rozhkova N. N., and Ōsawa E. (2007) Nanophase of water in nano-diamond gel, *J. Phys. Chem. C*, **111**, pp. 7330–7334.

28. Aleksenskii A. E., Osipov V. Y., Dideykin A. T., Vul' A. Ya., Adreaenssens G. J., and Afanasev V. V. (2000) Ultradisperse diamond cluster aggregation studied by atomic force microscopy, *Tech. Phys. Lett.*, **26**, pp. 819–821.

29. Vul' A. Ya., Golubev V. G., Grudinkin S. A., Krüger A., and Naramot H. (2002) Diamond films: initial CVD growth stage using nanodiamonds as nucleation centers, *Techn. Phys. Lett.*, **28**, pp. 787–789.

30. Chang L.-Y., Ōsawa E., and Barnard A. S. (2011) Confirmation of the electrostatic self-assembly of nanodiamonds, *Nanoscale*, **3**, pp. 958–962.

31. Aleksenskiy A., Baidakova M., Osipov V., and Vul' A. Ya. (2009) *Nanodiamonds: Applications in Biology and Nanoscale Medicine*, ed. Ho D. "The fundamental properties and characteristics of nanodiamonds" (Spinger, New York, Dordrecht, Heidelberg, London), pp. 55–77.

32. Ōsawa E. (2005) *Synthesis, Properties and Applications of Ultrananocrystalline Diamond*, NATO Science Series, Series II, Vol. 192, "Disintegration and purification of crude aggregates of detonation nanodiamonds" (Springer, the Netherlands), pp. 231–240.

33. Ōsawa E. (2008) Monodisperse single-nano diamond particulates, *Pure Appl. Chem.*, **80**, pp. 1365–1379.

34. Eidelman E. D., Siklitsky V. I., Sharonova L. V., Yagovkina M. A., Vul' A. Ya., Takahashi M., Inakuma M., Ozawa M., and Ōsawa E. (2005)

A stable suspension of single ultrananocrystalline diamond particles, *Diam. Relat. Mater.*, **14**, pp. 1765–1769.

35. Ōsawa E. (2007) Recent progress and perspectives in single-digit nanodiamond, *Diam. Relat. Mater.*, **16**, pp. 2018–2022.
36. Pentecost A., Gour S., Mochalin V., Knoke I., and Gogotsi Y. (2010) De-aggregation of nanodiamond powders using salt- and sugar-assisted milling, *ACS Appl. Mater. Interfaces*, **2**, pp. 3289–3294.
37. Mochalin V. N., Shenderova O., Ho D., and Gogotsi Yu. (2012) The properties and applications of nanodiamonds, *Nat. Nanotechnol.*, **7**, pp. 11–23.
38. Kang X., and Qunji X. (2004) A new method for deaggregation of nanodiamond from explosie detonation: graphitization-oxidation method, *Phys. Solid State*, **46**, pp. 649–650.
39. Kang X., and Qunji X. (2007) Deaggregation of ultradispersed diamond from explosive detonation by a graphitization–oxidation method and by hydroiodic acid treatment, *Diam. Relat. Mater.*, **16**, pp. 277–282.
40. Osswald S, Yushin G., Mochalin V., Kucheyev S. O., Gogots Y. (2006), Control of sp2/sp3 carbon ratio and surface chemistry of nanodiamond powders by selective oxidation in air, *J. Am. Chem. Soc.*, **128**, pp. 11635–11642.
41. Shenderova O., Petrov I., Walsh J., Grichko V., Tyler T., and Cunningham G. (2006) Modification of detonation nanodiamonds by heat treatment in air, *Diam. Relat. Mater.*, **15**, pp. 1799–1803.
42. Larionova I., Kuznetsov V., Frolov A., Shenderova O., Moseekov S., and Mazov I. (2006) Properties of individual fractions of detonation nanodiamond, *Diam. Relat. Mater.*, **15**, pp. 1804–1808.
43. Aleksenskiy A. E., Eydelman E. D., and Vul' A. Ya. (2011) Deagglomeration of detonation nanodiamonds, *Nanosci. Nanotechnol. Lett.*, **3**, pp. 68–74.
44. Williams O., Hees A., Dieker C., Jäger W., Kirste L., and Nebel C. (2010) Size-dependent reactivity of diamond nanoparticles, *ACS Nano*, **4**, pp. 4824–4830.
45. Ōsawa E. (2009) *Nanodiamonds: Applications in Biology and Nanoscale Medicine*, ed. Ho D. "Single-nano buckydiamind particles" (Spinger, New York, Dordrecht, Heidelberg, London), pp. 1–34.
46. Vul' A. Ya., Eydelman E. D., Sharonova L. V., Aleksenski, A. E., and Konyakhin S. V. (2011) Absorption and scattering of light in nanodiamond hydrosols, *Diam. Relat. Mater.*, **20**, pp. 279–284.

47. Ozerin A. N., Kurkin T. S., Ozerina L. A., and Dolmatov V. Yu. (2008) X-ray diffraction study of the structure of detonation nanodiamonds, *Crystallogr. Rep.*, **53**, pp. 60–67.

48. Kruger A., Kataoka F., Ozawa M., Fujino T., Suzuki Y., Aleksenskii A. E., Vul' A. Ya., and Ōsawa E. (2005) Unusually tight aggregation in detonation nanodiamond: Identification and disintegration, *Carbon*, **43**, pp. 1722–1730.

49. Barnard A. S. (2008) Self-assembly in nanodiamond agglutinates, *J. Mater. Chem.*, **18**, pp. 4038–4041.

50. Reich K. V. (2011) Optical properties of nanodiamond suspensions, *JETP Lett.*, **94**(1), pp. 22–26.

51. Aleksenskii E. A., Vul' A. Ya., Konyakhin S. V., Reich K. V., Sharonova L. V., and Eidel'man E. D. (2012) Optical properties of detonation nanodiamond hydrosols, *Phys. Sol. State*, **54**(3), pp. 578–585.

52. Pandey K. C. (1982) New dimerized-chain model for the reconstruction of the diamond (111)-(2×1) surface, *Phys. Rev. B*, **25**, pp. 4338–4346.

53. Kuznetsov V. L., Aleksandrov M. N., Zagoruiko I. V., Chuvilin A. L., Moroz E. M., Kolomiichuk V. N., Likholobov V. A., Brylyakov V. A., and Sakovitch G. V. (1991) Study of ultra disperse diamond obtained using explosion energy, *Carbon*, **29**, pp. 665–668.

54. Kulakova I. I. (2004) Surface chemistry of nanodiamonds, *Phys. Sol. State*, **46**, pp. 636–643.

55. Baidakova M., and Vul' A. Ya. (2007) New prospects and frontiers of nanodiamond clusters, *J. Phys. D: Appl. Phys.*, **40**, pp. 6300–6311.

56. Hawelek L., Brodka A., Dore J. C., Honkimaki V., Tomita S., and Burian A. (2008) Structural studies of nanodimond by high-energy X-ray diffraction, *Diam. Relat. Mater.*, **17**, pp. 1186–1193.

57. Palosz B., Pantea C., Grzanka E., Stelmakh S., Proffen Th., Zerda T. W., and Palosz W. (2006) Investigation of relaxation of nanodiamond surface in real and reciprocal spaces, *Diam. Relat. Mater.*, **15**, pp. 1813–1817.

58. Dolmatov V. Y. (2007) Detonation synthesis nanodiamonds: synthesis, structure, properties and applications, *Russ. Chem. Rev.*, **76**, pp. 339–360.

59. Danilenko V. V. (2005) *Synthesis, Properties and Applications of Ultrananocrystalline Diamond*, NATO Science Series, Series II, Vol. 192, "Nanocarbon phase diagram and condition for detonation nanodiamond Formation (Springer, the Netherlands), pp. 181–198.

Chapter 4

Optical and Rheological Properties of Nanodiamond Suspensions

Alexander Ya. Vul'[a] and Evgeniy D. Eydelman[a,b]
[a] *Ioffe Physical-Technical Institute, Solid State Electronics Department, 26 Polytechnicheskaya, St. Petersburg 194021, Russia*
[b] *St. Petersburg Chemical Pharmaceutical Academy, Physics Department, St. Petersburg 197022, Russia*
alexandervul@mail.ioffe.ru

In many cases, it is not diamond nanoparticles but their suspensions become the subject of a study. Moreover, it is suspensions of detonation nanodiamonds (DNDs) that are considered in many cases as the final commercial product.

This chapter addresses briefly the optical and rheological properties of hydrosols of diamond nanoparticles, as well as similar characteristics of nanoparticle suspensions in other liquid media.

We note that the optical and rheological properties of suspensions of diamond macro- and microparticles with a perfect surface are obvious because diamond is hydrophobic and transparent throughout the visible part of the spectrum. Now suspensions of diamond nanoparticles reveal radically different properties; indeed, the particles are at least partially graphitized, that is, coated by regions with sp^2-hybridized carbon atoms (see Chapter 3).

Detonation Nanodiamonds: Science and Applications
Edited by Alexander Ya. Vul' and Olga A. Shenderova

ISBN 978-981-4411-27-1 (Hardcover), 978-981-4411-28-8 (eBook)
www.panstanford.com

This accounts for noticeable absorption in the visible part of the spectrum and a change in their viscosity.

We start with an analysis of the conditions favoring formation of sp^2 regions on the surface of a diamond nanoparticle.

4.1 Modeling Graphitization of the Surface of Diamond Nanoparticles

A mechanical and/or thermal attack on the surface of detonation nanodiamond (DND) particles culminates in their partial graphitization[a] [1–3].

It is as well known that a decrease of surface energy is accompanied in crystals by reconstruction of their surface. In particular, on the (111) diamond face the neighboring carbon atoms regroup to form dimmers—the so-called Pandey chains [4]. Both these features initiate formation on the surface of diamond nanoparticle regions with sp^2-hybridized carbon atoms.

Consider first the simplest case of formation of sp^2 regions, assuming the particle to be of spherical shape, and their appearance as being initiated by a random factor (Fig. 4.1).

Mathematical modeling of particle surface graphitization treats this process as the appearance on the surface of a set of sp^2-hybridized carbon atoms N_p, which combine to form a graphite-like phase.

As the number N_p of these regions increases, they merge to produce irregular spots. Although our particular case cannot be identified with a discrete problem of graph treatment, we may conveniently use the terminology of this theory [5] and call a set of joined regions the "connectivity component."

Location of a new region was selected by a pseudorandom number generator employed by default, which provides uniform distribution with subsequent normalization. A possibility was also envisaged to specify the constants in the Gaussian distribution, which controls randomly the radii of the sp^2 regions. The most

[a]Similar graphitization is observed, for instance, in deagglomeration by the method of Ōsawa and fabrication of onion-like carbon from DNDs (see Chapter 3).

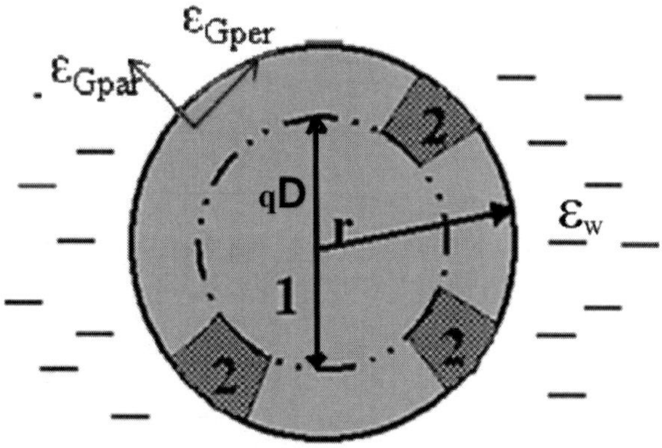

Figure 4.1 Model of a DND particle in suspension. (1) Crystalline diamond core; (2) regions of sp^2-hybridized carbon with weakly coupled π electrons; r and D, radius and diameter of the particle; qD, diamond core size; δ_D, thickness of graphite-like region; ε_D and ε_W, dielectric permittivities of diamond and water, respectively; ε_{Gpar} and ε_{Gper}, radial and angular components of the dielectric permittivity of graphite.

important characteristic is the ratio of the most probable size of the region to the radius of the sphere, r. The radii of the regions varied from 0 to $2r_0$.

After a new region has been randomly added, the already present graphitized spots overlapping with the new region were identified. If the new region did not border on the already available spots, it was identified with a separate connectivity component. If, however, the region contacted spots belonging to one particular connectivity component only, it was assigned the number of this component and it was assumed to belong to it. If the new region was found to overlap simultaneously several connectivity components, they were combined. All the spots belonging to the combined components were assigned the number equal to the smallest of the numbers of these connectivity components.

This mathematical model did not reveal such a clearly pronounced sign of the onset of percolation as, for instance, activation of a path over the lattice edges from one point to another [6]. Therefore the onset of percolation was derived from an analysis of the spot field.

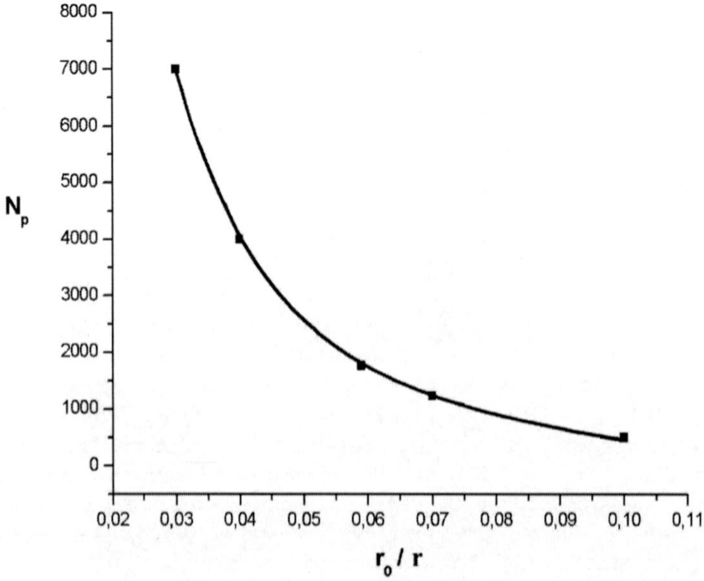

Figure 4.2 Number N_p of sp^2 graphitization regions at the onset of percolation plotted vs. the ratio of the most probable radius of the region, r_0, to that of the particle, r.

It was found that there exists such a number of spots, equal usually to 10–15, that adding one more sp^2 region makes almost all of the spots merge to form one large spot. This moment was taken to be the time of the onset of percolation. Figure 4.2 visualizes the times of the onset of percolation as derived from 30 observations for each point.

A common feature combining all visualizations was that at the moment of the onset of percolation the fraction of the surface covered by sp^2 regions was close to 0.7. For the model used, the value 0.7 has the same meaning as the well-known values of the percolation threshold for different lattices [6].

It is well known (see, e.g., Refs. [6, 7]) that percolation (the Foucault current) set in when more than one half of the surface becomes conducting, which in our case translates into the formation of one large spot dominated by sp^2 hybridization. The critical fraction of the bonds (connectivity components) for an arbitrary system is $x_c = (1/Z)[d/(d-1)]$, where Z is the number of atomic bonds and d is the dimension of the system. We may recall that for the graphite surface $d = 2$ and $Z = 3$, whence $x_c \approx 0.67$.

This model, as this will be demonstrated below, permits a satisfactory enough correlation between the optical (black color) and rheological (high viscosity) properties of hydrosols of diamond nanoparticles prepared by the method described in Ref. [2].

We note that the presence on the surface of diamond nanoparticles weakly coupled by π electron bonds is suggested also by computer modeling of stable diamond nanoparticles [8] and follows from direct nuclear magnetic resonance measurements [9].

4.2 Viscosity of Diamond Nanoparticle Suspensions

We are turning now to a consideration of the viscosity of water-based suspensions of DND particles whose surface is coated by a graphite-like layer composed of sp^2-hybridized carbon atoms.

This surface layer of a particle in suspension is surrounded by polar water molecules combined in a "loose cloud." As particles of this suspension move relative to one another (see Fig. 4.3), electrostatic interaction of the "clouds" enveloping each particle can, as this will be shown below, account for the high viscosity of the suspension. The corresponding relevant experimental data [2] is displayed in graphical form in Fig. 4.4.

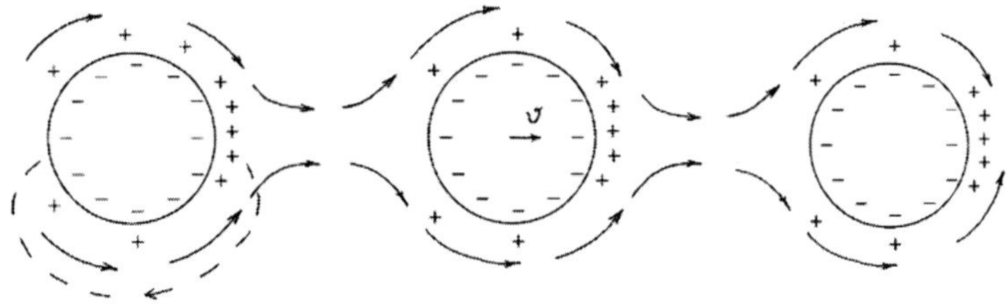

Figure 4.3 "Dumbbell-shaped flow." A model of electrostatic interaction of diamond nanoparticles flowing in suspension (hydrosol). v-relative velocity of a DND particle; solid lines liquid flow lines; dashed lines, ion motion in a double electric layer.

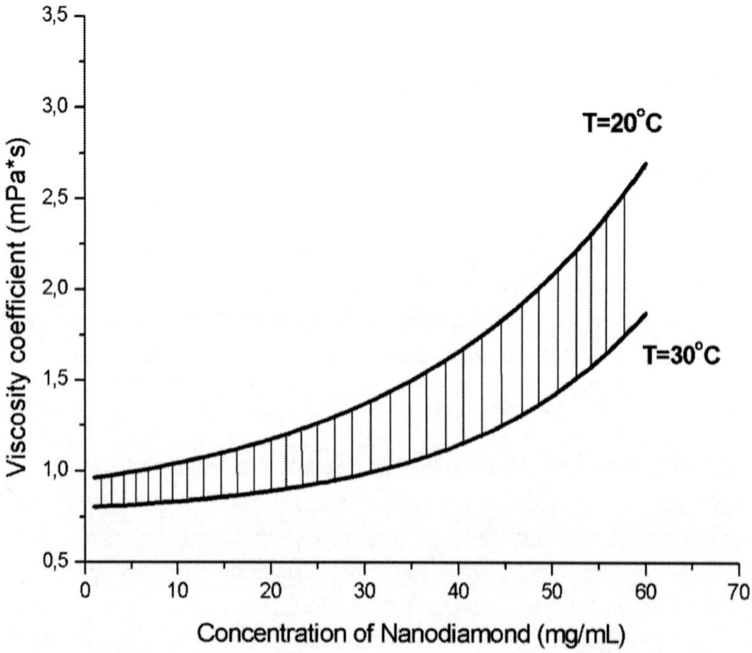

Figure 4.4 Viscosity of a water suspension plotted vs. DND concentration. The shaded region identifies the values corresponding to the temperature interval specified.

There are three factors that should be taken into account in viscosity calculations.

First, it is the viscosity of the dispersing medium, η_0. This liquid is usually water, whose viscosity in neutral conditions is approximately $\eta_0 \approx 1$ mP·s. The second factor is the effect exerted by particles immersed in a liquid on the flow. The presence of such particles affects the viscosity of the main (in a medium without particles) flow. This contribution was first calculated by Einstein (see, e.g., Ref. [10], §22) and can be written as

$$\frac{2}{5}\eta_0\frac{N\vartheta}{V} = \frac{2}{5}\eta_0\frac{c}{\rho}, \tag{4.1}$$

where V is the suspension volume and N and ϑ are the number of nanodiamond particles dispersed in the suspension and the volume of such a particle, respectively.

Obviously enough, the mass of diamond in the suspension will be mN (m is the particle mass). Then for the concentration of the particles dispersed in the suspension we obtain $c = mN/V$, with c being the mass of the particles referred to the volume of the suspension. The density of a nanodiamond particle is $\rho = m/\vartheta$. The nondimensional parameter c/ρ specifies the volume fraction of the suspension occupied by the particles. This is supported by the observation that at low concentrations (up to 40 mg/mL or $\sim$5 wt.%; see Fig. 4.4) the coefficient depends linearly on concentration.

At high particle concentrations, one should take into account the fraction of the liquid; similar to the way employed in the transition from the equation for an ideal to that for a real gas, the quantity (c/ρ) will become replaced (see, e.g., Ref. [11], §76) by

$$\frac{c}{\rho} \cdot \frac{1}{1 - \frac{c}{\rho}}. \tag{4.2}$$

Significantly, it is this effect that leads to the concentration dependence of the suspension viscosity becoming nonlinear, the conclusion supported by experiment.

The third and, as we believe, the most important factor interfering with the flow of nanoparticle suspension is the electrical interaction of the particles, together with the "clouds" surrounding them, that is, the electroviscous effect. This interaction can be characterized by three parameters, more specifically, dielectric permittivity ε of the medium, the electrokinetic potential ζ characterizing the double electric layer around the particle, and the suspension conductivity γ associated with the motion of the liquid.

The contribution of the electroviscous effect to the total viscosity of a suspension of charged dispersed particles was first calculated by Smolukhovskii (see Refs. [11, 12]). This contribution increases the viscosity by Einstein by a factor:

$$1 + \frac{1}{\eta_0 \gamma r^2}(2\varepsilon\varsigma)^2. \tag{4.3}$$

Here r is the radius of the particle with a double electric layer, 4 nm $< r <$ 10 nm.

The mass of the dispersed particle, m, and its density ρ should be taken accounting for the "cloud" surrounding it. The "cloud" itself may be considered to consist of two regions, a fixed layer coating the nanoparticle and moving together with it and a loose one, a diffuse part entrained by the flow.

Thus, the viscosity η of a suspension of DND particles in water or any other polar liquid, which precludes their immediate aggregation, should have the form

$$\frac{\eta}{\eta_0} = 1 + \frac{5}{2}\frac{c}{\rho}\frac{1}{1-\frac{c}{\rho}}\left[1 + \frac{1}{\eta_0\gamma r^2}(2\varepsilon\varsigma)^2\right]. \tag{4.4}$$

Judicious approximation of the curves in Fig. 4.4 suggests that the dimensionless parameter

$$E = \frac{10\varepsilon^2\zeta^2}{\eta_0\gamma r^2} \approx 1. \tag{4.5}$$

Estimation of the viscosity and its comparison with experiment can be simplified by neglecting the effect of the second factor, that is, the influence of the perturbations in the flow introduced by the particles

$$\eta = \eta_0 + \frac{5}{2}\frac{c}{\rho}\frac{1}{(1-\frac{c}{\rho})}\frac{(2\varepsilon\varsigma)^2}{\gamma r^2}. \tag{4.6}$$

The numerical estimates made using this expression and presented in Ref. [1] were found to fit well the experiment.

4.3 Qualitative Analysis of the Optical Properties of Nanodiamond Hydrosols

The size and mutual arrangement of the regions containing π electrons (Fig. 4.1) define the possibility of generation of Foucault currents on the surface (see Section 4.1 of this chapter). Thus the presence of π electrons not only accounts for the anomalous viscosity but should initiate absorption of optical radiation as well. A careful analysis of the laws forming the basis of the theory of percolation suggests a conclusion that conductivity and, accordingly, resonant absorption (plasmons) depend strongly on the structure of the sp^2 layer on the particle surface.

In contrast to earlier publications [13], we are going to present here the calculations of the cross sections of scattering and absorption by DND particles made under the assumption that π electrons behave similarly to electrons in a metal. This permits one to express the cross sections through the parameters of the sp^2 phase.

In our consideration of hydrosols we are going to assume the water to be absolutely transparent, in which case the processes of interest to us here will in actual fact represent scattering and absorption of electromagnetic waves on DND particles small compared to the wavelength. The cross sections of such processes are well known.

We write the expressions for the scattering cross section (Rayleigh law) as

$$\sigma_{\text{sca}} = \frac{8\pi^2}{3}\frac{\omega^4}{c^4}V^2|\alpha|^2, \tag{4.7}$$

and absorption cross section as

$$\sigma_{\text{abs}} = 4\pi\frac{\omega}{c}\ V\,\text{Im}\,(\alpha)\,. \tag{4.8}$$

In these expressions, $\omega = 2\pi c/\lambda$ is the frequency of incident electromagnetic radiation with a wavelength λ, $c = 3{\cdot}10^8$ m/s is the velocity of light, V is the particle volume, and α is the coefficient of polarization. We shall assume the light to fall and be scattered uniformly and isotropically.

The calculation of the polarization coefficient is based on the above model of the DND particle structure used to explain the viscosity of the suspension.

While it is quite possible that graphite-like spots do not totally coat the particle and are of different thickness, in our analysis of the optical properties we are going to consider a DND particle to be of spherical shape and coated by a continuous layer of sp^2-hybridized carbon.

The calculation of the polarization coefficient α for small particles reduces to calculation of the dipole term in the distortion of a uniform electric field introduced by a particle. The solution of this problem is well known (see Refs. [12, 13] and references therein).

The coefficient of polarizability of a multilayer spherical carbon particle produced from DND can be found in Ref. [14].

In these calculations, it was assumed that the dielectric permittivity of water $\varepsilon_{\text{w}} = 1.7$ and of the diamond (diamond-like) core

$\varepsilon_D = 5.7$ and that these quantities do not depend on the incident wavelength. The dielectric permittivity (the tensor of dielectric permittivity) of the surface spot of graphite-like material, ε_G, was assumed to be anisotropic. At each point of the sp^2-hybridized region the tensor of dielectric permittivity was taken to be diagonal with components ε_{Gpar} and ε_{Gper} (see Fig. 4.1). Each of these components has, respectively, a real and an imaginary part. All the four quantities obtained in this way depend on wavelength [15].

An experimental study was conducted [16] of the spectral relations of optical density $y = \ln(I_0/I)$; I_0 and I are the intensities of the incident light and of the light that traversed the DND hydrosol, respectively. A typical dependence of optical density on wavelength $y(\lambda)$ is shown graphically in Fig. 4.5.

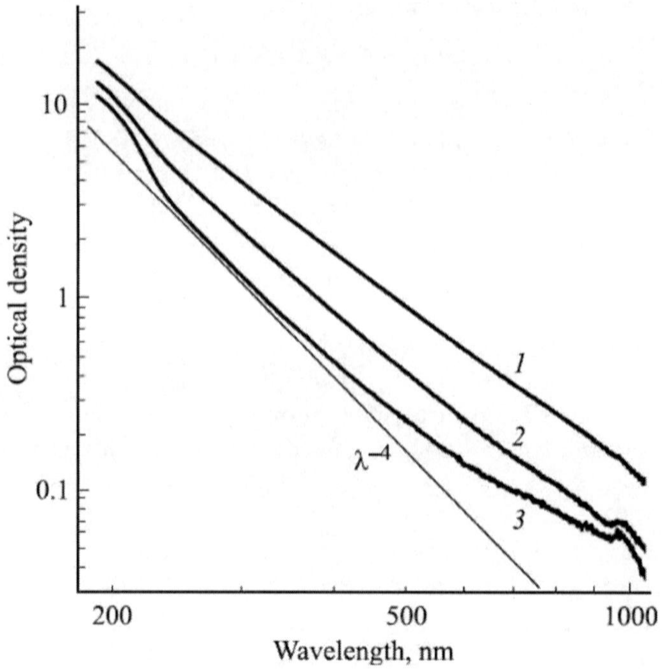

Figure 4.5 Spectral response of the optical density of a DND hydrosol, $y = \ln(I_0/I)$ [17]. The maximum of particle size distribution shifts successively from 200 nm (curve 1) to 4 nm (curve 3). Thin line, Rayleigh scattering law [17].

For transition from calculations made using Eqs. (4.7) and (4.8) to the measured optical density, one makes use of the well-known relation

$$y = \sigma_{\text{ext}}\, nl, \tag{4.9}$$

in which n is the concentration of DND particles in the suspension and l is the optical length equal to the product of the path length passed by light in the suspension by its refractive index.[a] The extinction cross section can be expressed through the corresponding absorption, σ_{abs}, and scattering, σ_{sca}, cross sections as

$$\sigma_{\text{ext}} = \sigma_{\text{abs}} + \sigma_{\text{sca}}. \tag{4.10}$$

The calculated data treated in terms of the model of a spherical diamond particle coated by a continuous layer of sp^2-hybridized carbon atoms corresponds best with the experiment for suspensions with an average DND particle in the suspension larger than 15 nm [16]. The same model was used in treating the hydrosol viscosity (see Section 4.2).

A typical example of such a fitting is presented in Fig. 4.6.

We note that to fit the measurements with calculations within this simplest model [16] one had to assume the thickness of a graphite-like region to be on the order of or even smaller than the distances in the diamond crystal lattice. This means that not all of the particle surface is really graphitized, a conclusion supporting the calculations presented earlier in this chapter (Section 4.1).

Let us turn now to the features observed in the spectral behavior of the optical density of suspensions with DND particles less than 15 nm in size (see Fig. 4.5, curves 2 and 3).

The observed absorption can be related within a classical approach to the presence of weakly bonded π electrons in sp^2-hybridized regions on the particle surface.

The characteristics of DND suspensions employed to estimate the viscosity (see Section 4.2) can be used readily to evaluate the concentration n of the weakly bonded π electrons. Indeed, the simple relation developed for the potential of a spherical particle (e is the electronic charge) $\zeta = nve/(4\pi\varepsilon r)$, where v is the volume

[a] The concentration of nanoparticles in suspensions is determined, as a rule, by the dynamic light scattering (DLS) technique.

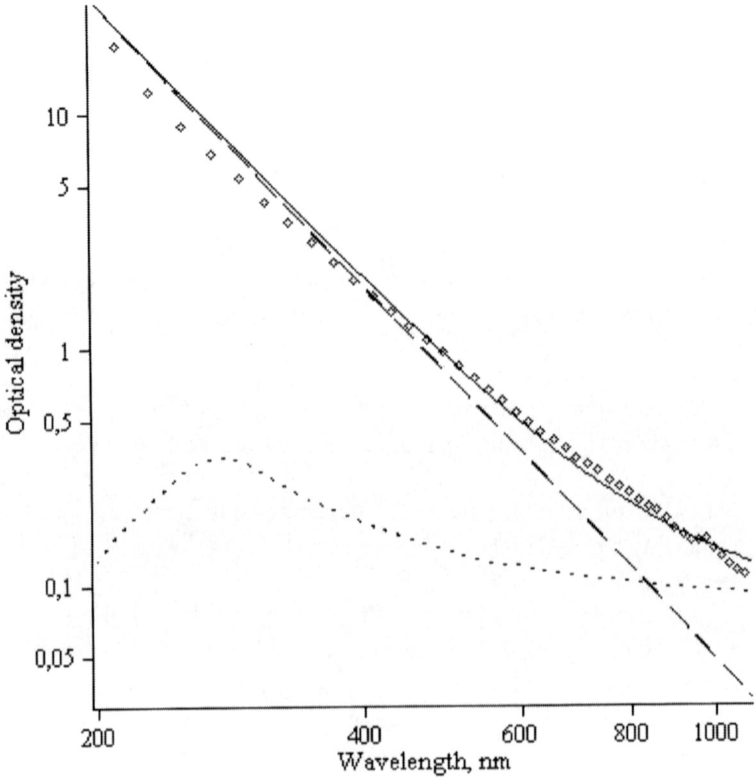

Figure 4.6 Optical density of a DND hydrosol, $y = \ln(I_0/I)$, plotted vs. wavelength λ for nanoparticles with an average size (maximum of the size distribution) above 15 nm. Rhombuses, experimental data. Calculations: Dashed line, scattering; dotted line, absorption; solid line, resultant extinction [16].

of a bare diamond nanoparticle (without the "cloud" of water molecules), yields for 4 nm particles $n = (1 - 10)\cdot 10^{26}$ m^{-3}. For such a concentration of free π electrons the resonance frequency of vibrations and, hence, the wavelength λ_p of the collective vibrations of the electron plasma (plasmon) will be (see, e.g., Ref. [18])

$$\lambda_p = 2\pi c\sqrt{\frac{m\varepsilon_0}{ne^2}}. \tag{4.11}$$

Substitution of the corresponding numerical values yields $\lambda_p =$ 300–500 nm.

We readily see that this straightforward estimation matches well with the data obtained from an analysis of the absorption curve (see Fig. 4.6), both qualitatively and quantitatively. Estimates suggest also a possible plasmon absorption resonance in the long-wavelength part of the spectrum (close to 1,000 nm).

The dependence of dielectric permittivity on frequency can be expressed with a good enough accuracy through the Lorentz relation

$$\varepsilon = \varepsilon_{\mathrm{d}} + \frac{\omega_{\mathrm{p}}^2}{\omega_0^2 - \omega^2 - i\Gamma\omega}. \tag{4.12}$$

Here ε_{d} is the dielectric permittivity at frequencies low compared with the excitation frequency of the system ω_{p} (plasmon frequency).

For a system of simple harmonic oscillators, $\varepsilon_{\mathrm{d}} = 1$. The plasmon frequency is related to the wavelength, Eq. (4.11). Through the plasmon frequency is expressed the damping coefficient

$$\Gamma = \varepsilon_0 \omega_{\mathrm{p}}^2 \rho = \frac{\varepsilon_0 \omega_{\mathrm{p}}^2}{\gamma}, \tag{4.13}$$

which is directly proportional to the resistivity of the material, ρ, and therefore, the damping coefficient is inversely proportional to γ.

One more characteristic of a material is the internal frequency ω_0 of vibrating systems.

In the case under consideration here, Eq. (4.12) should be applied separately to the diamond core (ε_{D}), the *sp*^{2} shell (ε_{G}), and the water surrounding the particle (ε_{W}). The uniqueness of the problem under analysis consists of coexistence of diamond-like *sp*^{3} and graphite-like *sp*^{2} regions at nanoscale distances from one another.

It is well known (see Chapter 1) that diamond is a wide-band semiconductor with a low free-electron concentration $n_{\mathrm{D}} \cong 10^{12}$ -10^{14} m^{-3} and internal and plasmon frequencies $\omega_{0\mathrm{D}} = 8{\cdot}10^{15}$ s^{-1}; $\omega_{\mathrm{pD}} = (0.5 - 1.5){\cdot}10^9 \mathrm{s}^{-1}$.

Whence it follows immediately that $\omega_{\mathrm{pD}} << \omega$, that is, that the plasmon frequency of a diamond core is always lower than that of incident light in the visible region and, thus, cannot cause absorption. Such estimates suggest that the plasmon frequency for graphite-like regions matches the wavelengths in the long-wavelength region of the spectrum, $\lambda \geq 1{,}000$ nm. This all brings us to a qualitative conclusion that the spectra of DND suspensions in water are determined, throughout the visible range, by scattering, with absorption appearing at the edges of the range.

4.4 Numerical Calculations of the Optical Properties of Diamond Nanoparticle Hydrosols

The above qualitative consideration of the optical properties of DND suspensions was supported by rigorous quantum-chemical calculations [19], which evolve from the model of diamond surface reconstruction (Section 4.1) involving formation of the Pandey dimer chains [4] (Fig. 4.7). Such a chain may be considered as a periodic one-dimensional structure with a basal vector **a** and a unit cell consisting of two atoms, A and B. Each atom in such a chain is coupled both to the other atoms in the chain and atoms of carbon C in the nanodiamond core. Each individual chain may be considered separately from the others because the distance between chains is twice that between atoms in a chain.

While in real conditions the surface of DND particles is coated by various functional groups [20], nevertheless, the dimers formed of carbon atoms should invariably be present. The only difference from a clean surface of diamond can apparently consist of the dimer assembly in the form of long chains being less probable.

Numerical calculations of the energy bands and of the densities of states were conducted for wave vectors **k** near the Brillouin zone edge being $\mathbf{k} = \pi/\mathbf{a}$, the conditions in which bulk states are not admixed to surface states, thus permitting one to solve analytically the Schrödinger equation and derive the dependence of energy on the wave vector for an electron on the diamond surface, $E(\mathbf{k}) = \pm|\varepsilon(\mathbf{k})|$. The equation was solved over the whole region of the wave vectors **k**.

Mainly the density of states, D, which for one-dimensional systems is given by the following relation, determines the optical

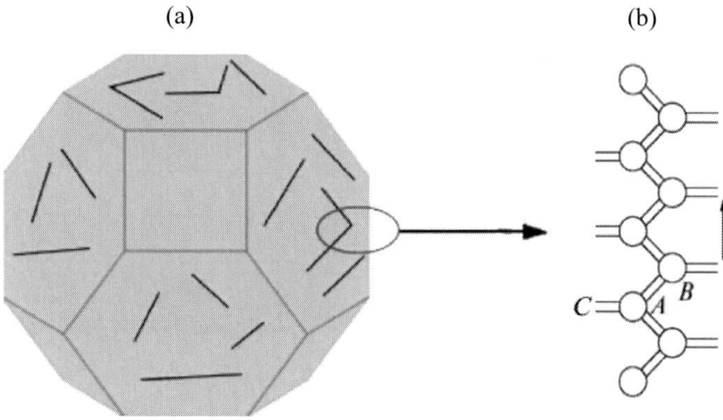

Figure 4.7 Schematic of the surface of a diamond nanoparticle coated by dimer chains. (a) Neighboring carbon atoms on the reconstructed (111) diamond surface form a dimer chain. (b) Schematic of a chain, a periodic one-dimensional structure with a unit cell A–B and a basis vector **a**. Each carbon atom in such a chain is coupled to bulk atoms (atom C).

absorption coefficient

$$D\left(\varepsilon\right)=\frac{1}{\varepsilon'\left(k_0\right)'}=\varepsilon\left(k_0\right)=\varepsilon. \tag{4.14}$$

The results of the calculations can be found in Refs. [17, 19]. The dependence of electron energy on the wave vector displayed graphically in Fig. 4.8 permits one to draw a conclusion concerning the absorption characteristic of a nanodiamond surface. As seen from Fig. 4.8, electrons can transfer from the bulk to surface states in the region where the D_2 and D_3 densities of states cross. These transitions, E_1 and E_2, are specified in the figure.

The origin of these peaks can be identified from purely qualitative considerations as well (see Section 4.3). The energy of the first transition, E_1, correlates with the width of the bandgap separating the surface states and is about 1.5 eV (measured in Ref. [21]).

The optical transition at an energy $E_2 \approx 5$ eV coupling the lower valence band edge with the conduction band is affected by several factors.

First, as seen from Fig. 4.8, the upper valence band edge of surface states is located fairly far from the characteristic energies

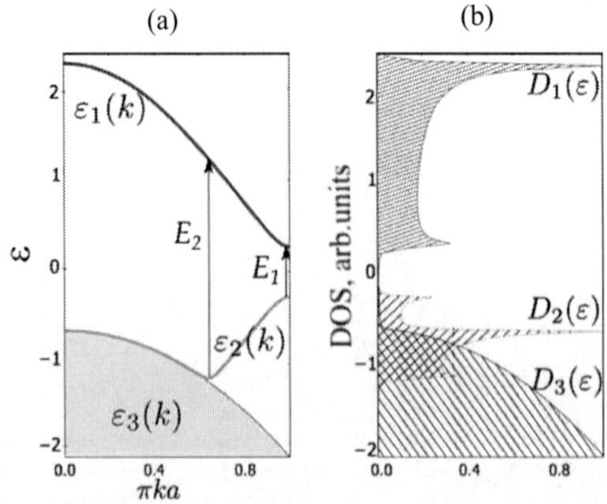

Figure 4.8 Energy diagram for a diamond nanoparticle. Dependence of energy ε in units of the energy characteristic of the Pandey chain: (a) on wave vector **k** and (b) on density of states. In addition to the band–band transition typical of bulk diamond, in a DND particle one observes electronic transitions E_1 and E_2 in the surface states, dimers formed on the reconstructed diamond surface. Indices 1 and 2 specify surface zones and 3 the bulk zone.

for an electron in the bulk of diamond. This is not so, however, for the lower valence band edge. The energy of electrons on the surface overlaps that of the bulk electrons. Said otherwise, we witness the formation of resonance surface states, which is also observed to occur in diamond [22]. Therefore, the optical transition E_2 should be noticeably stronger in amplitude than the E_1 optical transition because of the admixture of bulk states.

Second, the E_2 transition practically coincides in energy with the valence conduction band transition for bulk diamond. Here, however, a significant contribution to the absorption edge should come only from electronic transitions on the surface because they have a strong dependence on the incident light energy: $E \sim 1/\sqrt{E_2 - E_1}$, as is the case with any absorption in a one-dimensional case.

To sum up, although a nanodiamond surface should absorb throughout the spectral region studied, the spectrum should reveal two characteristic maxima with wavelengths near 300 nm and 1,000 nm, with the latter one being significantly larger in amplitude.

Thus, the noticeable absorption of radiation in the visible region of the spectrum, which originates from the dimer chains (the Pandey chains), should account for the dark coloring of the hydrosol at a high enough concentration of DND particles, an effect observed regularly in experiments.

4.5 Specific Features of the Optical Properties of DND Hydrosols

The results of the studies presented above (Sections 4.3 and 4.4) suggest the following conclusion.

The core of a diamond nanoparticle is virtually nonabsorbent in the visible region. Absorption occurs primarily on the surface of diamond nanoparticles in the suspension, and it is therefore Rayleigh scattering that plays an essential part in the extinction coefficient. As the particle size decreases, the fraction of scattered light should decrease too, whereas that of absorbed light will grow with increasing fraction of the surface.

It is known that the scattering cross section is proportional to the squared particle volume V—see Eq. (4.7)—and that the concentration at a fixed total amount of material is inversely proportional to volume, whence it follows that larger particles provide a substantially larger contribution to scattering. For instance, the contribution to scattering due to 10^6 particles is the same as from a single particle 10 times larger in size. This means that purification of a hydrosol from large particles should reduce efficiently the scattering.

At the same time, absorption on the particle surface is proportional to $V^{2/3}$. An obvious conclusion therefrom is that removal of large particles from a hydrosol should not affect, noticeably, light absorption.

This opens up a possibility of making a rough judgment of the size distribution of diamond particles in a DND sample from comparatively simple optical measurements.

The above quantitative and qualitative considerations suggest an essential conclusion that without due allowance for the presence of sp^2-hybridized carbon atoms on the surface of a DND particle, it is not possible to correlate the calculation with experiment, either in determination of the viscosity or in optical absorption.

An analysis of light absorption and scattering in DND hydrosols provides explanation for some phenomena observed in the course of a thorough purification of the industrial detonation soot from which these hydrosols are obtained.

The assumption of the formation of photonic crystals in DND hydrosols [23] should possibly be attributed to the effect associated with separation of particles into fractions according to their size. It is light scattering from particles of different size stacked in layers that can account for the rainbow coloring of DND hydrosols. The same considerations can be invoked to explain [24] the variation of suspension color with pH, because aggregation of particles can entail a strong increase of the fraction of scattered light even at a small variation of pH.

References

1. Eidelman E. D., Siklitsky V. I., Sharonova L. V., Yagovkina M. A., Vul' A. Ya., Takahashi M., Inakuma M., Ozawa M., and Ōsawa E. (2005) A stable suspension of single ultrananocrystalline diamond particle, *Diam. Relat. Mater.*, **14**, pp. 1765–1769.
2. Vul' A. Ya., Eydelman E. D., Inakuma M., and Ōsawa E. (2007) Correlation between viscosity and absorption of electromagnetic waves in an aqueous UNCD suspension, *Diam. Relat. Mater.*, **16**(12), pp. 2023–2028.
3. Osipov V., Baidakova M., Enoki T., Takai K., and Vul' A. Ya. (2006) Magnetic properties of hydrogen-terminated surface layer of diamond nanoparticles, *Fullerenes Nanotubes Carbon Nanostruct.*, **14**, pp. 565–572.
4. Pandey K. C. (1982) New dimerized-chain model for the reconstruction of the diamond (111)-(2×1) surface, *Phys. Rev. B*, **25**, pp. 4338–4346.
5. Tutte W. T. (1988) *Graph Theory* (Cambridge Mathematical Library).
6. Efros A. L. (1986) Physics and Geometry of Disorder: Percolation Theory (Science for Everyone) (Mir, Moscow).
7. Shklovskii B. I., and Efros A. L. (1984) *Electronic Properties of Doped Semiconductors* (Springer-Verlag, Berlin, Heidelberg, New York, Tokyo).

8. Raty J. Y., Galli G., Bostedt C., van Buuren T. W., and Terminello L. J. (2003) Quantum confinement and fullerenelike surface reconstructions in nanodiamonds, *Phys. Rev. Lett.*, **90**, pp. 037401–037404.
9. Panich A. M., Shames A. I., Vieth H. M., Ōsawa E., Takahashi M., and Vul' A. Ya. (2006). Nuclear magnetic resonance study of ultrananocrystalline diamonds, *Eur. Phys. J.*, **52**, pp. 397–403.
10. Landau L. D., and Lifschitz E. M. (2010) *Fluid Mechanics* (Pergamon Press).
11. Landau L. D., and Lifschitz E. M. (1980) *Statistical Physics*, Part 1 (Pergamon Press, Oxford).
12. Chernoberezhskii Yu. M., Kuchuk V. I., Klochkova O. V., and Golikova E. V. (1987) Temperature dependence of the stability of natural diamond dispersions in electrolyte solutions, *Langmuir*, **3**(5), pp. 654–659.
13. Tomita S., Fujii M., and Hayashi S. (2002) Optical extinction properties of carbon onions prepared from diamond nanoparticles, *Phys. Rev. B*, **66**, pp. 245424(1–7).
14. Lucas A. A., Henrard L., and Lambin Ph. (1992) Computation of the ultraviolet absorption and electron inelastic scattering cross section of multishell fullerenes, *Phys. Rev. B*, **49**, pp. 2888–2896.
15. Drain B. T., and Lee H. M. (1994) Optical properties of interstellar graphite and silicate grains, *Astrophys. J.*, **285**, pp. 89–108.
16. Vul' A. Ya., Eidelman E. D., Sharonova L. V., Aleksenskiy A. E., and Konyakhin S. V. (2011) Adsorption and scattering of light in nanodiamond hydrosols, *Diam. Relat. Mater.*, **20**(3), pp. 279–284.
17. Aleksenskii A. E., Vul' A. Ya., Konyakhin S. V., Reich K. V., Sharonova L. V., and Eidel'man E. D. (2012) Optical properties of detonation nanodiamond hydrosols, *Phys. Sol. State*, **54**(3), pp. 578–585.
18. Platzman P. M., and Wolff P. A. (1973) *Waves and Interactions in Solid State Plasmas* (Academic Press, New York, London).
19. Reich K. V. (2011) Optical properties of nanodiamond suspensions, *JETP Lett.*, **94**(1), pp. 22–26.
20. Baidakova M., and Vul' A. Ya. (2007) New prospects and frontiers of nanodiamond clusters, *J. Phys. D: Appl. Phys.*, **40**, pp. 6300–6311.
21. Bussetti G., Goletti C., Chiarada P., and Derry T. (2007) Optical gap between dangling-bond states of a single-domain diamond C(111) - 2×1 by reflectance anisotropy spectroscopy, *Eur. Phys. Lett.*, **79**, pp. 570021–570025.

22. Graupner R., Hollering M., Ziegler A., Ristein J., and Ley L. (1997) Dispersions of surface states on diamond (100) and (111), *Phys. Rev. B*, **55**, pp. 10841–10847.

23. Grichko V., Tyler T., Grishko V. I., and Shenderova O. (2008) Nanodiamond particles forming photonic structures, *Nanotechnology*, **19**(22), pp. 2252011–2252015.

24. Pentecost A., Gour S., Mochalin V., Knoke I., and Gogotsi Y. (2010) Deaggregation of nanodiamond powders using salt- and sugar-assisted milling, *ACS Appl. Mater. Interfaces*, **2**, pp. 3289–3294.

Chapter 5

Raman and Photoluminescence Spectroscopy of Detonation Nanodiamonds

Igor I. Vlasov[a] **and Olga A. Shenderova**[b]

[a] *General Physics Institute, Russian Academy of Sciences, 38 Vavilov Street, Moscow 119991, Russia*
[b] *International Technology Center, 8100 Brownleigh Dr., Raleigh, NC 27617, USA*
vlasov@nsc.gpi.ru

Raman scattering (RS) spectroscopy enjoys broad application in diagnostics of carbon-based materials. This method appears particularly attractive because each allotropic form of carbon is easily identifiable by its Raman spectra, is nondestructive, and requires a small amount of material for analysis (in its confocal version, the volume of the sample can be as small as 1 μm^3). When applied to analysis of detonation nanodiamonds (DNDs), Raman spectroscopy can be used to advantage in determination of the perfection of crystal structure, phase purity, and crystallite size. The high quantum yield of luminescence of some impurity centers embedded in diamond and their absolute photostability have made nanodiamond (ND) one of the most promising materials for fabrication of single-photon emitters and photoluminescent biomarkers. This accounts for the pursuit of photoluminescent

Detonation Nanodiamonds: Science and Applications
Edited by Alexander Ya. Vul' and Olga A. Shenderova

ISBN 978-981-4411-27-1 (Hardcover), 978-981-4411-28-8 (eBook)
www.panstanford.com

properties of ND, including DND, having become a major goal in present-day nanophotonics.

5.1 Interpretation of DND Raman Spectra

A Raman spectrum of highly purified detonation diamond in the 1,000–2,000 cm^{-1} range exhibits typically four lines at 1,250 cm^{-1}, 1,326–1,332.5 cm^{-1}, 1,630–1,640 cm^{-1}, and 1,730 cm^{-1} (Fig. 5.1).

The line observed in the 1,326–1,332.5 cm^{-1} interval can be identified with the RS-active diamond lattice vibrations [1]. The line position and shape depend on the actual size of diamond nanocrystallites. This relation, attributed to phonon confinement in nanosized diamond crystallites, is discussed at some length in the next section. The broader line at 1,250 cm^{-1}, which resembles

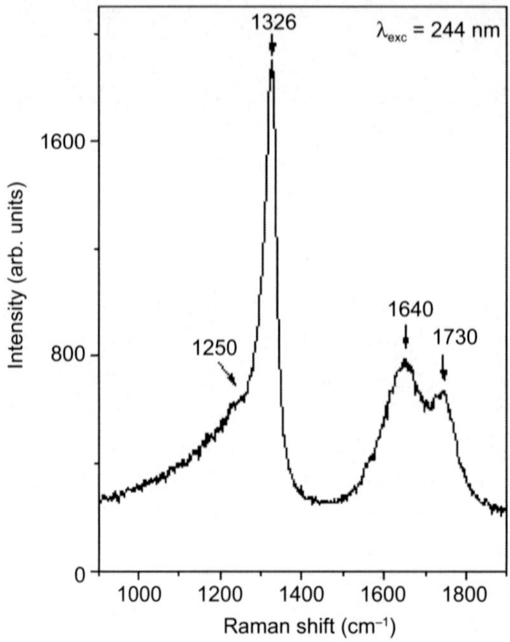

Figure 5.1 Typical Raman spectrum of highly purified DND measured under excitation by laser UV radiation at a wavelength of 244 nm in normal conditions. *Abbreviation*: UV, ultraviolet.

a low-frequency shoulder of the fundamental diamond line in the Raman spectrum, likewise evolves from phonon vibrations of the diamond lattice corresponding to the maximum in the distribution of the phonon density of states in the diamond. These vibrations are identified with the presence of small (<1 nm) crystallites in DND [2] or with a high concentration of extended defects in DNDs, which confine phonon vibrations to a distance of <3 nm [3]. The 1,730 cm^{-1} line present in DND Raman spectra is unambiguously attributed to valence vibrations of the carboxyl groups on the DND surface [2, 3]. The assignment of the line at 1,630–1,640 cm^{-1} is in no way unanimous. Different studies trace it to valence vibrations of sp^2-hybridized amorphous carbon on the surface of ND crystallites (*G* line) [4, 5], deformation vibrations of OH groups on the crystallite surface [3], or local lattice vibrations close to interstitial carbon atoms in the bulk of the crystallites [6].

5.2 Phonon Confinement

In the one-phonon region of the spectrum, Raman scattering (RS) in a perfect bulk diamond crystal produces a 2 cm^{-1} wide narrow line at 1,332.5 cm^{-1} [1]. Of the rich diversity of phonon modes existing in the diamond lattice, wave vector (**k**) selection rules show only zone-center phonons, that is, phonons with $q \approx 0$, to be Raman active. And it is the vibration frequencies of such phonons that lie close to 1,332.5 cm^{-1} [1].

As one approaches nanosized diamond crystallites, additional lines associated with the pronounced increase of the surface area to volume in ND material may appear in the Raman spectrum. Structural surface defects and various functional groups adsorbed on the surface render the Raman spectrum of NDs more complicated and dependent on the methods of their preparation used, surface chemistry, and other factors. The position and shape of the diamond line itself change. This finds the following explanation [7]. RS in the confined space of a diamond grain involves an uncertainty in the wave vectors of the phonons taking part in scattering equal to $2\pi/L$, where L is the characteristic grain dimension. This follows directly from the Heisenberg uncertainty principle or from Fourier analysis of spatially confined vibrations. Thus, a signal in the Raman spectrum of a small enough crystal will contain phonon modes from the uncertainty interval $\Delta\mathbf{k}$ in the Brillouin zone, whose width depends on the crystal dimensions. This effect of crystal size on the Raman spectrum has come to be known in literature as phonon confinement. For diamond, this effect becomes appreciable at crystallite dimensions of less than 30 nm [8, 9].

The size effect associated with phonon confinement was studied in most detail on silicon nanostructures. Richter was the first to propose a model for calculation of the low-frequency shift and the asymmetric shape of the silicon Raman line, which was based on the size of the Si crystallite under study [10]. By Richter's model, the Raman line of nanocrystalline silicon represents essentially superposition of phonon modes, taken with the corresponding weights and wave vectors q in the $2\pi L$ interval from the Brillouin zone center. The phonon modes are described by Lorentzians of width Γ and the maximum at the frequency $\omega(q)$. Three-dimensional integration in the Brillouin zone reduces to one-dimensional in the spherical Brillouin zone. Finally, the dependence of the silicon Raman line intensity on frequency is given by the relation

$$I(\omega) = \int_0^1 \frac{\exp(-q^2 L^2/4/)\, 4\pi q^2}{[\omega - \omega(q)]^2 - (\Gamma_0/2)^2}\, dq, \tag{5.1}$$

where the value of Γ_0 is assumed equal to the intrinsic silicon Raman linewidth and various branches of the dispersion relation coupling the frequency of phonons with their wave vector, $\omega(q)$, are approximated by one averaged dispersion relation

$$\omega(q) = A + B\cos(q\pi). \tag{5.2}$$

The model of Richter was successfully used in interpretation of the Raman line shape and estimation of the size of Si [11–13] and GaAs [14–15] grains.

The lattices of diamond and silicon have the same cubic symmetry, and lattice dynamics of the two crystals reveals the same pattern over the two crystallographic directions, with phonon frequency decreasing with increasing wave vector. The dispersion curve in diamond exhibits a specific behavior only in the [001] direction. At a certain distance from the zone center, the phonon frequency reaches a maximum corresponding to the frequency of

Table 5.1 Dependence of the diamond Raman line maximum position on a grain size

Grain size, L (nm)	∞	20	15	12	10	7	5
Diamond line position (cm^{-1})	1,332.50	1,331.88	1,331.50	1,331.13	1,330.75	1,329.38	1,327.05

$\sim$1,338 cm^{-1}, to fall off thereafter with increasing wave vector. Ager et al. [8] demonstrated that the specific dispersion curve affects only weakly the dispersion curve for ND averaged over all directions and that when calculating the diamond Raman line shape in terms of the phonon confinement model, the dispersion relation may be approximated by Eq. (5.2). Table 5.1 contains positions of the diamond Raman line maximum calculated [8] for different diamond crystallite sizes.

Attempts at observing the phonon confinement effect in Raman spectra did not meet with success for a long time. The observations were performed primarily on thin films of nanocrystalline diamond produced by chemical vapor deposition (CVD) on a silicon substrate. It was shown that the position and shape of the Raman spectral line in CVD diamond are affected predominantly by the high local lattice strains [8, 16, 17].

ND powder and suspensions produced by detonation synthesis (detonation nanodiamond [DND]) appear to be a more suitable material for probing the phonon confinement effect, although measurement of Raman spectra in DND is notoriously more difficult than that in the CVD diamond because the heat conductivity of a powder sample is substantially lower than that of aggregated material (a solid film). To exclude the impact of the shift of the diamond Raman line in position caused by overheating by the probing laser beam, the incident laser power density employed in the Raman analysis of DND powder samples is usually an order of magnitude lower than that used traditionally in analysis of CVD-fabricated DND powders.

Shenderova et al. [18] studied the position of the Raman line in spectra of samples of DNDs, which differed in the size of individual grains, coherent-scattering regions (CSRs) in the grains, and content of the nitrogen impurity. The size of the original diamond grains was

Table 5.2 Crystallite size characteristics and nitrogen concentration for three types of NDs obtained by detonation synthesis

Sample no.	Grain size (nm) (SAXS)	CSR size (nm) (XRD)	Nitrogen concentration (N/C) (%)	Diamond line position ($\pm 0.3\ cm^{-1}$)
DND-1	6	4	3	1,328.5
DND-2	6	4	<1	1,328.5
DND-3	34	9	<1	1,332.0

determined by two methods, more specifically by small-angle X-ray scattering (SAXS) and X-ray diffraction (XRD). In the first method, determination of the size is based essentially on the difference between the electron densities of the scattering particle and of the surrounding medium, and in the second, one probes the size of the region revealing the CSRs of X-rays in the material. It thus follows that the size found by XRD is, as a rule, smaller than or equal to that derived from the SAXS measurements. The size of the diamond grains, the concentration of nitrogen in them, and the position of the diamond lines in Raman spectra of the DND samples reported in Ref. [18] are presented in Table 5.2.

The Raman spectra of the samples reported in Table 5.2 are plotted in Fig. 5.2.

The positions of the diamond line presented in Table 5.2 for three samples fit nicely to the data calculated by Ager et al. [8] if we assume that the size of the phonon confinement region in the samples under study can be identified with the SAXS data. Indeed, by Ager et al., in diamond grains larger than 20 nm the shift of the Raman diamond line relative to its standard position at the frequency of 1,332.5 cm^{-1} should be less than 0.6 cm^{-1} (Table 5.1), exactly what is observed in the DND-3 sample with a characteristic grain size of 34 nm. The diamond lines in the Raman spectra of DND-1 and DND-2 with 6 nm crystallites are seen to be down-shifted in frequency to 1,328.5 cm^{-1}, a figure that likewise correlates well with the calculated data of Table 5.1, by which for crystallites 5 nm and 7 nm in size the diamond lines are shifted, respectively, to 1,327.05 cm^{-1} and 1,329.38 cm^{-1}. It appears appropriate to note here that despite the extremely high concentration of the nitrogen impurity in the bulk of DND [19], it does not affect the diamond line position

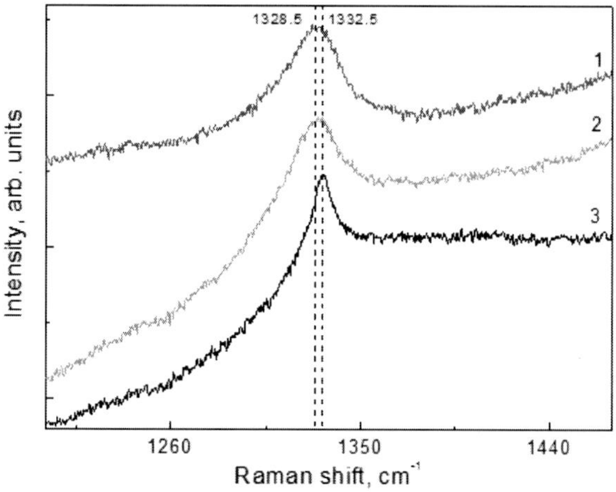

Figure 5.2 Raman spectra of DND-1, DND-2, and DND-3 samples measured at room temperature. Raman spectra were excited by 488 nm argon laser radiation. The Raman spectra were taken from Shenderova O. A., Vlasov I. I., Turner S., Van Tendeloo G., Orlinskii S. B., Shiryaev A. A., Khomich A. A., Sulyanov S. N., Jelezko F., and Wrachtrup J. (2011) Nitrogen control in nanodiamond produced by detonation shock-wave-assisted synthesis, *J. Phys. Chem. C*, **115**, pp. 14014–14024.

in DND, because the line shift in the spectra of DND-1 and DND-2 samples is the same, although the nitrogen contents in them are notably different.

In defect-free NDs, the characteristic dimensions of the phonon confinement region are determined by those of the crystallite. For crystallites with extended defects (twins, stacking faults, dislocation walls, etc.), the size of the confinement region will be determined, however, by the characteristic distances between defects, so in the case of anisotropic distribution of defects it may also be found related with crystallographic orientation [20]. DNDs exhibit typically the presence of narrow stacking faults [19] confined to the (111) plane (see Fig. 5.3).

In this case, phonon confinement will depend both on the transverse and on the longitudinal dimensions of individual faults (CSRs) rather than on their averaged size (presented, e.g., in Table 5.2).

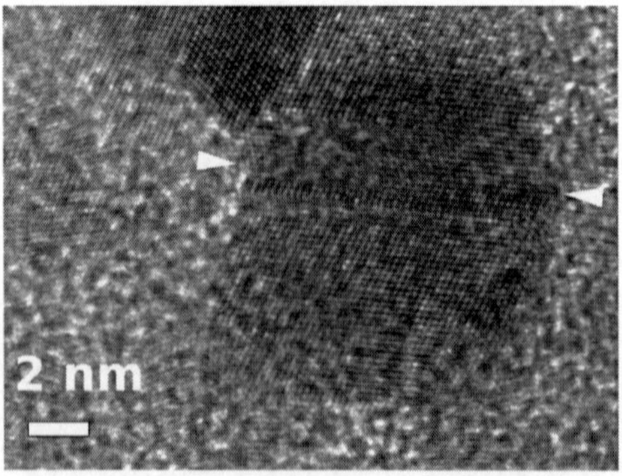

Figure 5.3 HRTEM image of a DND grain. Arrows denote boundaries of narrow twins. *Abbreviation*: HRTEM, high-resolution transmission electron microscopy.

As follows from Fig. 5.3, phonon vibrations propagating in the plane of the stacking faults perpendicular to the [111] direction are confined to the region that coincides in size with the diamond crystallite. Now in the [111] direction, spatial confinement of phonon vibrations is defined by the characteristic thickness of the stacking faults, which is ~1 nm.

For such a small region, phonons with wave vectors of about $\pi/0.5\ \text{nm}^{-1}$, that is, practically all possible phonon vibrations of the diamond lattice, become Raman active. This is what accounts for the observation in Raman spectra of DNDs, besides the comparatively narrow band close to 1,332.5 cm^{-1}, of a wide band in the region of 1,250 cm^{-1}, which can be identified with one of the maxima in the distribution of the density of phonon states of the diamond lattice in frequency (see above). Studies of phonon confinement in DND Raman spectra [21, 22] invoked data on the crystallite size obtained by XRD measurements.

The models employed usually in this method to estimate the CSR dimensions disregard the size anisotropy characteristic of this material. As a result, the shift of the diamond line in the Raman spectra of the samples studied deviated slightly from the figures obtained with Ager's model. The authors include the contribution to the Raman spectrum due to larger particles, which are shown by transmission electron microscopy (TEM) and SAXS measurements to be present in some amounts in DND in order to make them correspond better to the observed shape of the diamond line in DND spectra.

5.3 Photoluminescence of DND

5.3.1 *Single-Photon Emitters Based on Luminescent Nanodiamonds*

Progress in the technology of nanosystems and nanomaterials stimulated the development of devices whose operation, rather than obeying the laws of classical physics, is based on quantum mechanical principles. One of the most promising directions pursued in the creation of quantum mechanical systems in the beginning of the 21st century has become a new approach in informatics, more specifically quantum processing of information (QPI) [23–25]. An ever-growing demand for increasing computer power, the data transmission rate, and protection of transferred data (cryptography) has stimulated a search for radically new approaches to information processing and transfer, which has culminated in the development of QPI.

The development of the principles underpinning quantum mechanical information processing [26] translated into a search for systems that could serve as a platform for realization of QPI. Among a fairly limited array of solid materials (superconductors, semiconductor quantum dots), which are potentially suitable for engineering realization of QPI, point defects in diamond are the most promising candidates because they are presently the only optically stable room-temperature sources of emission [27]. The negatively charged "nitrogen vacancy" (NV) defect occupies a prominent place in this context. This defect has an electron spin (i.e., is paramagnetic) and consists of a nitrogen atom at a site of the diamond lattice and a vacancy in the neighboring lattice site, as shown in Fig. 5.4.

Two different energy states of the spin are ideally suited for realization of a two-level quantum system. With the C^{13} isotope having no nuclear spins that would interact with the spin of the NV center, the diamond lattice is essentially an ideal matrix for the spin

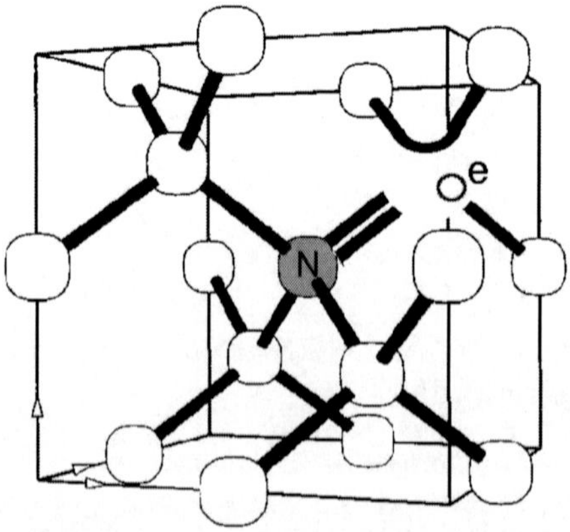

Figure 5.4 Structure of the NV^- center in the diamond lattice.

qubit, thus permitting operation with it in normal conditions due to its coherence time (on the order of a few millisecond) being the longest among solids at room temperature [28–30].

The single spin in the diamond can be used also as a nanomagnetometer with atomic-scale spatial resolution [31]. Conventional magnetic resonance methods fail in the detection of a single spin. The presence of an excited electronic state and the concomitant efficient photoluminescence (PL) of the defect in one of its spin states offer a possibility of initializing and recording single-spin states by hypersensitive optical methods [23]. This requires the development of selectively detectable single NV centers (single-photon photoemitters) in the diamond lattice.

Two main approaches to the formation of arrays of single-photon emitters of the desired architecture involving NV centers are presently pursued. The first of them assumes formation of NV centers in the bulk of nitrogen-free diamonds by implantation of nitrogen into the diamond lattice. The second way lies in creating NV centers in isolated diamond nanoparticles.

In the first approach, NV centers are formed in a spin-free diamond crystal, thus excluding the undesirable interaction of NV spins with other spins and increasing their coherence time. The use of a focused nitrogen ion beam to produce NV centers of a specified geometry was reported [32]. Ions slowing down in the diamond lattice knock carbon atoms out of their sites and form in this way vacancies. Annealing combines the implanted nitrogen atom with the vacancy into an NV complex. The efficiency of formation of NV defects was not high, about 2.5%, presumably because of the small number of vacancies capable of interacting efficiently with the nitrogen. This approach to the formation of NV defects was improved by incorporating additional implantation of carbon ions into the diamond lattice [33]. In this way one succeeded in reaching a more than 50% yield in the formation of NV defects.

Figure 5.5 visualizes the luminescent surface of a diamond. At the top one can see a part of the surface implanted by nitrogen ions only and at the bottom an area with implanted ions of nitrogen and carbon. In the bottom part, the density of luminescent NV defects is markedly higher. The diamond plate used in the experiment was grown by chemical deposition from the gas phase (CVD method) with an extremely low nitrogen concentration (less than 1 ppb).

The second approach to the formation of NV centers in diamond involves preparation of ND particles containing one or two such centers. One makes use here of doping the diamond with nitrogen in the stage of its synthesis. Here NV centers form in the natural way, also in the course of synthesis, and the process allows intensification by irradiating the material with electrons and annealing after completion of the synthesis. This approach contains important aspects that need to be clarified, namely, (1) the effect of ND size on the efficiency of NV center production and (2) the optimum conditions of synthesis favoring formation of the NV centers with a high probability in each nanoparticle of the diamond.

Thus far, emphasis has been placed primarily on confronting the first problem by using NDs prepared by the "high pressure–high temperature" (HPHT) method. The dimensions of the starting HPHT diamonds lying in the micrometer range, production of NDs from them requires application of the labor-intensive procedure of grinding and fractionation of this material [34, 35]. The crystallites of HPHT diamond, typically 7 nm in size, prepared in this way demonstrated a remarkable stability of single NV centers under illumination (Fig. 5.6).

Significantly, the efficiency of NV center production by electron bombardment was found to be independent of crystallite size.

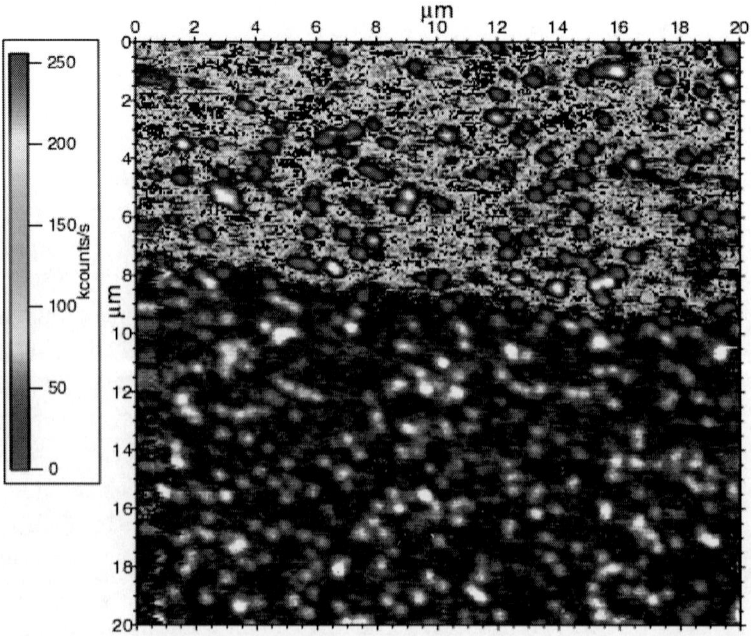

Figure 5.5 Image of two parts of a luminescent diamond surface. The top part was implanted by nitrogen ions only and the bottom part by nitrogen and carbon ions. The bright dots are the single luminescent NV centers. Reproduced with permission from Naydenov B., Richter V., Beck J., Steiner M., Neumann Ph., Balasubramanian G., Achard J., Jelezko F., Wrachtrup J., and Kalish R. (2010) Enhanced generation of single optically active spins in diamond by ion implantation, *Appl. Phys. Lett.*, **96**, pp. 163108–163110. Copyright 2010, American Institute of Physics.

The PL lifetime of NV centers in nanocrystals, about 17 ns, did not practically differ from that measured on bulk crystals. After the corresponding treatment of the surface of ND crystallites the relaxation time of the electron spins associated with the NV centers was a few microseconds, a figure that likewise does not differ much from the values characteristic of bulk diamond. Thus one can say that the properties of NV centers in HPHT NDs vary little compared with those in bulk diamond material.

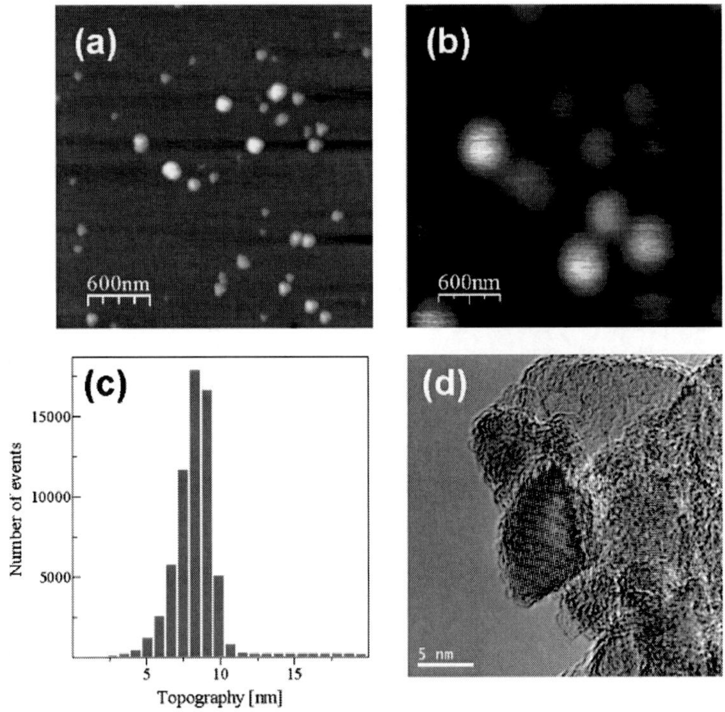

Figure 5.6 (a) Topography of diamond nanocrystals dispersed on a quartz substrate, which was obtained by AFM. (b) Image of luminescent diamond nanoparticles in the same fragment of the substrate as in (a). (c) Size distribution histogram of the nanoparticles, demonstrating an average size of 7-8 nm. (d) A high-resolution electron microscope image of a diamond nanoparticle. Reproduced with permission from Tisler J., Balasubramanian G., Naydenov B., Kolesov R., Grotz B., Reuter R., Boudou J.-P., Curmi P. A., Sennour M., Thorel A., Borsch M., Aulenbacher K., Erdmann R., Hemmer Ph. R., Jelezko F., and Wrachtrup J. (2009) Fluorescence and spin properties of defects in single digit nanodiamonds, *ACS Nano*, **3**, pp. 1959–1965. Copyright 2009, American Chemical Society. *Abbreviation*: AFM, atomic force microscopy.

5.3.2 *Optical Markers Based on Luminescent Nanodiamonds*

The need for development of fluorescent markers of a new type for biology and medicine arises from the fact that the fluorescent indicators in wide use presently, which are based on various fluorescent dyes, are plagued by a number of shortcomings; indeed, they depend to a great extent on solution acidity, are optically unstable, and are toxic for animal systems.

The fluorescent nanoparticles called quantum dots, which proved their usefulness as markers, cannot be broadly employed in vivo because of their toxicity. Using NDs as markers in biosystems and in medicine exhibits a number of obvious advantages [36, 37]. The core of a DND is 4–6 nm in size, which accounts for the highly developed polyfunctional surface of the material. The various functional groups formed on the surface of NDs in their synthesis or purposefully grown by chemical modification of the DND surface in solution provide a possibility of various compounds attaching to the diamond surface, including amino acids and proteins.

Experiments suggest that NDs do not exert harmful action on biological objects; indeed, they are mechanically, chemically, and thermally stable; exhibit a high hydrolytic stability within a broad range of acidity variation; possess remarkable sorption properties to various compounds of biological and nonbiological nature; and possess an array of unique mechanical characteristics, such as rigidity, high strength of the matrix, and absence of swelling in solvents. The above properties make NDs attractive adsorbents for excretion of undesirable and toxic compounds (products of metabolism, heavy metals, radionuclides) from the organism and convenient platforms for drug delivery in medicine. The extremely high fluorescence intensity achievable in NDs makes them attractive as optical markers in biomedical diagnostics [36].

Biomedical experiments making use of luminescent NDs employ usually the high-efficiency luminescence of the NV centers and the intense broadband PL, which presumably originates from structural defects on the ND surface. It should be stressed that nanosized HPHT particles of static diamond generate bright PL originating from the NV centers, whereas the content of these centers in typical DND is very low. Therefore the results of cellular studies presented below are those obtained with HPHT ND. The potential of DND as markers is discussed below.

A study of the possibility of PL intensity enhancement of NV centers in NDs with a characteristic size of 100 nm synthesized

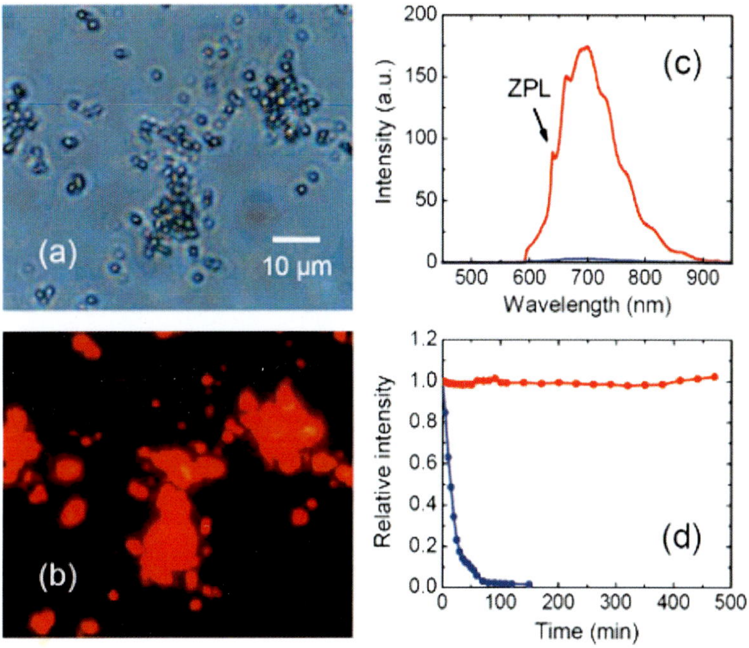

Figure 5.7 (a) Image of 100 nm diamond powder as viewed with an optical microscope and (b) PL distribution pattern obtained in the 590-900 nm range for the same region of the sample. (c) PL spectra of the proton-irradiated (red) and original (blue) ND. (d) Optical stability test: evolution of the PL signal with the time of illumination by a mercury lamp for ND (red curve) and polystyrene (blue curve). Reproduced with permission from Yu S. J., Kang M. W., Chang H. C., Chen K. M., and Yu Y. C. (2005) Bright fluorescent nanodiamonds: no photobleaching and low cytotoxicity, *J. Am. Chem. Soc.*, **127**, pp. 17604–17605. Copyright 2005, American Chemical Society.

by the HPHT method [38] (Fig. 5.7) showed such diamonds to be nontoxic in human liver cells. In this study, diamond nanoparticles were irradiated by 3 MeV protons to a dose of 5×10^{15} ions/cm^2 to create a high concentration of vacancies forming optically active complexes with a nitrogen impurity in the course of thermally stimulated migration over the diamond crystal lattice. Annealing of the irradiated sample resulted in amplification of the NV center luminescence signal peaking at about 700 nm by two orders of magnitude (Fig. 5.7c). The optical stability of the centers was tested

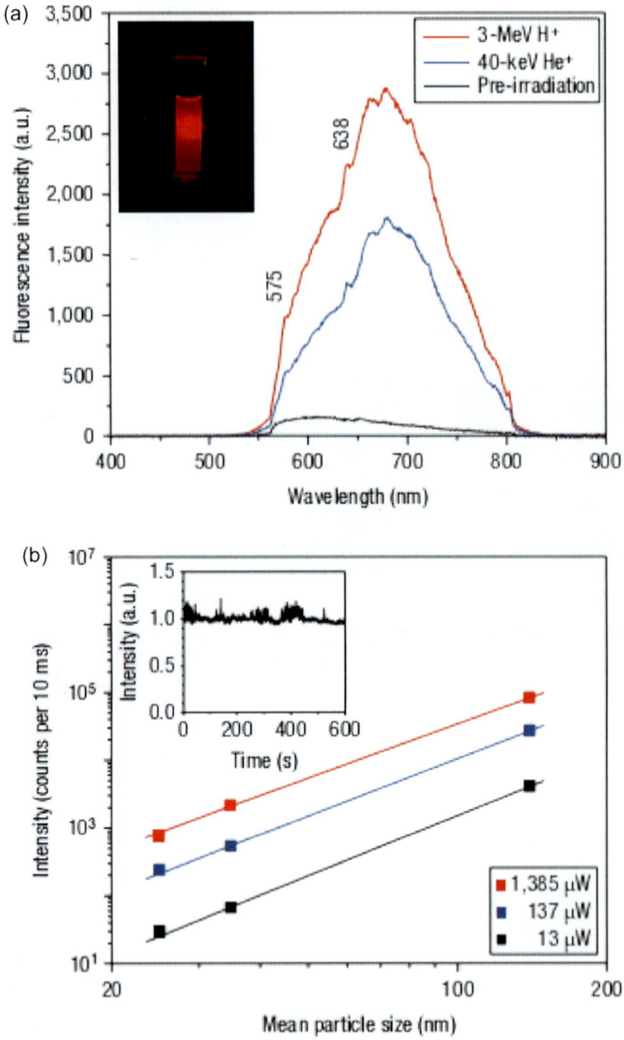

Figure 5.8 (a) Fluorescence spectra of 35 nm DND powder dispersed in water (1 mg/mL) before and after irradiation with hydrogen and helium ions. Inset shows an image of a suspension of 35 nm DND powder under illumination with a laser operating at a wavelength of 532 nm. (b) PL intensity plotted vs. DND size for different laser power levels. Reproduced with permission from Chang Y. R., Lee H. Y., Chen K., Chang C. C., Tsai D. S., Fu C. C., Lim T. S., Tzeng Y. K., Fang C. Y., Han C. C., Chang H. C., and Fann W. (2008) Mass production and dynamic imaging of fluorescent nanodiamonds, *Nat. Nanotech.*, **3**, pp. 284–288. Copyright 2008, Nature Publishing Group.

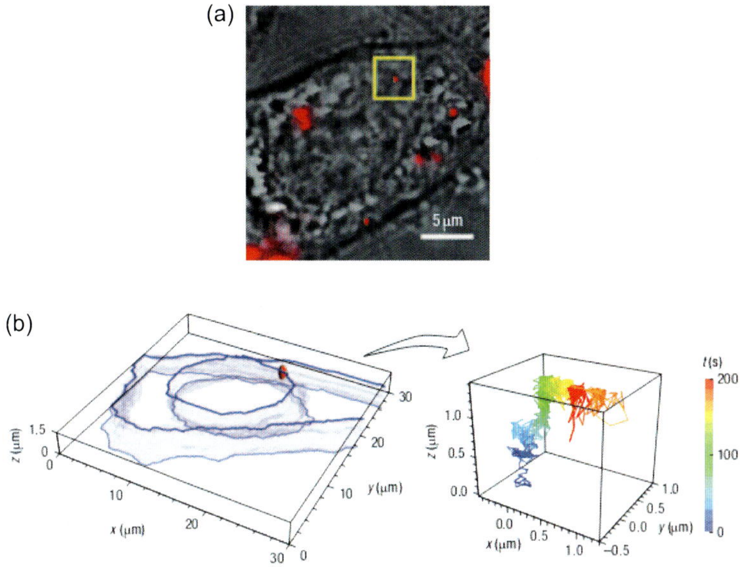

Figure 5.9 Three-dimensional monitoring of a 35 nm diamond particle in a living HeLa cell. (a) Optical image of a living cell matched with point PL emitted by particulate NDs (red points). (b) Three-dimensional reconstruction (left part) of the boundaries of the cellular nucleus and cytoplasm. Three-dimensional trajectory (right part) of displacement of a diamond particle inside a cell during 200 s. Reproduced with permission from Chang Y. R., Lee H. Y., Chen K., Chang C. C., Tsai D. S., Fu C. C., Lim T. S., Tzeng Y. K., Fang C. Y., Han C. C., Chang H. C., and Fann W. (2008) Mass production and dynamic imaging of fluorescent nanodiamonds, *Nat. Nanotech.*, **3**, pp. 284–288. Copyright 2008, Nature Publishing Group.

by illuminating the sample under study with a powerful mercury lamp for eight hours, with its subsequent comparison with that of the organic dye polystyrene subjected to a similar treatment. This treatment was found not to affect the PL intensity of the NV centers, whereas the intensity of the polystyrene signal dropped to zero after 30 minutes of such processing (Fig. 5.7d).

A later study [39] showed other light ions (of helium) to be as efficient for preparation of bright fluorescent NDs (Fig. 5.8). The efficiency of formation of NV centers as a function of ND size was also studied in the 25–150 nm range. It was demonstrated that the efficiency of PL intensity enhancement by bombardment by helium

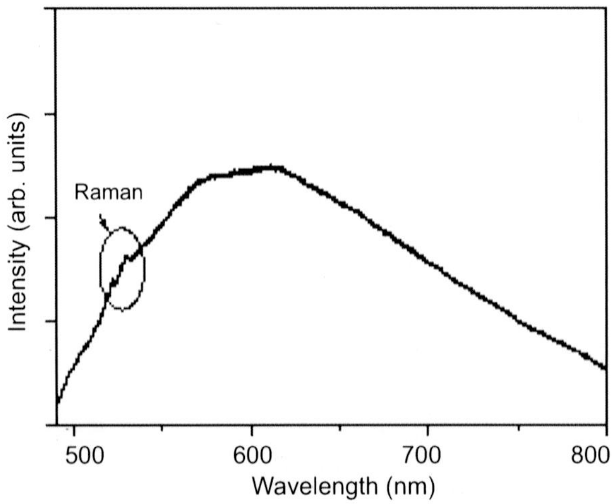

Figure 5.10 Typical PL spectrum of DND measured under excitation with 488 nm laser radiation in normal conditions.

ions does not decrease with nanoparticle size decreasing down to 25 nm.

It was demonstrated that the bright and stable PL of diamond nanoparticles provides a possibility of three-dimensional monitoring of individual diamond particles migrating in a living cell (Fig. 5.9).

5.3.3 *Photoluminescence of DND*

For a long time, studies of DND spectra demonstrated only wide-band PL within the 400–700 nm interval, with a maximum depending on the excitation wavelength (Fig. 5.10). Its origin is tentatively assigned to structural defects present on the surface of DND particulates [40, 41]. It was assumed that optically active defects (primarily those related with nitrogen) either are absent in the bulk of the crystallites or their concentration is extremely low. These assumptions were underpinned by theoretical predictions that the energy supporting the stability of nitrogen atoms in the diamond lattice core should decrease with a decrease of its size [42].

The first successive observation of nitrogen-containing defects consisting of a single nitrogen atom and a vacancy occupying

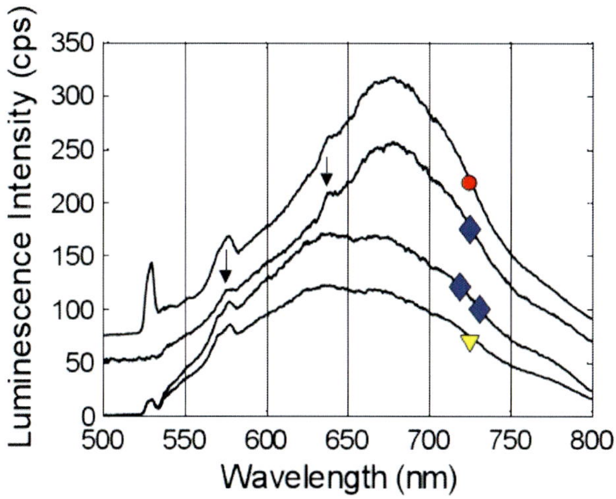

Figure 5.11 PL spectra measured under excitation by pulsed laser radiation at a wavelength of 531 nm. The spectra recorded before (yellow triangle) and after (two blue rhombuses) DND irradiation by protons feature only a broad structureless band in the 550-750 nm range. The narrow bands in the 572–578 nm interval belong to the DND Raman spectrum. The spectrum of irradiated DND measured with a 3.5 ns delay relative to the excitation pulse is dominated by lines of the neutral (NV^0) and negatively charged (NV^-) centers, with their zero-phonon components located at wavelengths of 575 nm and 637 nm (one blue rhombus). Also shown for comparison is a spectrum of proton-bombarded 55 nm diamonds produced by the HPHT method (red circle). The spectrum was measured in the ungated mode, and it is likewise dominated by lines of the NV^0 and NV^- centers. Spectra measured without delay following the excitation pulse also feature the 531 nm laser line. Reproduced with permission from Smith B. R., Inglis D. W., Sandnes B., Rabeau J. R., Zvyagin A. V., Gruber D., Noble C. J., Vogel R., Osawa E., and Plakhotnik T. (2009) Five-nanometer diamond with luminescent nitrogen-vacancy defect centers, *Small*, **5**, pp. 1649–1653. Copyright 2009, Wiley-VCH Verlag GmbH & Co. KGaA, Weinheim.

neighboring lattice sites (the NV centers) in DND was made by time-resolved pulsed PL spectroscopy [43].

To increase the concentration of vacancy-containing centers, in particular of the NV centers, in the samples under study, one invoked the standard approach of irradiation of crystals by high-energy particles (in our case, by protons). It was found that the PL lifetime

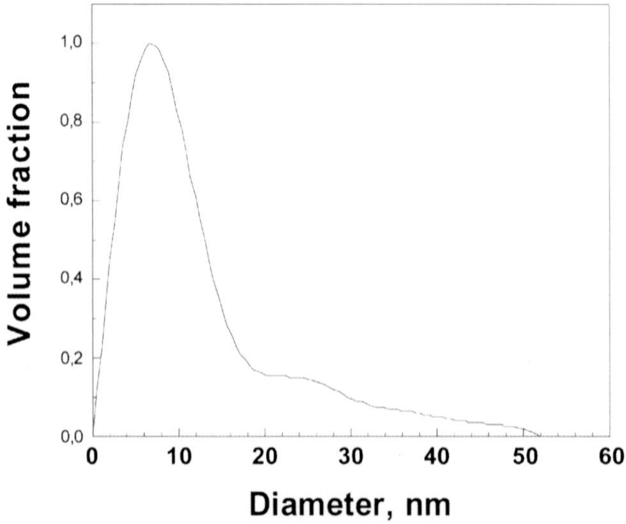

Figure 5.12 Characteristic polydisperse distribution of DND in size obtained by SAXS. The maximum of the distribution corresponds to a particle size of 6 nm. Reproduced with permission from Vlasov I. I., Shenderova O., Turner S., Lebedev O. I., Basov A. A., Sildos I., Rahn M., Shiryaev A. A., and Van Tendeloo G. (2010) Nitrogen and luminescent nitrogen-vacancy defects in detonation nanodiamond, *Small*, **6**, pp. 687–694. Copyright 2010, Wiley-VCH Verlag GmbH & Co. KGaA, Weinheim.

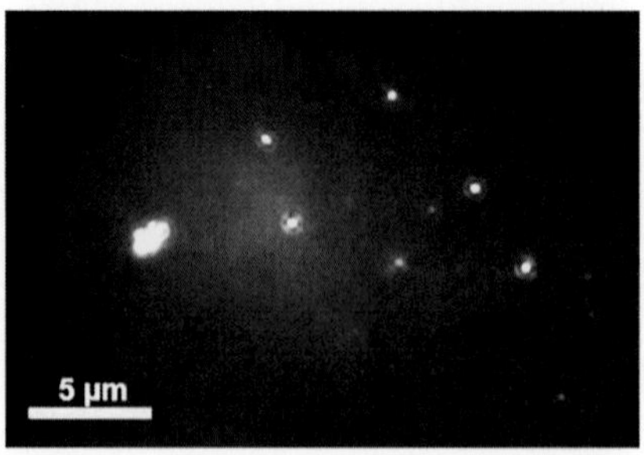

Figure 5.13 Image of the distribution of integrated PL intensity from DND particles dispersed on a silicon substrate. The luminescence was excited by a 532 nm laser radiation ion a ≈50 μm diameter spot on the substrate. Reproduced with permission from Vlasov I. I., Shenderova O., Turner S., Lebedev O. I., Basov A. A., Sildos I., Rahn M., Shiryaev A. A., and Van Tendeloo G. (2010) Nitrogen and luminescent nitrogen-vacancy defects in detonation nanodiamond, *Small*, **6**, pp. 687–694. Copyright 2010, Wiley-VCH Verlag GmbH & Co. KGaA, Weinheim.

of surface defects does not exceed 1 ns, whereas for the NV centers it is on the order of 10 ns. By properly using for the PL excitation a pulsed laser operating at 531 nm with a pulse length of 0.6 ns, one succeeded in suppressing efficiently the PL signal of surface defects to allow reliable detection of the PL of NV centers, with the measurements delayed by 3.5–6 ns relative to the laser pulse (Fig. 5.11).

Because DND crystallites exhibit, as evident from Fig. 5.12, polydisperse distribution in size extending from 0 nm to 50 nm, it remained unclear of what size were the particles that produced the dominant contribution to the NV center PL reported in Ref. [43].

The transition to studying the luminescence of DND particulates irradiated by electrons opened a way to radical suppression of the background luminescence generated by surface defects, which immediately revealed bright luminescence of NV centers in single crystallites or small aggregates of DND grains initiated by continuous wave (CW) laser excitation (Fig. 5.13) [44].

By combining the study of the photoemissive properties of DND with Raman spectroscopy, one succeeded in estimating the size of the diamond nanocrystallites that produce intense PL associated with NV centers. Raman spectra measured at bright luminescent points revealed the absence of any shift of the diamond line with respect to the 1,332.5 cm^{-1} frequency, a finding that, by the phonon confinement model (see Section 5.2), suggests that the intense local luminescence of the NV centers is initiated in crystallites >20 nm in size (Fig. 5.14).

Subsequent measurements performed with the use of PL detection systems of enhanced sensitivity revealed single NV centers in selected nonirradiated 5 nm DND grains [45] (Fig. 5.15).

It was demonstrated that the PL generated by a 5 nm grain blinks with a certain frequency [45]. Such blinking luminescence is characteristic of many fluorescent nanosystems, including fluorescent proteins and a variety of quantum dots [46], which interferes with their application. In the particular case of NDs, however, a way was found to stabilize the luminescence of the NV centers by embedding diamond nanocrystallites in a polymer matrix [45].

Quite recently scientists in St. Petersburg (Russia) have succeeded in producing DND with an anomalously high concentration

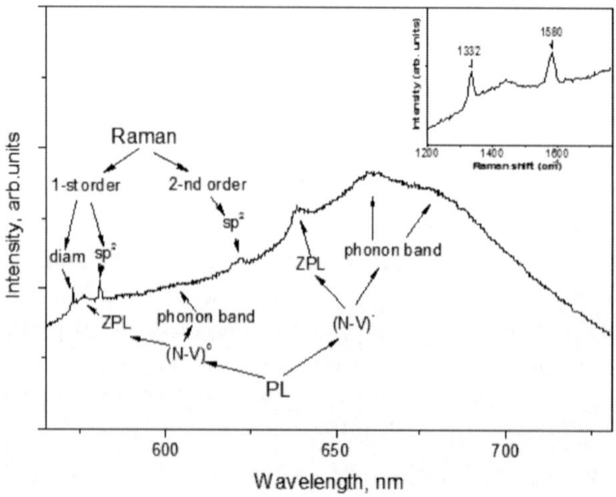

Figure 5.14 PL/Raman spectrum measured from one of the bright luminescent dots visualized in Fig. 5.13. The spectral range covering the Raman spectrum is shown in the inset in cm^{-1} units. The spectrum was measured under laser excitation at a wavelength of 532 nm in normal conditions. Reproduced with permission from Vlasov I. I., Shenderova O., Turner S., Lebedev O. I., Basov A. A., Sildos I., Rahn M., Shiryaev A. A., and Van Tendeloo G. (2010) Nitrogen and luminescent nitrogen-vacancy defects in detonation nanodiamond, *Small*, **6**, pp. 687–694. Copyright 2010, Wiley-VCH Verlag GmbH & Co. KGaA, Weinheim.

of NV centers [47]. This was reached by sintering the starting 4–5 nm DND crystallites at $T = 1,500°C$ and a pressure $P = 6$ GPa for 20 seconds. As follows from structural studies of the sintered DND powder, sintering initiates self-organization of the starting diamond nanocrystallites into oriented aggregates about 1–10 μm in size. Figure 5.16 shows an image of luminescent diamond particles obtained with an optical microscope under illumination of the sintered DND powder by laser radiation at a wavelength of 532 nm, together with a typical PL spectrum of one of the luminescent diamond clusters.

Estimates based on the magnitude of the electron paramagnetic resonance (EPR) signal from NV centers in sintered DND clusters yield $10^{20}–10^{21}$ cm^{-3} for the concentration of NV centers in this material [47]. This result holds forth considerable promise in the

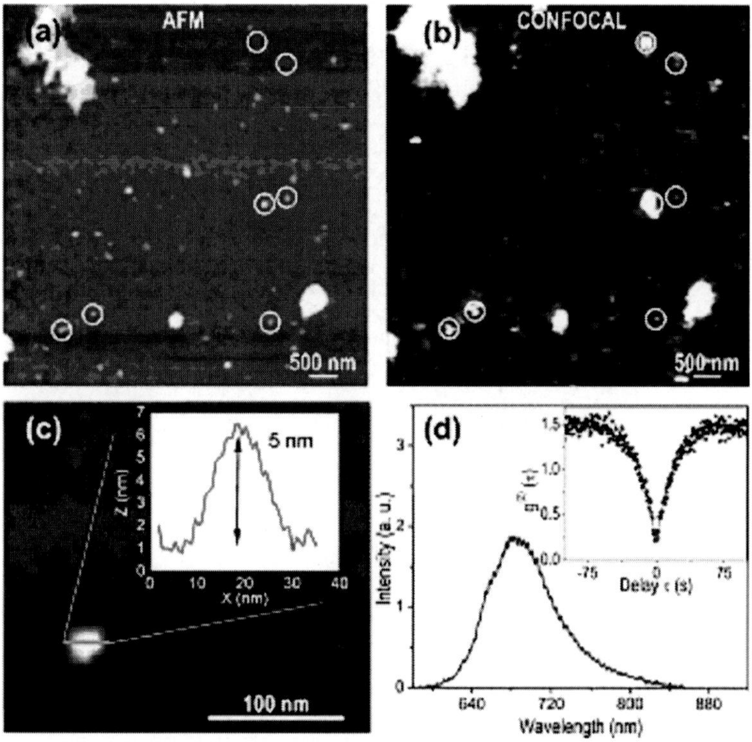

Figure 5.15 (a) AFM image of the topography of diamond nanocrystals dispersed on a glass substrate. (b) Image of luminescent diamond crystallites obtained with a confocal microscope on the same fragment of the substrate as that of panel (a). (c) Magnified AFM image and the corresponding surface profile of a 5 nm crystallite. (d) PL spectrum of the same crystallite and (inset) second-order correlation function of PL intensity, which suggests that the spectrum of the crystallite under study was produced by a single NV center. Reproduced with permission from Bradac C., Gaebel T., Naidoo N., Sellars M. J., Twamley J., Brown L. J., Barnard A. S., Plakhotnik T., Zvyagin A. V., and Rabeau J. R. (2010) Observation and control of blinking nitrogen-vacancy centres in discrete nanodiamonds, *Nat. Nanotech.*, **5**, pp. 345–349. Copyright 2010, Nature Publishing Group.

technology of production of NDs with a high concentration of NV centers.

In conclusion, one of the major achievements in the ND community is related to production of ND particles containing

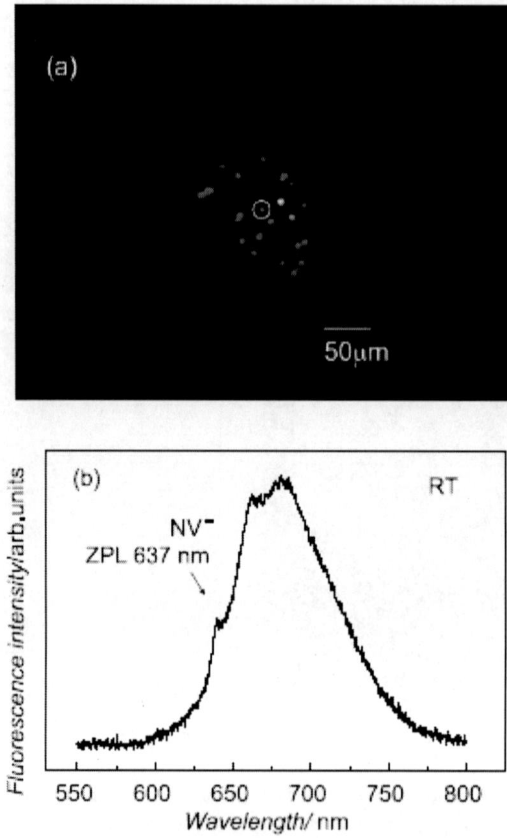

Figure 5.16 (a) Image of luminescent diamond particles obtained with an optical microscope under illumination of a sintered DND powder sample with laser radiation at a wavelength of 532 nm and a filter with a limited transmission band (>600 nm) for luminescence cutoff. (b) Room-temperature PL spectrum of a diamond cluster, circled in panel (a). The spectrum is dominated by PL lines of NV^- centers with a zero-phonon line at 637 nm. Reproduced with permission from Baranov P. G., Soltamova A. A., Tolmachev D. O., Romanov N. G., Babunts R. A., Shakhov F. M., Kidalov S. V., Vul' A. Ya., Mamin G. V., Orlinskii S. B., and Silkin N. I. (2011) Enormously high concentrations of fluorescent nitrogen-vacancy centers fabricated by sintering of detonation nanodiamonds, *Small*, **7**, pp. 1533–1537. Copyright 2011, Wiley-VCH Verlag GmbH & Co. KGaA, Weinheim.

specific impurity defects, resulting in NDs with outstanding photoluminescent properties and the ability to manipulate their spin state. NDs with color centers are aimed to revolutionize biological imaging and quantum optics applications. However, a central focus of many research groups remains the ability to control production of NV and other color centers in nanoscale diamond.

References

1. Solin S. A., and Ramdas A. K. (1970) Raman spectrum of diamond, *Phys. Rev. B*, **1**, pp. 1687–1698.
2. Mykhaylyk O. O., Solonin Y. M., Batchelder D. N., and Brydson R. (2005) Transformation of nanodiamond into carbon onions: a comparative study by high-resolution transmission electron microscopy, electron energy-loss spectroscopy, x-ray diffraction, small-angle x-ray scattering, and ultraviolet Raman spectroscopy, *J. Appl. Phys.*, **97**, 074302.
3. Mochalin V., Osswald S., and Gogotsi Y. (2009) Contribution of functional groups to the Raman spectrum of nanodiamond powders, *Chem. Mater.*, **21**, pp. 273–279.
4. Vul' A. Ya. (2006) *Ultrananocrystalline Diamond*, eds. Shenderova O., and Gruen D., "Characterization and physical properties of UNCD particles" (William Andrew, Norwich, New York, USA), pp. 379–404.
5. Aleksenskii A. E., Osipov V. Y., Vul' A. Ya., Ber B. Y., Smirnov A. B., Melekhin V. G., Adriaenssens G. J., and Iakoubovskii K. (2001) Optical properties of nanodiamond layers, *Phys. Solid State*, **43**(1), pp. 145–150.
6. Prawer S., Nugent K. W., Jamieson D. N., Orwa J. O., Bursill L. A., and Peng J. L. (2000) The Raman spectrum of nanocrystalline diamond, *Chem. Phys. Lett.*, **332**(1–2), pp. 93–97.
7. Prawer S., and Nemanich R. J. (2004) Raman spectroscopy of diamond and doped diamond, *Phil. Trans. R. Soc. Lond. A*, **362**, pp. 2537–2565.
8. Ager III J. W., Veirs D. K., and Rosenblatt G. M. (1991) Spatially resolved Raman studies of diamond films grown by chemical vapor deposition, *Phys. Rev. B*, **43**, pp. 6491–6499.
9. Sun Z., Shi J. R., Tay B. K., and Lau S. P. (2000) UV Raman characteristics of nanocrystalline diamond films with different grain size, *Diam. Relat. Mater.*, **9**, pp. 1979–1983.
10. Richter H., Wang Z. P., and Ley L. (1981) The one phonon Raman spectrum in microcrystalline silicon, *Solid State Commun.*, **39**, pp. 625–629.

11. Gonzalez-Hernandez J., Azarbayejani G. H., Tsu R., and Pollak F. H. (1985) Raman, transmission electron microscopy, and conductivity measurements in molecular beam deposited microcrystalline Si and Ge: a comparative study, *Appl. Phys. Lett.*, **47**, pp. 1350–1352.
12. Campbell H., and Fauchet P. M. (1986) The effects of microcrystal size and shape on the one phonon Raman spectra of crystalline semiconductors, *Solid State Commun.*, **58**, pp. 739–741.
13. Fauchet and P. M., and Campbell I. H. (1988) Raman spectroscopy of low-dimensional semiconductors, *Crit. Rev. Solid State Mater. Sci.*, **14**, pp. S79–S101.
14. Tiong K. K., Amirtharaj P. M., Pollak F. H., and Aspnes D. E. (1984) Effects of As^+ ion implantation on the Raman spectra of GaAs: "spatial correlation" interpretation, *Appl. Phys. Lett.*, **44**, p. 122.
15. Braunstein G., Tuschel D., Chen S., and Lee S. T. (1989) Raman scattering study of lattice disorder in 1 MeV Si-implanted GaAs, *J. Appl. Phys.*, **66**, pp. 3515–3522.
16. Bergman L., and Nemanich R. J. (1995) Raman and photoluminescence analysis of stress state and impurity distribution in diamond thin films, *J. Appl. Phys.*, **78**, pp. 6709–6719.
17. Le Grice Y. M., Nemanich R. J., Glass J. T., Lee Y. H., Rudder R. A., and Markunas R. J. (1990) *Domain Size Determination in Diamond Thin Films*, Vol. 162, eds. Glass J. T., Messier R. F., and Fujmori N., "Diamond, silicon carbide and related wide-bandgap semiconductors" (Materials Research Society, Warrendale, Pennsylvania, USA), p. 219.
18. Shenderova O. A., Vlasov I. I., Turner S., Van Tendeloo G., Orlinskii S. B., Shiryaev A. A., Khomich A. A., Sulyanov S. N., Jelezko F., and Wrachtrup J. (2011) Nitrogen control in nanodiamond produced by detonation shock-wave-assisted synthesis, *J. Phys. Chem. C*, **115**, pp. 14014–14024.
19. Turner S., Lebedev O. I., Shenderova O., Vlasov I. I., Veerbeck J., and Tendeloo G. Van (2009) Determination of size, morphology, and nitrogen impurity location in treated detonation nanodiamond by transmission electron microscopy, *Adv. Funct. Mater.*, **19**, pp. 2116–2124.
20. Gouadec G., and Colomban Ph. (2007) Raman spectroscopy of nanomaterials: how spectra relate to disorder, particle size and mechanical properties, *Prog. Cryst. Growth Charact. Mater.*, **53**, pp. 1–56.
21. Obraztsova E. D., Korotushenko K. G., Pimenov S. M., Ralchenko V. G., Smolin A. A., Konov V. I., and Loubnin E. N. (1995) Raman

and photoluminescence investigations of nanograined diamond films, *Nanostruct. Mater.*, **6**, pp. 827–830.

22. Osswald S., Mochalin V., Havel M., Yushin G., and Gogotsi Y. (2009) Phonon confinement effects in the Raman spectrum of nanodiamond, *Phys. Rev. B*, **80**, 075419.
23. Gruber A., Drabenstedt A., Tietz C., Fleury L., Wrachtrup J., and von Borczyskowski C. (1997) Scanning confocal optical microscopy and magnetic resonance on single defect centers, *Science*, **276**, pp. 2012–2014.
24. Wang C., Kurtsiefer C., Weinfurter H., and Burchard B. J. (2006) Single photon emission from SiV centres in diamond produced by ion implantation, *Phys. B*, **39**, pp. 37–41.
25. Gaebel T., Popa I., Gruber A., Domhan M., Jelezko F., and Wrachtrup J. (2004) Stable single-photon source in the near infrared, *New J. Phys.*, **6**, p. 98.
26. Jelezko F., and Wrachtrup J. (2006) Single defect centres in diamond: a review, *Phys. Stat. Sol.(a)*, **203**, pp. 3207–3225.
27. Rabeau J. R., Stacey A., Rabeau A., Prawer S., Jelezko F., Mirza I., and Wrachtrup J. (2007) Single nitrogen vacancy centers in chemical vapor deposited diamond nanocrystals, *Nano Lett.*, **7**, pp. 3433–3437.
28. Gurudev Dutt M. V., Childress L., Jiang L., Togan E., Maze J., Jelezko F., Zibrov A. S., Hemmer P. R., and Lukin M. D. (2007) Quantum register based on individual electronic and nuclear spin qubits in diamond, *Science*, **316**, pp. 1312–1316.
29. Balasubramanian G., Neumann P., Twitchen D., Markham M., Kolesov R., Mizuochi N., Isoya J., Achard J., Beck J., Tisler J., Jacques V., Hemmer P. R., Jelezko F., and Wrachtrup J. (2009) Ultralong spin coherence time in isotopically engineered diamond, *Nat. Mater.*, **8**, pp. 383–387.
30. Childress L., Dutt M. V. G., Taylor J. M., Zibrov A. S., Jelezko F., Wrachtrup J., Hemmer P. R., and Lukin M. D. (2006) Coherent dynamics of coupled electron and nuclear spin qubits in diamond, *Science*, **314**, pp. 281–285.
31. Balasubramanian G., Chan I. Y., Kolesov R., Al-Hmoud M., Tisler J., Shin C., Kim C., Wojcik A., Hemmer P. R., Krueger A., Hanke T., Leitenstorfer A., Bratschitsch R., Jelezko F., and Wrachtrup J. (2008) Nanoscale imaging magnetometry with diamond spins under ambient conditions, *Nature*, **455**, pp. 648–651.
32. Meijer J., Burchard B., Domhan M., Wittmann C., Gaebel T., Popa I., Jelezko F., and Wrachtrup J. (2005) Generation of single color centers

by focused nitrogen implantation, *Appl. Phys. Lett.*, **87**, pp. 261909–261911.

33. Naydenov B., Richter V., Beck J., Steiner M., Neumann Ph., Balasubramanian G., Achard J., Jelezko F., Wrachtrup J., and Kalish R. (2010) Enhanced generation of single optically active spins in diamond by ion implantation, *Appl. Phys. Lett.*, **96**, pp. 163108–163110.
34. Boudou J.-P., Curmi P. A., Jelezko F., Wrachtrup J., Aubert P., Sennour M., Balasubramanian G., Reuter R., Thorel A., and Gaffet E. (2009) High yield fabrication of fluorescent nanodiamonds, *Nanotechnology*, **20**, p. 235602.
35. Tisler J., Balasubramanian G., Naydenov B., Kolesov R., Grotz B., Reuter R., Boudou J.-P., Curmi P. A., Sennour M., Thorel A., Borsch M., Aulenbacher K., Erdmann R., Hemmer Ph. R., Jelezko F., and Wrachtrup J. (2009) Fluorescence and spin properties of defects in single digit nanodiamonds, *ACS Nano*, **3**, pp. 1959–1965.
36. Schrand A. M., Ciftan Hens S. A., and Shenderova O. A. (2009) Nanodiamond particles: properties and perspectives for bioapplications, *Crit. Rev. Solid State Mater. Sci.*, **34**, pp. 18–74.
37. Chao J. I., Perevedentseva E., Chung P.-H., Liu K.-K., Chang H.-H., and Cheng C.-L. (2007) Nanometer-sized diamond particle as a probe for biolabeling, *Biophys. J.*, **93**, pp. 2199–2208.
38. Yu S. J., Kang M. W., Chang H. C., Chen K. M., and Yu Y. C. (2005) Bright fluorescent nanodiamonds: no photobleaching and low cytotoxicity, *J. Am. Chem. Soc.*, **127**, pp. 17604–17605.
39. Chang Y. R., Lee H. Y., Chen K., Chang C. C., Tsai D. S., Fu C. C., Lim T. S., Tzeng Y. K., Fang C. Y., Han C. C., Chang H. C., and Fann W. (2008) Mass production and dynamic imaging of fluorescent nanodiamonds, *Nat. Nanotech.*, **3**, pp. 284–288.
40. Zhao F. L., Gong Z., Liang S. D., Xu N. S., Deng S. Z., Chen J., and Wang H. Z. (2004) Ultrafast optical emission of nanodiamond induced by laser excitation, *Appl. Phys. Lett.*, **85**, pp. 914–916.
41. Chung P. H., Perevedentseva E., and Cheng C. L. (2007) The particle size-dependent photoluminescence of nanodiamonds, *Surf. Sci.*, **601**, pp. 3866–3870.
42. Barnard A. S., and Sternberg M. (2007) Can we predict the location of impurities in diamond nanoparticles? *Diam. Relat. Mater.*, **16**, pp. 2078–2082.
43. Smith B. R., Inglis D. W., Sandnes B., Rabeau J. R., Zvyagin A. V., Gruber D., Noble C. J., Vogel R., Osawa E., and Plakhotnik T. (2009) Five-nanometer

diamond with luminescent nitrogen-vacancy defect centers, *Small*, **5**, pp. 1649–1653.

44. Vlasov I. I., Shenderova O., Turner S., Lebedev O. I., Basov A. A., Sildos I., Rahn M., Shiryaev A. A., and Van Tendeloo G. (2010) Nitrogen and luminescent nitrogen-vacancy defects in detonation nanodiamond, *Small*, **6**, pp. 687–694.
45. Bradac C., Gaebel T., Naidoo N., Sellars M. J., Twamley J., Brown L. J., Barnard A. S., Plakhotnik T., Zvyagin A. V., and Rabeau J. R. (2010) Observation and control of blinking nitrogen-vacancy centres in discrete nanodiamonds, *Nat. Nanotech.*, **5**, pp. 345–349.
46. Mahler B., Spinicelli P., Buil S., Quelin X., Hermier J. P., and Dubertret B. (2008) Towards non-blinking colloidal quantum dots, *Nature Mater.*, **7**, pp. 659–664.
47. Baranov P. G., Soltamova A. A., Tolmachev D. O., Romanov N. G., Babunts R. A., Shakhov F. M., Kidalov S. V., Vul' A. Ya., Mamin G. V., Orlinskii S. B., and Silkin N. I. (2011) Enormously high concentrations of fluorescent nitrogen-vacancy centers fabricated by sintering of detonation nanodiamonds, *Small*, **7**, pp. 1533–1537.

Chapter 6

Study of Detonation Nanodiamonds by Electron Paramagnetic Resonance

Alexander I. Shames
Department of Physics, Ben-Gurion University of the Negev, Be'er Sheva 84105, Israel
sham@bgu.ac.il

A characteristic feature of detonation nanodiamonds (DNDs), as of most nano-objects in which the number of atoms on the surface is comparable to that in the bulk, is the pronounced effect of the surface on the structural and electronic properties of material. This is why investigation of the differences in the structure and behavior of defects and impurities between nano- and bulk diamond is attracting particular interest. Where defects and impurities are located in nanodiamond, and how do they form? Can the specific effects originating from the nanosized scale of carbon objects be revealed in DND? Answers to these questions can be derived from present-day magnetic resonance methods widely used in modern research. One of such powerful nondistracting methods is electron paramagnetic resonance (EPR), which provides a possibility of studying the crystalline and electronic-nuclear structure of materials and their electronic and magnetic properties.

Detonation Nanodiamonds: Science and Applications
Edited by Alexander Ya. Vul' and Olga A. Shenderova

ISBN 978-981-4411-27-1 (Hardcover), 978-981-4411-28-8 (eBook)
www.panstanford.com

This chapter outlines the physical basis underlying the method and the associated techniques, with subsequent summing up of the results obtained with EPR in studies of DND in the recent years.

6.1 Introduction to Electron Paramagnetic Resonance

6.1.1 *EPR and its Applications in Physics, Chemistry, and Biology*

The phenomenon of electron paramagnetic resonance (EPR) discovered in 1944 by Soviet scientist Evgeny Konstantinovich Zavoisky may be considered as a remarkable development of the experiments of Stern-Gerlach, who found that the total electron magnetic moment of an atom in a magnetic field can assume only some definitely fixed discrete values. The magnetic moments originating from electrons derive from the presence of an uncompensated spin or orbital moment, or with their combination. In the vast majority of cases, more than 99% of the magnetic moment is generated by the spin, with the orbital moment accounting only for a small contribution. It is the resonant absorption of electromagnetic radiation by such systems placed in an external static magnetic field that one calls "paramagnetic resonance," "electron paramagnetic resonance" (EPR), and "electron spin resonance" (ESR). It has become common practice presently to combine the EPR/ESR phenomenon with phenomena that are close to it in their nature and involve absorption of electromagnetic radiation by magnetically ordered systems, namely, ferromagnets and antiferromagnets; these phenomena are ferromagnetic resonance (FMR) and antiferromagnetic resonance (AFMR), which likewise originate from the resonance in the electron subsystem. The present scientific literature becomes increasingly used to the term "electron magnetic resonance" (EMR), which covers all types of magnetic resonance involving electrons. In this chapter we are going to employ the term "EPR," which is presently accepted in worldwide practice. For more details of the EPR method, its theoretical background, and application to studies of a large variety of systems, the reader can be referred to remarkable books by Altshuler and Kozyrev [1], Wertz and Bolton [2], and Abragam and Bleaney [3].

Application of the EPR method to spectroscopy provided scientists with unique possibilities in identification of the electron-nuclear structure of various substances, from gases to complex biological molecules. Indeed, one could hardly recall another method that would permit as clear an illustration of the fundamentals of quantum mechanics as EPR. Unlike many other spectroscopic and analytical methods, EPR offers a researcher important information on the structure and dynamics of a system, including the chemical and physical processes at work in it, without destruction of the sample or application of chemical reactions. To sum up, EPR is an ideal method capable of complementing and refining all the other approaches employed in research.

6.1.2 *Fundamentals of the Theory of EPR*

By spectroscopy one commonly understands measurement of the differences between the energies of discrete states of atoms or molecules, as well as their interpretation. Knowledge of these differences allows one, as it were, to look into a quantum system and determine the characteristic features, structure, and dynamics of the system under study.

By Planck's law, an electromagnetic wave can be absorbed by an atom or a molecule only if the relation

$$\Delta E = h\upsilon, \tag{6.1}$$

is satisfied, where h is Planck's constant (6.620693 (11) $\times$ 10^{-34} J·s) and v is the frequency of electromagnetic radiation (Fig. 6.1).

The dependence of the intensity of radiation absorption on the frequency v of electromagnetic radiation defines the EPR spectrum. The frequency range employed in EPR spectroscopy lies in the gigahertz wave band (1–100 GHz). The energy difference, which is the object of the study by EPR spectroscopy, is the result of interaction of unpaired electrons in the material of the sample placed in a magnetic field.[a] Because the electron has an intrinsic magnetic moment, the spin, it can, in the process of orientation in the external magnetic field $\mathbf{B_0}$, occupy one of the two states with the spin

---

[a]A point to be reminded: The splitting of energy levels of atoms or molecules placed in an external magnetic field is called the Zeeman effect.

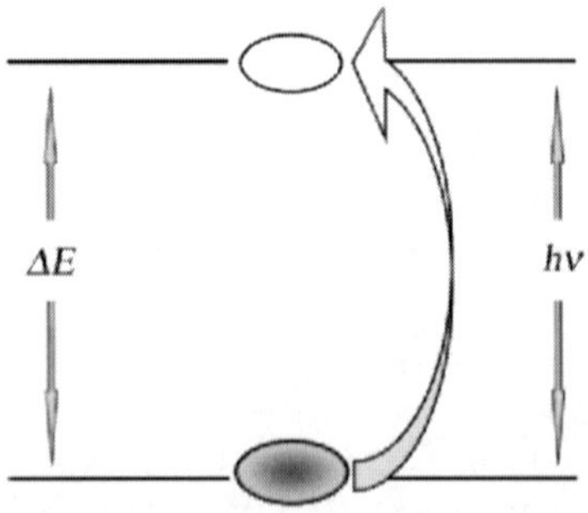

Figure 6.1 Transitions between discrete energy levels of a quantum system initiated by absorption of electromagnetic radiation energy by the system.

vector directed along and opposite to the vector $\mathbf{B_0}$. These two states can be specified in accordance with the projections of the electron spin vector M_s on the direction of the external magnetic field. The electron spin **S** has in absolute value $S = 1/2$, the direction along the field is identified with $M_s = -1/2$, and that opposite to the magnetic field, $M_s = +1/2$.

The main equations describing EPR are

$$E = g\mu_B B_0 M_s = (+1/2)g\mu_B B_0 \tag{6.2}$$

and

$$E = hv = g\mu_B B_0, \tag{6.3}$$

where g is the g factor, or the spectroscopic splitting factor, a proportionality constant, $g \cong 2$ for most samples, which varies only weakly with the electron configuration of a radical or ion (for free electrons $g_e = 2.0023193043718$ [75]), and μ_B is the Bohr magneton, a unit of the electron magnetic moment, $\mu_B = 9.27400949\,(80) \times 10^{-24}\ \mathrm{J{\cdot}}T^{-1}$.

As evident from Eqs. (6.2) and (6.3) and Fig. 6.2:

- With no external magnetic field applied, both spin states have the same energy.
- The energy gap between two spin state levels split by a magnetic field increases linearly with the magnetic field.

The possibility of varying the difference between the energy levels of spin states (energy level splitting) by proper adjusting,

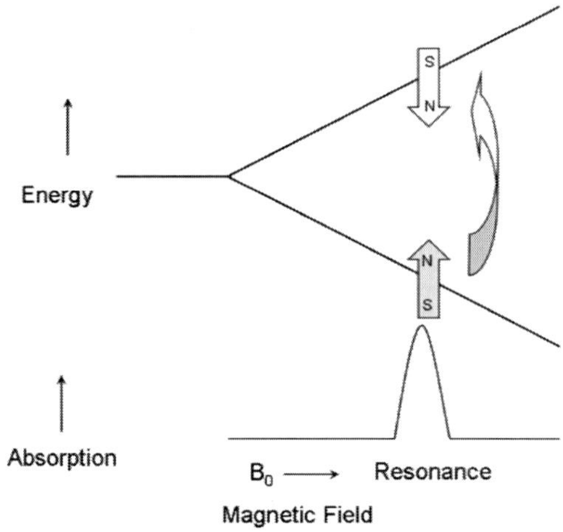

Figure 6.2 Spin state energy plotted vs. magnetic field.

the magnetic field strength offers alternate ways for spectral measurements:

(i) by scanning the frequency of electromagnetic radiation in a constant magnetic field, the approach employed in traditional spectroscopic methods; and
(ii) by maintaining a constant frequency of electromagnetic radiation and scanning the magnetic field (Fig. 6.2).

The absorption of electromagnetic radiation is maximal where the magnetic field is "tuned," as it were, to the frequency of the transition between two energy states at which the energy difference between the split levels is exactly equal to the radiation energy. Such a magnetic field is called "resonant." The specifics associated with electronic components employed in the microwave (gigahertz) band imposes certain constraints on the possibilities of frequency scanning, while scanning the magnetic field does not meet with any difficulties. Therefore it is the method of magnetic field scanning at a constant frequency of microwave radiation incident on the sample that underlies the design of the most modern commercial EPR spectrometers.

Interestingly, because a spectrum can be measured at different frequencies and, accordingly, in different magnetic field ranges, the

Table 6.1 Resonant fields B_{res} for signals with $g = 2.00$ presented for different microwave bands

Microwave band	Frequency (GHz)	B_{res} (T)
L	1.1	0.0392
S	3.0	0.107
X	9.75	0.348
K	24.0	0.857
Q	34.0	1.2
W	94.0	3.4
D	130	4.6

resonant field is not a unique characteristic of the material under study (its "fingerprint," as it were). On the other hand, the g factor

$$g = h\nu/\mu_B B_0, \tag{6.4}$$

which is independent of microwave frequency and is much better suited for identification of a material. Significantly, in EPR spectra obtained at a fixed frequency, larger values of the g factor appear in weaker magnetic fields and vice versa. Table 6.1 lists resonant fields for signals with $g = 2.00$ for the microwave frequency bands usually employed in modern EPR spectrometers.

While measurement of g factors provides very useful information on the material under study, the knowledge of the g factor alone is far from being adequate for forming a comprehensive idea of the molecular structure of a material. Fortunately, the unpaired electron involved in EPR is very sensitive to its local environment (other electrons and atomic nuclei). Indeed, atomic nuclei in a molecule or a complex possess quite frequently a nonzero magnetic moment generating a local magnetic field, which acts upon the electron. The interaction between an electron and a nucleus causes a change in the EPR spectrum and formation of a hyperfine structure in this spectrum, which provides important information on the sample under study. It is the hyperfine structure that makes possible identification of the atoms composing a molecule or a cluster and their location with respect to the unpaired electron.

The nature of hyperfine interaction can be readily unraveled if we assume the nucleus of an atom to act as a small magnet generating a magnetic field $\mathbf{B_I}$ in which the electron resides. This field $\mathbf{B_I}$ is capable of increasing or lowering the external magnetic field in which a maximum in EPR is observed. In an EPR spectrum obtained for a substance containing an unpaired electron interacting with a nucleus with spin $I = 1/2$ (such nuclear spins have, for instance, 100% of fluorine nuclei, 99.98% of hydrogen nuclei, 1.7% of carbon nuclei, or 2.12% of iron nuclei), the only absorption signal (a singlet) will split into two signals (a doublet), each shifted by $\mathbf{B_I}$ relative to the original singlet signal.

If a molecule contains one more nucleus with spin 1/2, each of the signals of the doublet will be split once more into a pair of signals to form a hyperfine four-line (quadruplet) hyperfine structure. As the number of magnetic nuclei increases, the number of signals in the hyperfine structure grows sharply by complex rules, which have to account for the fact that even nuclei with the same spin can be nonequivalent, that is, located differently with respect to the electron (e.g., at different distances).

Significantly, the magnitude of the EPR signal contains important information concerning the number of EPR-active centers (radicals, transition metal ions, defects, or unpaired charge carriers) in the object under study. In the language of spectroscopy, the "magnitude of the signal" is defined as the integrated intensity, that is, the area under the absorption curve.

The integrated intensity of the EPR signal is proportional to the concentration of paramagnetic centers or, termed otherwise, to the magnetic susceptibility of a sample at the corresponding microwave frequency. It should be borne in mind, however, that the intensities of observed signals depend not only on the concentration of EPR-active centers but also on the power of the microwave radiation. If one does not operate at very high levels of microwave power, the intensity of the observed signal grows proportional to the square root of power. At high levels of microwave power one observes a broadening and drop of the EPR peak intensity. To ensure correct measurement of the linewidth and shape, as well as of weakly separated hyperfine lines, one should avoid this saturation effect.

Although the Zeeman effect was known in optical spectroscopy long before the discovery of EPR, the first direct EPR demonstration became possible only after progress in the development of high-

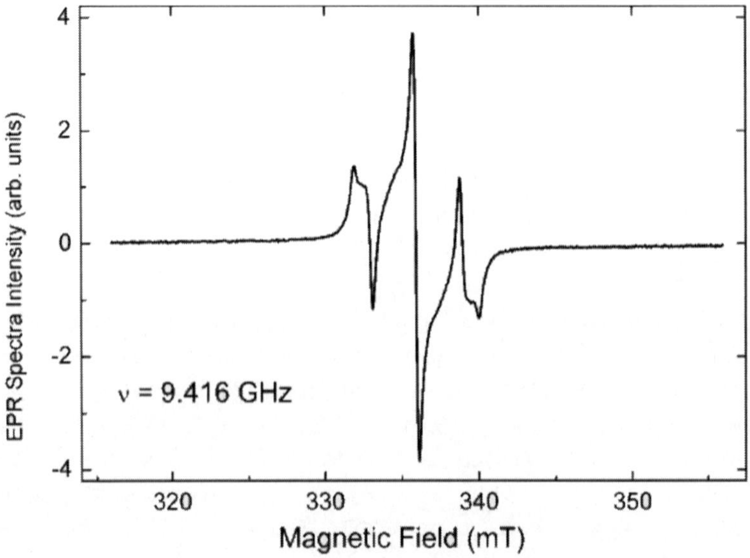

Figure 6.3 EPR spectrum of a powder of micron-size synthetic diamonds obtained in the X-band ($\nu = 9.416$ GHz) at room temperature.

frequency radar technology during the Second World War. By that time the electronics needed for the design and construction of EPR spectrometers were already commercially available.

A characteristic feature of most of modern EPR spectrometers is the specific technique employed in recording EPR signals; indeed, in contrast to the optical and nuclear magnetic resonance (NMR) spectrometers, EPR signals are presented in the form of the first derivative of the absorption signal. A typical X-band EPR spectrum is shown in Fig. 6.3.

6.2 EPR Studies of DND Powders

The first EPR studies of nanodiamonds produced by detonation nanodiamond (DND) synthesis appeared in the mid-1970s [4–6]. It was established already at that time that DND powders are strongly paramagnetic, an effect originating from unpaired electrons that are localized at dangling bonds between carbon atoms. DND samples available from different manufacturers exhibited similar

EPR characteristics, more specifically the absence of multiplet (triplet, quintuple, etc.) EPR signals typical of unpaired electrons localized in the vicinity of nitrogen impurity atoms and the presence of a strong singlet signal with $g = 2.003$ and linewidth $\Delta H_{\rm pp} \approx 0.7$–0.9 mT, which could be assigned to paramagnetic defects localized at carbon atoms. The total concentration of paramagnetic centers coupled to defects amounts to 10^{18}–10^{19} spins/g. The EPR signals observed in DND approach most closely those identified with the signals produced by defects originating from dangling C–C bonds on the surface of bulk (not nanosized) diamonds ($g = 2.0027$, $\Delta H_{\rm pp} =$ 0.55 mT) [7].

Subsequent EPR studies of DND obtained by both wet and dry synthesis [8–10] (see Chapter 2) confirmed that the above main parameters of EPR signals characteristic of DND (with the exception of the linewidth $\Delta H_{\rm pp}$) are independent of the method by which a sample was prepared.

The results accumulated in the most complete investigation of DND powders of different degrees of purification by methods of magnetic resonance (EPR, NMR) and high-resolution transmission electron microscopy are summarized in Ref. [11].

It was demonstrated [11] that EPR is a powerful and reliable method allowing one to monitor the content of para- and ferromagnetic impurities in DND powder, as well as to follow promptly the nanodiamond purification processes.

Figure 6.4a illustrates graphically the difference between the EPR spectra of detonation carbon obtained at room temperature (detonation soot), DND-1, and the spectrum of a DND sample subjected to high-level purification with nitric acid at 200°C, DND-2. In the first case, magnetic impurities (Fe, Ni, Co, etc.) present in the sample produce an intense broad ($\Delta H_{\rm pp} \approx 200$ mT) line, while in the second one observes only one intense narrow ($\Delta H_{\rm pp} \approx 0.97$ mT) Lorentzian line with $g \approx 2.00$, which can be assigned to paramagnetic centers of the free radical-type center (RC), that is, unpaired electrons localized at various atoms and molecules.

As seen from Fig. 6.4b, the specific density of RCs in detonation soot (before the isolation of DND) is lower by an order of magnitude than that in purified DND, which was estimated to be $\sim 10^{20}$ spins/g.

Precise measurements of the RC g factor showed that $g =$ 2.00282 ± 0.00003 within the temperature range of 4.2–300 K.

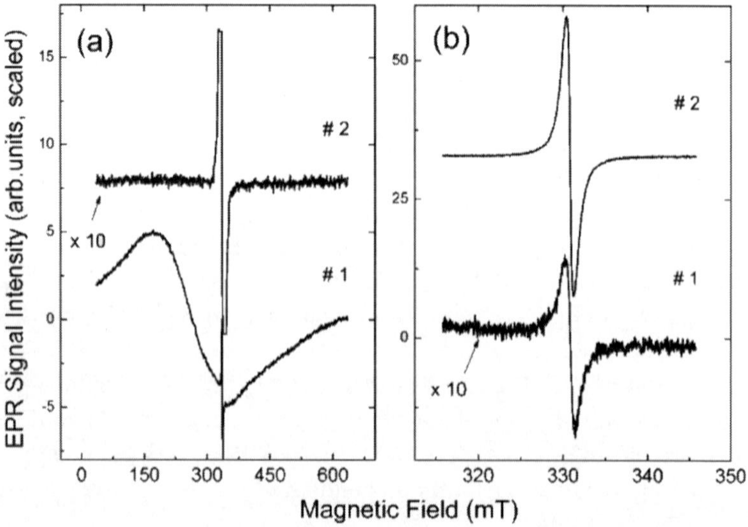

Figure 6.4 EPR spectra obtained from a ~1 g sample of the original detonation soot, no. 1, and of the same amount of a well-purified DND sample, no. 2. (a) General view, $\upsilon = 9.38$ GHz; (b) $g = 2.00$ region, $\upsilon = 9.27$ GHz. The spectra are plotted in the same scale. The gain is specified in the left-hand part of the spectrum. $T = 300$ K. The spectra were taken with kind permission from Springer Science+Business Media: Shames A. I., Panich A. M., Kempinski W., Baidakova M. V., Osipov V. Yu., Enoki T., and Vul' A. Ya. (2005) *Synthesis, Properties and Applications of Ultrananocrystalline Diamond*, NATO Science Series II: Mathematics, Physics and Chemistry, Vol. 192, eds. Gruen D. M., Shenderova O. A., and Vul' A. Ya, "Magnetic resonance study of nanodiamonds" (Springer), pp. 271–282.

The temperature dependence of the RC EPR line intensity measured within the same temperature range shows it to follow the Curie law. The data obtained gives one grounds to attribute RCs observed in nanodiamond powders to noninteracting (or weakly interacting) unpaired spins originating from defects in a diamond-like carbon sp^3 system (so-called dangling bonds).

Moreover, the observed shift of the EPR g factor of the RC EPR line from that of the free electron ($g_e = 2.0023$) may be interpreted in terms of the model of unpaired spin localized at a mesoscopic object (to which nanodiamonds also belong). This interpretation permits us to estimate the size of the crystallite (DND) as 20–60 Å, which correlates well with the DND dimensions derived from X-ray diffraction and high-resolution transmission electron microscopy.

The electron spin-lattice relaxation times T_{1e} were evaluated for different DND samples by successive saturation of the EPR signal at $T = 300$ K and found to be fairly short, $T_{1e} < 10$ μs.

6.3 EPR of Ultrananocrystalline Diamond Films

EPR signals were observed also from ultrananocrystalline diamond (UNCD) films with crystal grains 3–5 nm in size, which were grown by chemical deposition from the gas phase using Ar/CH_4 plasma with/without addition of various amounts of nitrogen (Fig. 6.5d,e) [12, 13]. Similar to DND EPR spectra (Fig. 6.5a–c), EPR signals of UNCD films were found to be at all temperatures Lorentzian singlets with a close g factor of 2.0027 ± 0.0001.

All this points at carbon as the origin of these signals while telling us nothing about the RC localization. Thus, the only conclusion these EPR spectra permit is that the spin states in UNCD films are localized far away from nitrogen atoms. The RC spin densities in UNCD films do not depend on nitrogen content in the original gas and vary in the interval of $(1.3–2.0) \times 10^{20}$ spin/cm^3, a figure close to that for DND.

The main difference from the EPR signals of DND powders lies in the width of the lines, which turned out to be much narrower than those of DND powders and vary from 0.2 mT to 0.4 mT. This may offer additional information on internal interactions, both in DND powders and in UNCD films.

The EPR linewidth in a solid is governed primarily by interaction among the spins of the system. The observed linewidth depends on three main factors: (i) magnetic dipole interaction, (ii) the averaging resulting from spin motion, and (iii) exchange interaction.

An analysis of the evolution with temperature of the spin–spin and spin–lattice relaxation times in UNCD films suggested the conclusion that exchange interaction between the unpaired spins of defects in UNCD films is stronger compared with that in DND powders. In view of the spin and mass densities of RCs in UNCD films, and assuming the spins to be distributed uniformly over the film volume, we come to 1.0–1.5 nm for the average spin separation.

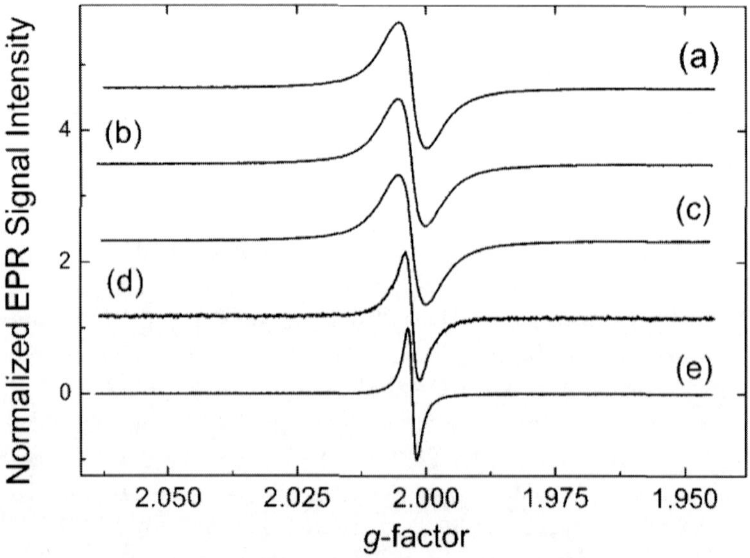

Figure 6.5 EPR spectra of (a) original powder of aggregated nanoparticles from a well-purified DND sample, (b) wet, (c) DMSO suspension of deaggregated original DND powder, (d) UNCD film grown without addition of nitrogen to the Ar/CH_4 plasma, and (e) UNCD film grown with 18% nitrogen added to the Ar/CH_4 plasma. $T = 300$ K. The EPR spectra were taken from Shames A. I., Panich A. M., Porro S., Rovere M., Musso S., Tagliaferro A., Baidakova M. V., Osipov V. Yu., Vul' A. Ya., Enoki T., Takahashi M., Osawa E., Williams O. A., Bruno P., and Gruen D. M. (2007) Defects localization and nature in bulk and thin film utrananocrystalline diamond, *Diam. Relat. Mater.*, **16**, pp. 1806–1812. *Abbreviation*: DMSO, dimethylsulfoxide.

But exchange interactions are short ranged; indeed, the atomic-like functions of localized spin states decay at distances of the order of 0.1 nm. This gives us grounds to assume that in UNCD films, where one observes strengthening of exchange interaction, the unpaired spins of defects are located much closer than this would follow from the model of arbitrary distribution of paramagnetic defects throughout the film volume. This inference suggests as a logically consistent the model of spin localization in UNCD films in which paramagnetic defects combine in groups (clusters) arranged in similar environmental conditions, such as, for instance, grain boundaries.

6.4 EPR of Multilayer "Onion-Like" Nanographites Obtained from DND

EPR spectroscopy was successfully employed to study the transformation of nanodiamond subjected to high-temperature treatment (HTT) in an argon atmosphere [14, 15]. It is well known that the surface of the original material, DND, is covered by various functional groups (see Chapters 2 and 3). HTT removes readily functional groups, which initiates rearrangement of the diamond (111) planes into graphitic (001) planes. Both X-ray diffraction and transmission electron microscopy, as well as Raman scattering spectra, demonstrate progressing development of the graphitic phase with increasing HTT temperature, starting to become evident at $T = 1,200^{\circ}$C. Complete transformation of nanodiamond to graphite occurs at $T = 1,600^{\circ}$C. EPR spectra obtained at room temperature reflect these changes and provide valuable information on the transformations occurring in the electronic system of nanographite objects. As DND is heated, all parameters of the DND RC EPR line undergo noticeable changes; indeed, the g factor decreases from 2.0028 (the value for the original DND and a sample heated to 900°C) to 2.0013 in the sample heated to 1,600°C. Accordingly, the linewidth H_{pp} increases from 0.9 mT to 3.5 mT, while the electron spin–lattice relaxation time becomes shorter by an order of magnitude.

This evolution of EPR signals, combined with data derived from magnetic measurements, demonstrates graphically that the original DND, similar to the DND heated to 900°C, belongs to diamond-like materials dominated by carbon sp^3 bonds in which all unpaired electrons are localized and the orbital magnetic susceptibility $|\mathrm{X}_{\mathrm{orb}}|$ is negligible, whereas a sample heated to 1,600°C is a fully graphitized system with predominantly carbon sp^2 bonds.

The broad ($\Delta H_{pp} = 3.5$ mT) line with $g = 2.0013$ seen in EPR spectra of a sample prepared under heating to 1,600°C can be assigned to the spins localized at nonbonding *n* electronic edge states (the so-called Fugita states). These states are formed by zigzag-shaped boundaries of graphene sheets making up a nanographitic particle, randomly shifted and turned about the azimuthal axis. Intermediate samples prepared at 1,200°C and 1,400°C are systems with displaced sp^2:sp^3 phases. The EPR signal from edge spins in nanographites turned out to be very sensitive to external factors, more specifically to partial oxygen pressure. Indeed, in nanographite samples subjected to prolonged evacuation (down to 10^{-6} millibar) at temperatures of 300–400°C, the broad signal ($\Delta H_{pp} \approx 1$–3 mT in air) narrows by a few times. Subsequent exposure of nanographite to oxygen forces the EPR signal to broaden practically to the original values. Further studies of the magnetic properties and EPR of nanographites prepared by temperature treatment of DND can be found in Refs. [16–18].

6.5 EPR Study of DND with the Surface Modified by Transition Metal Ions

Because DND particles 4–5 nm in size exhibit a unique combination of a mechanically stable and chemically inert diamond core with a chemically active surface with controllable properties, targeted modification of DND surface by various atoms, ions, and molecular groups appears to be of practical significance. Such modification offers a possibility of controlling the chemical, electronic, and physical characteristics of diamond nanoparticles. For instance, embedding ions of paramagnetic and rare-earth metals into DND or bonding them to its surface changes its magnetic properties and opens up a way to using DND as a stable carrier of metallic and metal oxide catalysts with improved characteristics. Application of the EPR method to DND with the surface modified by transition metal ions made it possible to establish the nature and geometry of the metal-containing complexes forming on the surface and their stability in various conditions.

An earlier paper [19] announced observation in EPR spectra of DND particles with surfaces modified by iron, nickel, cobalt, and copper ion additional signals (besides those due to the RCs, defects in the diamond-like carbon sp^3 system discussed above), which originated from unpaired electrons of ions of transition ions. It was shown also that the ion most appropriate for such EPR studies is that of divalent copper, with the most informative part of its EPR spectrum located aside from the main intensive RC signal, thus providing a fruitful possibility for a careful study of new complexes and for numerical estimates of the number of such complexes per diamond particle. A comprehensive study of superclean DND particulates whose surface was modified by divalent ions of copper and cobalt by ion exchange in water suspensions was reported in Ref. [20]. Two series of dry surface-modified DND powder were studied: Series A contained samples modified by copper in different starting concentrations (A1 is the original clean DND; A2 and A3 refer to modifications with 0.06 wt.% and 0.6 wt.% copper); in series B, sample Bl was the same starting-clean DND, while sample B3 was modified by 0.6 wt.% cobalt. Each series contained also two additional samples, A3-1, A3-2 and B3-1, B3-2, prepared by annealing samples A3 and B3 in a hydrogen flow for two hours at temperatures of 550°C and 900°C, respectively.

The EPR spectrum of the original sample A1 reveals a strong singlet RC signal with a g factor of 2.0028 described above in Section 6.2. The copper-modified samples A2 and A3, as well as A3-1 and A3-2, exhibit additional signals that are prominent in the low-field part of the EPR spectrum, that is, to the left of the strong RC signal. As evident from Fig. 6.6a, the intensity of the additional signals reduced to unit sample mass increases with increasing concentration of added copper, whereas after annealing in a hydrogen flow the intensity of these signals falls off (see Fig. 6.6b).

The recorded additional signals correlate clearly with the specific EPR spectra observed with powder samples containing paramagnetic complexes of divalent copper ($3d^9$, electron spin $S = 1/2$, nuclear spin $I = 3/2$), in which the Cu^{2+} ion resides in the ligand field of axial symmetry.

Such EPR spectra are described by a spin Hamiltonian that covers both interaction of electrons of a copper ion with an external

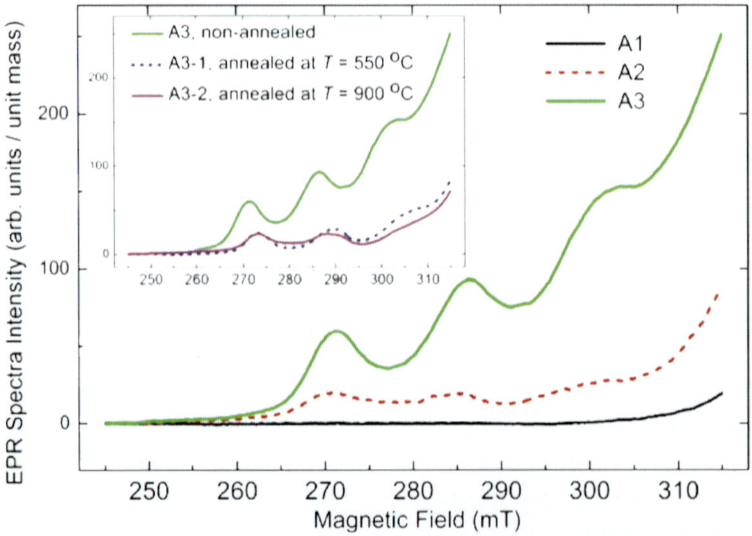

Figure 6.6 EPR spectra (υ = 9.464 GHz) of DND samples of series A modified by copper ions and plotted (a) vs. copper concentration (in units of 10^{18} spin/g): Al, 0; A2, 4.2; A3, 16.74; and (b) vs. annealing temperature: A3-1, 550°C ($4.1 \cdot 10^{18}$ spin/g) and A3-2, 900°C ($3.9 \cdot 10^{18}$ spin/g). T = 300 K. The EPR spectra were taken with kind permission from Springer Science+Business Media: Panich A. M., Shames A. I., Medvedev O., Osipov V. Yu., Aleksenskiy A. E., and Vul' A. Ya. (2009) Magnetic resonance study of detonation nanodiamonds with surface chemically modified by transition metal ions, *Appl. Magn. Reson.*, **36**, pp. 317–329.

magnetic field and hyperfine interaction (see Section 6.2) between the electron and nuclear spins of the Cu^{2+} ion

$$\begin{aligned}\mathbf{H} &= g_{||} S_z B_z + g_{|}(S_x B_x + S_y B_y) \\ &\quad + A_{||} S_z I_z + A_{|}(S_x I_x + S_y I_y) \qquad (6.5)\end{aligned}$$

with the parameters $g_{||} = 2.308 \pm 0.005$ and $A_{||} = 15.1 \pm 0.03$ mT.[a] Significantly, in the case under consideration we observe in spectra of modified DND only the part of the spectrum that is accounted for by "parallel" parameters of the spin Hamiltonian—three of the four components of the "parallel" hyperfine structure of the spectrum. Observation of the complete EPR spectrum of ions of divalent

[a] For more details on the spin Hamiltonian and EPR spectra of transition and rare-earth metals, see the books referred to above [1–3].

copper is impossible because the part of the spectrum located in the region of the g factor 2.00 is masked by the much more intense narrow RC line. This does not allow precise determination of the "perpendicular" parameters of the spin Hamiltonian and interferes markedly with estimation of the content of divalent copper ions in these samples. Nevertheless, a special technique based on computer simulation of the spectrum of Cu^{2+} ions and comparison with the spectrum of a $CuCl_2 \cdot 2H_2O$ standard specimen [20, 21] permitted evaluation of the copper concentration in the samples studied. For instance, in sample A2 divalent copper is contained in a concentration of $4.2 \cdot 10^{18}$spin/g, in A3 $1.67 \cdot 10^{19}$ spin/g, in A3-1 $4.1 \cdot 10^{18}$ spin/g, and in A3-2 $3.9 \cdot 10^{18}$ spin/g. Interestingly, the content of copper modifying the DND surface turned out to be much lower than that of copper introduced into the solutions used in modification. This implies that even if all of the copper introduced reacts with DND a sizable fraction of the copper ions does not form the paramagnetic mononuclear Cu^{2+} complexes.

A comprehensive analysis of the shape of the RC EPR signal showed that this singlet line with $g = 2.0028$, close to Lorentzian in shape, is best fitted by a sum of two true Lorentzian lines, a "broad" and a "narrow" one, with close (practically indistinguishable in X-band) g factors. The algorithm of the deconvolution employed is described in Ref. [21]. Application of this procedure to the RC EPR signal of series A samples revealed the widths $\Delta H_{\rm pp}$ of the lines of both Lorentzian components of the main signal to be sensitive to the number of Cu^{2+} ions determined by EPR for each sample of this series. Figure 6.7a displays a typical RC EPR signal (solid black curve) and its simulation (green circles) by a sum of two lines, a broad blue one and a narrow red one, obtained by the least-squares technique.

The data obtained in a series of simulations is displayed in Figs. 6.7b and 6.7c. Figure 6.7b visualizes a gradual broadening of both components of the RC EPR signal observed to occur with an increasing amount of copper ions that participated in DND modification.

On the other hand, the decrease of the amount of the copper modifier recorded to occur under sample annealing brings about narrowing of the corresponding EPR line components (see Fig. 6.7c).

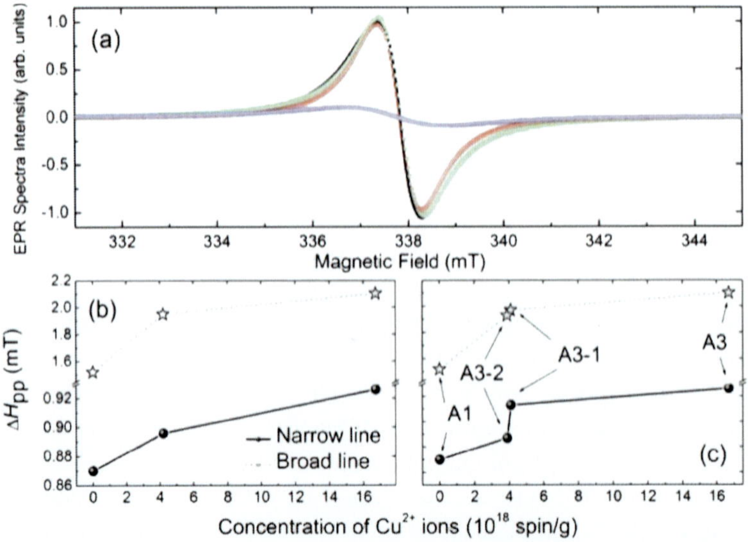

Figure 6.7 (a) Experimental RC EPR spectrum obtained on the A3 sample (solid black line) and its simulation by a sum (green circles) of two Lorentzian lines: a narrow one (red curve) and a broad one (blue curve), $\upsilon =$ 9.464 GHz. (b) Linewidth ΔHpp of RC EPR signals plotted vs. concentration of the modifying copper (in units of 10^{18} spin/g): Al, 0; A2, 4.2; A3, 16.74. (c) Linewidth ΔHpp plotted vs. annealing temperature: A3-1, 550°C ($4.1 \cdot 10^{18}$ spin/g) and A3-2, 900°C ($3.9 \cdot 10^{18}$ spin/g). Circles: narrow component; stars: broad component. The lines were drawn for the sake of convenience. T = 300K. The figure was taken with kind permission from Springer Science+Business Media: Panich A. M., Shames A. I., Medvedev O., Osipov V. Yu., Aleksenskiy A. E., and Vul' A. Ya. (2009) Magnetic resonance study of detonation nanodiamonds with surface chemically modified by transition metal ions, *Appl. Magn. Reson.*, **36**, pp. 317–329.

The estimated concentrations of the paramagnetic ions of the copper modifier turned out to be fairly low. Even in the case of the maximum modification (sample A3), the number of Cu^{2+} ions for each DND particle is not more than four. At the highest copper content (sample A3) one still observes well-resolved copper fine structure lines. This suggests that copper ions present in these samples are "magnetically diluted" and interact with one another and with the RC spins as elementary magnets.

This gives one grounds to assume magnetic dipole–dipole interaction among the spins of dangling bonds and copper ions to lie at the base of the observed broadening of the RC EPR lines. But the dipole–dipole interaction is a fairly short-range factor, with its efficiency falling off with spin separation as r^{-6}. Recalling that RC spins reside in the inner near-surface layer of a DND particle [19], one comes to the following assumption on the location of Cu^{2+} ions in modified DND: copper ions reside on or close to the surface of a nanoparticle.

Displayed in Fig. 6.8 are EPR spectra obtained at $T = 300$ K of DND samples modified by divalent cobalt ions, more specifically a sample modified by 0.6 wt.% Co (B3) and B3 samples annealed in a hydrogen flow at 550°C (B3-1) and 900°C (B3-2).

The spectrum of sample B3 contains, besides the strong RC signal for $g > 2$, a group of signals overlapping the main peak, omitted in Fig. 6.8. These signals can be assigned to the observed part of the EPR spectrum typical of polycrystalline samples containing deformed octahedral complexes of divalent cobalt ions ($3d^7$, $S = 1/2$, $I = 7/2$). After annealing of sample B3 at 550°C, these additional signals disappear, to leave in their place a new strong broad signal shifted toward low fields—see the spectrum of sample B3-1 in Fig. 6.8. Such broad signals are the signature of FMR spectra of powdered ferromagnetic metallic cobalt. Annealing at 900°C results in a noticeable strengthening of the FMR signal of metallic cobalt (see Fig. 6.8, spectrum of sample B3-2). Unlike the samples of series A, the components of the EPR line of the RC signal broaden not only because of the DND having been modified by cobalt ions but also as a result of annealing, due to the change in magnetic phase state of modifying cobalt ions (see inset of Fig. 6.8).

Thus, EPR demonstrated that modification of DND by divalent ions of copper and cobalt initiates formation of paramagnetic complexes of these ions on the surface of DND particles. Annealing of modified samples appears to produce both narrowing (in the case of Cu^{2+}) and broadening (for Co^{2+}) of the RC EPR lines. Both these effects evidence destruction of the surface paramagnetic centers, with the metal ions moving away from the surface and, possibly, forming metallic nanoclusters. In the case of copper ions, these clusters are not magnetic, which brings about weakening of the

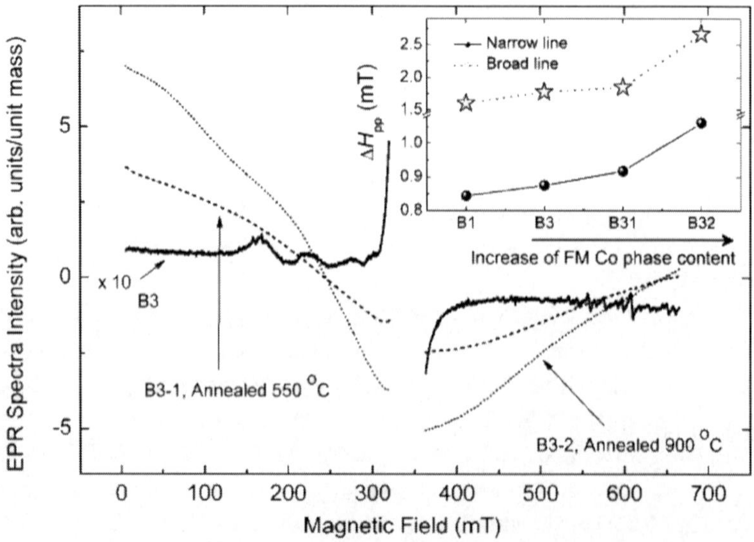

Figure 6.8 EPR spectra (υ = 9.464 GHz) of DND samples of series B modified by cobalt ions: solid black curve, sample B3; dashed and dotted curves, samples B3-1 and B3-2 annealed at 550°C and 900°C, respectively. The spectrum of sample B3 is scaled up by a factor 10. The inset visualizes the dependence of the linewidths ΔHpp of RC EPR signals on cobalt modification and subsequent annealing. Black circles, the narrow component; red stars, the broad component. The lines are drawn for clarity. T = 300 K. The figure was taken with kind permission from Springer Science+Business Media: Panich A. M., Shames A. I., Medvedev O., Osipov V. Yu., Aleksenskiy A. E., and Vul' A. Ya. (2009) Magnetic resonance study of detonation nanodiamonds with surface chemically modified by transition metal ions, *Appl. Magn. Reson.*, **36**, pp. 317–329.

effect of RC line broadening induced by magnetic interactions. As for the cobalt ions, the forming clusters are ferromagnetic, resulting in an additional broadening of all EPR lines, which is initiated by the influence of the external inhomogeneous magnetic field generated by a ferromagnetic particle in the vicinity of DND grains. This field turns out to be strong enough even when the cobalt ferromagnetic clusters are located not directly on the surface but at a distance from DND particles.

6.6 Localization of RCs in DND by EPR

As demonstrated in the preceding section, various magnetic (both para- and ferromagnetic) particles, molecules, or molecular fragments added to clean DND samples interact with the RC spins intrinsic to the original DND particles. This interaction becomes manifest in some changes revealed in the RC EPR signal. Thus one acquires a possibility of purposefully introducing various magnetic systems to act as probes to derive structural information both on the DND particles themselves and on the location of RCs inside a DND particle.

One of such natural and frequently employed magnetic probes is oxygen of the air. Indeed, air contains molecules of oxygen O_2, with unpaired electrons of two oxygen atoms forming a ground state with spin $S = 1$, which is magnetic and called a "triplet state." Spins of the triplet oxygen molecule are capable of coupling with those of RCs both by the dipole–dipole mechanism (see Section 6.5) and through exchange interaction that is efficient in collisions of molecules of gaseous oxygen with spins localized at particles.

A recent publication [19] reported on a very simple in situ EPR experiment in which one measured the peak intensity of the RC signal of purified DND as a function of time, with the sample fixed in the cavity being evacuated by a pump down to a vacuum of 10^{-4} Torr. Shown in Fig. 6.9a are the variations of the RC EPR signal recorded at the turning on of a high-vacuum pump and its subsequent turning off, after which access of air to the test tube was opened again. As seen clearly from this experiment, pumping out of air brings about a sharp increase of the peak intensity of the RC EPR signal, while subsequent access of the air to the sample results in an as fast a drop of this intensity to a level intermediate between the original and maximum ones.

Besides this curve demonstrating the kinetics of the process, one recorded also complete EPR spectra of the RC of the starting DND sample. A fragment of such a spectrum is displayed in Fig. 6.9b, with the black line related to the air-containing sample before evacuation, the red line to the same sample evacuated down to 10^{-4} Torr, and the blue line to the sample after the pumping out, in air again.

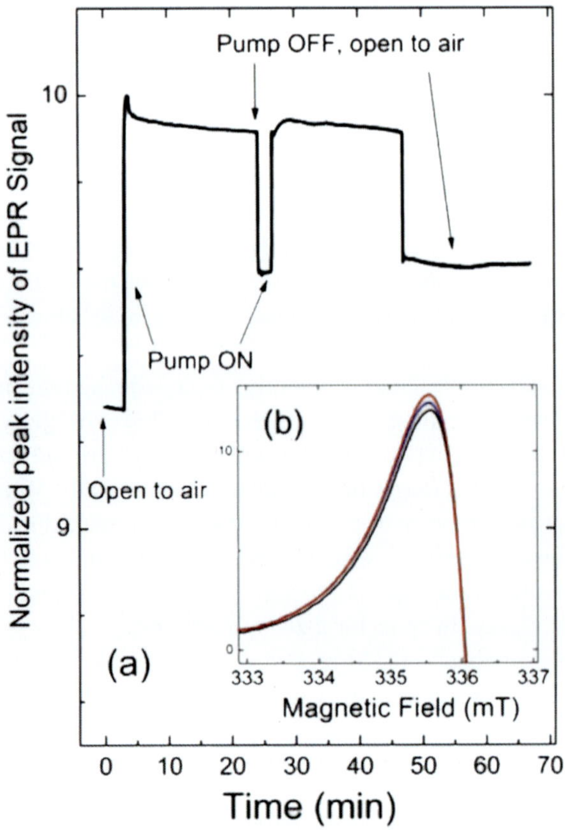

Figure 6.9 (a) Variation of the RC EPR signal of purified DND recorded at the turning on of a high-vacuum pump and its subsequent turning off, after which access to air into the test tube was open again. (b) A fragment of the RC EPR spectrum of purified DND: black line, air-containing sample before evacuation; red line, same sample evacuated to 10^{-4} Torr; blue line, sample after evacuation, again in air. The RC EPR signal was taken with kind permission from Springer Science+Business Media: Shames A. I., Panich A. M., Kempinski W., Baidakova M. V., Osipov V. Yu., Enoki T., and Vul' A. Ya. (2005) *Synthesis, Properties and Applications of Ultrananocrystalline Diamond*, NATO Science Series II: Mathematics, Physics and Chemistry, Vol. 192, eds. Gruen D. M., Shenderova O. A., and Vul' A. Ya, "Magnetic resonance study of nanodiamonds" (Springer), pp. 271–282.

Thus, it was demonstrated that pumping air out from the sample does not broaden the EPR signal, neither does it shift or change its shape. The only observed effect consists of an increase of the peak (and, hence, integral) intensity of this signal by about 10%. If we allow air contact with the sample again, the signal drops rapidly in intensity to regain slowly its original level again.

A reasonable explanation for these metamorphoses of the EPR signal can be formulated if we assume that only some RC spins, out of the totality of RC spins, can interact, by magnetic dipole coupling and exchange, with molecules of triplet oxygen, whereas the other RC spins are insensitive to these interactions.

We may recall (see Section 6.5) that dipole–dipole interaction falls off very fast with distance, while exchange interaction becomes realized only in direct collisions of triplet oxygen molecules with spins. Thus, the above changes in the intensity of the RC EPR signal (and, accordingly, the number of observed RCs per particle) derive from the interaction of spins of oxygen molecules with those only of the RCs that oxygen can approach to short enough (<1 nm) distances. As a result, not more than 10% of spins are able to interact with magnetic oxygen, and their signal will become broadened because of the presence of oxygen to the extent where it will not be detectable altogether in an air-saturated sample.

But this is possible only if these readily interacting spins occupy sites that can be approached by oxygen closely enough, that is, on the DND surface. The other spins are located so deep in the bulk of a DND particle as to practically not interact with oxygen at all.

Superconducting quantum interference device (SQUID) measurements (see Chapter 8) refined the RC concentration in DND, which was estimated by EPR as $\sim 10^{20}$ spin/g. The RC concentration was estimated [22] as $N_s = 6.3 \cdot 10^{19}$ spins/g, which gives 13–14 RC spins per DND particle. This estimate suggests that one to two spins are located on (or close to) the DND surface, with the others "isolated" from oxygen in the diamond core of the particle and the near-surface layer.

The question of the precise depth of RC location in a DND particle remains open. As for oxygen, it is not suitable as a probe for spins of this type—RC spins simply do not respond to this paramagnetic agent. As shown earlier, in Section 6.5, the RC EPR signal becomes effectively broadened in interaction of RC spins with spins of divalent copper. In Section 6.5, this broadening effect was used to demonstrate that the spins of DND-modifying copper are firmly localized on the DND surface.

To sum up, by varying properly the amount of Cu^{2+} ions forming paramagnetic complexes on the DND surface and knowing the pattern of line broadening in dipole–dipole interaction, one can readily calculate the distances, from the DND surface, at which RCs of different types (i.e., responsible for both the broad and the narrow components of the RC EPR signal) are located. This experiment and the associated calculations are described in detail in Ref. [23].

The most probable mutual arrangement of the paramagnetic probes (Cu^{2+} ions) and RC spins is shown schematically in Fig. 6.10. In this experiment, to the three samples of series A described in Section 6.5 were added two DND samples modified by copper ions in higher concentrations. The copper ion concentrations in the series, as established by EPR, were A2, $4.1 \cdot 10^{18}$ spin/g; A3, $1.67 \cdot 10^{19}$ spin/g; A4, $2.08 \cdot 10^{19}$ spin/g; and A5, $3.85 \cdot 10^{19}$ spin/g,

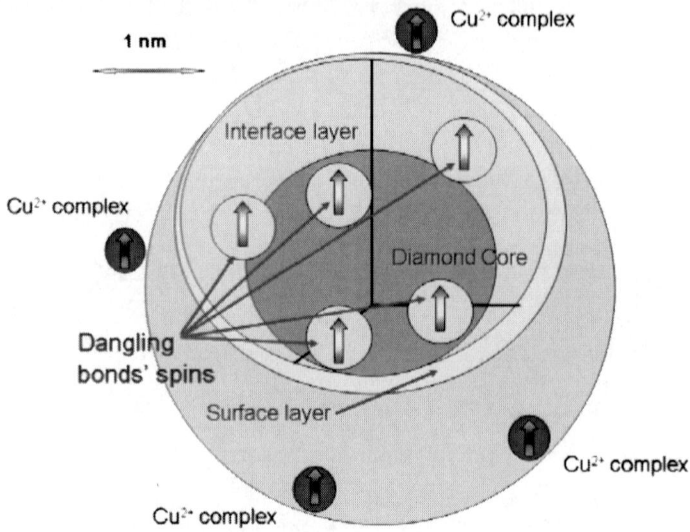

Figure 6.10 Schematic of a DND particle in section and of a possible mutual arrangement of the paramagnetic probes, i.e., divalent copper ions and RC spins.

which corresponds to 1, 4, 5, and 9 Cu^{2+} ions per DND particle. The spectra of all samples of this series were treated by the same technique described in Section 6.5, namely, deconvolution of the RC EPR line into two Lorentzian components, a broad and a narrow one. Figure 6.11a shows graphically the evolution of the EPR linewidths of these Lorentzian components with increasing concentration of the paramagnetic copper modifying the DND surface, as estimated by the EPR method, Figure 6.11b plots the absolute (reduced to the linewidth of DND sample A1, which was not modified by copper) increment of the widths of the corresponding lines resulting from an increase in the number of copper complexes per RC electron spin. We readily see that as the number of the paramagnetic probes per DND particle increases, the broad component of the RC EPR signal reveals a larger increase in width than the narrow one does. This gives one grounds to assume that the RC spins responsible for the broad component interact with the probes stronger than those accounting for the narrow one.

As demonstrated above, the RC spins revealed in a nonevacuated sample reside inside a DND particle, whereas the copper ions couple to its surface. Therefore the shortest-range exchange interaction can be excluded from possible broadening factors. Thus, the broadening dipole–dipole interaction varying inversely with the sixth power of distance exerts a stronger influence on a group of close-located RC spins responsible for the broad component, whereas the spins contributing to the narrow component are located farther away from the paramagnetic probes.

The method of spectral line moments developed by van Vleck (more details on the method can be found in Ref. [24]) establishes the dependence between the increment of the width of the EPR line under study, which originates from dipole–dipole interaction with a specific broadening agent (in our case, with paramagnetic probes), and the average distance between these agents and the spins responsible for this EPR line. Application of this method to Lorentzian curves is described in Ref. [23]. It is shown that the absolute increments in EPR linewidth in Fig. 6.11b can be fitted with the use of a simple relation

$$\Delta H_{\mathrm{pp}}\,(\text{modif.Cu}) - \Delta H_{\mathrm{pp}}\,(\text{non-modif.Cu}) = \frac{CN_{\mathrm{Cu}}\overline{r_{\mathrm{Cu}}^{-6}}}{a}, \quad (6.6)$$

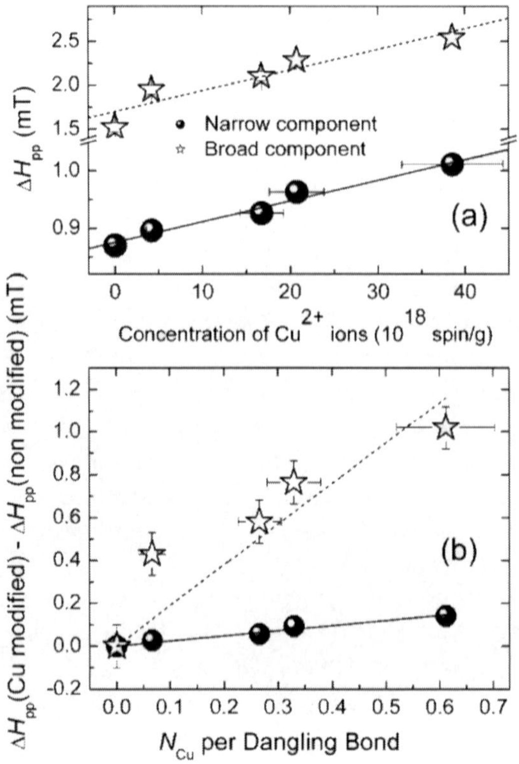

Figure 6.11 (a) RC EPR linewidth plotted vs. concentration of Cu^{2+} ions: circles, narrow component; stars, broad component. The lines are drawn for clarity. (b) Increase in the RC EPR linewidth resulting from an increasing number of the surface copper complexes per RC electron spin: circles, narrow component; stars, broad component. The lines are best fits to the results obtained with Eq. (6.5). The figure was taken from Shames A. I., Osipov V. Yu., Aleksenskiy A. E., Osawa E., and Vul' A. Ya. (2011) Locating inherent unpaired orbital spins in detonation nanodiamonds through the targeted surface decoration by paramagnetic probes, *Diam. Relat. Mater.*, **20**, pp. 318–321.

where C is a fitting constant, N_{Cu} is the number of copper ions per RC spin, $\overline{r_{Cu}{}^{-6}}$ is the average distance between a copper ion and the RC in nanometers, and $a = 10\Delta H_{pp}^{max}$ is the distance at which the wings of the Lorentzian line are cut off, which is derived from the maximum width of the EPR line of the broad component. The solid lines in Fig. 6.11b are the least-square fits obtained with the use of Eq. (6.6). The fittings performed yielded the following values for the distances: $\overline{r_{\text{Cu}}^{-6}} = 1.1 \pm 0.1$ nm for the RC spins responsible for the broad component and $\overline{r_{\text{Cu}}^{-6}} = 1.8 \pm 0.1$ nm for those accounting for the narrow component.

Investigation of the chemical processes involved in formation of a copper complex on the DND surface revealed that copper ions are located, on average, at a distance of 0.31 nm from the DND surface.

To sum up, application of divalent copper ions as paramagnetic probes arranged purposefully on the surface of a DND particle established that it contains carbon-inherited RC centers with an electron spin $S = 1/2$ of (at least) two types, more specifically spins responsible for the broad component of a Lorentzian-like RC center and located at a depth of 0.8 nm from the DND surface and those accounting for the narrow component of this signal, which lie at a depth of 1.5 nm from the surface.

References

1. Altshuler S. A., and Kozyrev B. M. (1964) *Electron Paramagnetic Resonance* (Academic Press, New York, London).
2. Wertz J. E., and Bolton J. R. (1972) *Electron Spin Resonance. Elementary Theory and Practical Applications* (McGraw-Hill, New York).
3. Abragam A., and Bleaney B. (1970) *Electron Paramagnetic Resonance of Transition Ions*, Vol. 1 (Clarendon Press, Oxford).
4. Belyankina A. V., Nachal'naya T. A., Sozin Yu. I., and Shul'man L. A. (1975) X-ray diffraction data and EPR spectra of explosion-formed diamonds, *Sint. Almazy*, **5**, pp. 5–7 (in Russian).
5. Bakul V. N., and Andreev V. D. (1975) AV-type diamonds synthesized by explosion, *Sint. Almazy*, **5**, pp. 3–4 (in Russian).
6. Vlasova M. V., Kakazei N. G., and Sawakin G. I. (1979) Some properties of ultrafine diamond powders produced under high-temperature synthesis conditions, *Izv. Akad. Nauk SSSR, Neorg. Mater.*, **15**(7), pp. 1303–1304 (in Russian).
7. Samsonenko N. D. and Sobolev E. V. (1967). Electron paramagnetic resonance caused by diamond surface, *Pis'ma Zh. Eksp. Tear. Fiz.*, **5**(9), pp. 304–307 (in Russian).

8. Vereshchagin A. L., Komarov V. F., Mastikhin V. M., Novoselov V. V., Petrova L. A., Zolotukhina L. L., Vychin N. V., Baraboshin K. S., and Petrov A. E. (1991) Properties of a detonation diamond phase, *Proc. Vth All-Union Meeting Detonat.*, **1**, pp. 99–103.
9. Iakuobovskii K., Adriaenssens G. J., Meykens K., Nesladek M., Vul' A. Ya. and Osipov V. Yu. (1999) Study of defects in CVD and ultradisperse diamond, *Diam. Relat. Mater.*, **8**, pp. 1476–1479.
10. Iakuobovskii K., Baidakova M. V., Wouters B. H., Stesmans A., Adriaenssens G. J., Vul' A. Ya., and Grobet P. J. (2000) Structure and defects of detonationsynthesis nanodiamond, *Diam. Relat. Mater.*, **9**, pp. 861–865.
11. Shames A. I., Panich A. M., Kempinski W., Alexenskii A. E., Baidakova M. V., Dideikin A. T., Osipov V. Yu., Siklitski V. L., Osawa E., Ozawa M., and Vul' A. Ya. (2002) Defects and impurities in nanodiamonds: EPR, NMR and TEM study, *J. Phys. Chem. Solids*, **63**, pp. 1993–2001.
12. Rovere M., Porro S., Musso S., Shames A., Williams O., Bruno P., Tagliaferro A., and Gruen D. M. (2006) Low temperature electron spin resonance investigation of ultrananocrystalline diamond films as a function of nitrogen content, *Diam. Relat. Mater.*, **15**, pp. 1913–1916.
13. Shames A. I., Panich A. M., Porro S., Rovere M., Musso S., Tagliaferro A., Baidakova M. V., Osipov V. Yu., Vul' A. Ya., Enoki T., Takahashi M., Osawa E., Williams O. A., Bruno P., and Gruen D. M. (2007) Defects localization and nature in bulk and thin film utrananocrystalline diamond, *Diam. Relat. Mater.*, **16**, pp. 1806–1812.
14. Andersson O. E., Prasad B. L. V., Sato H., Enoki T., Hishiyama Y., Kaburagi Y., Yoshikawa M., and Bandow S. (1998) Structure and electronic properties of graphite nanoparticles, *Phys. Rev. B*, **58**, pp. 16387–16395.
15. Prasad B. L. V., Sato H., Enoki T., Hishiyama Y., Kaburagi Y., Yoshikawa M., Rao A. M., Eklund P. C., and Endo M. (2000) Heat-treatment effect on the nanosized graphite 7i-electron system during diamond to graphite conversion, *Phys. Rev. B*, **62**, pp. 11209–11218.
16. Osipov V. Yu., Enoki T., Takai K., Takahara K., Endo M., Hayashi T., Hishiyama Y., Kaburagi Y. and Vul' A. Ya. (2006) Magnetic and high resolution TEM studies of nanographite derived from nanodiamond, *Carbon*, **44**, pp. 1225–1234.
17. Osipov V. Yu., Shames A. I., Enoki T., Takai K., Endo M., Hayashi T., Kaburagi Y., Vul' A. Ya. (2009) Magnetic and EPR studies of edge-localized spin paramagnetism in multi-shell nanographites derived from nanodiamonds, *Diam. Relat. Mater.*, **18**, pp. 220–223.

18. Osipov V. Yu., Shames A. I., Enoki T., Takai K., Endo M., Kaburagi Y., Vul' A. Ya. (2010) Interaction between edge-localized spins and molecular oxygen in multishell nanographites derived from nanodiamonds, *Diam. Relat. Mater.*, **19**, pp. 492–495.

19. Shames A. I., Panich A. M., Kempinski W., Baidakova M. V., Osipov V. Yu., Enoki T., and Vul' A. Ya. (2005) *Synthesis, Properties and Applications of Ultrananocrystalline Diamond*, NATO Science Series II: Mathematics, Physics and Chemistry, Vol. 192, eds. Gruen D. M., Shenderova O. A., and Vul' A. Ya, "Magnetic resonance study of nanodiamonds" (Springer), pp. 271–282.

20. Panich A. M., Shames A. I., Medvedev O., Osipov V. Yu., Aleksenskiy A. E. and Vul' A. Ya. (2009) Magnetic resonance study of detonation nanodiamonds with surface chemically modified by transition metal ions, *Appl. Magn. Reson.*, **36**, pp. 317–329.

21. Shames A. I., Panich A. M., Osipov V. Yu., Aleksenskiy A. E., Vul' A. Ya., Enoki T., and Takai K. (2010) Structure and magnetic properties of detonation nanodiamond chemically modified by copper, *J. Appl. Phys.*, **107**, pp. 014318(1–10).

22. Osipov V. Yu., Shames A. I., Enoki T., Takai K., Baidakova M. V., and Vul' A. Ya. (2007) Paramagnetic defects and exchange coupled spins in pristine ultrananocrystalline diamonds, *Diam. Relat. Mater.*, **16**, pp. 2035–2038.

23. Shames A. I., Osipov V. Yu., Aleksenskiy A. E., Osawa E., and Vul' A. Ya. (2011) Locating inherent unpaired orbital spins in detonation nanodiamonds through the targeted surface decoration by paramagnetic probes, *Diam. Relat. Mater.*, **20**, pp. 318–321.

24. Slichter C. P. (1978) *Principle of Magnetic Resonance*, 2nd Ed. (Springer-Verlag, Berlin, Heidelberg, New York).

Chapter 7

Nuclear Magnetic Resonance in Nanodiamonds

Alexander M. Panich
Department of Physics, Ben-Gurion University of the Negev, Be'er Sheva 84105, Israel
pan@bgu.ac.il

7.1 Introduction

Most of the atomic nuclei possess an intrinsic angular moment, or spin $\hbar I$, measured in units of $\hbar = h/2\pi$, where h is Planck's constant, and a nuclear magnetic moment $\mu = \gamma \hbar I$, which is proportional to the magnitude of the spin. In an external magnetic field $\mathbf{B}_0$ the nuclear spin 1/2 has two energy levels corresponding to its "up" and "down" orientation (Fig. 7.1) and separated by an interval $\Delta E = \hbar\gamma\mathbf{B}_0$, where γ is the gyromagnetic ratio, which is a characteristic of each nucleus. An oscillating electromagnetic field of frequency ν_0 applied to the system such that $h\nu_0 = \Delta E$ induces a resonant transition between the energy levels. This phenomenon is called nuclear magnetic resonance (NMR), and ν_0 is referred to as the Larmor frequency.

In a real system, each nuclear spin is coupled to neighboring nuclear and electronic spins, an interaction that brings about a shift

Detonation Nanodiamonds: Science and Applications
Edited by Alexander Ya. Vul' and Olga A. Shenderova

ISBN 978-981-4411-27-1 (Hardcover), 978-981-4411-28-8 (eBook)
www.panstanford.com

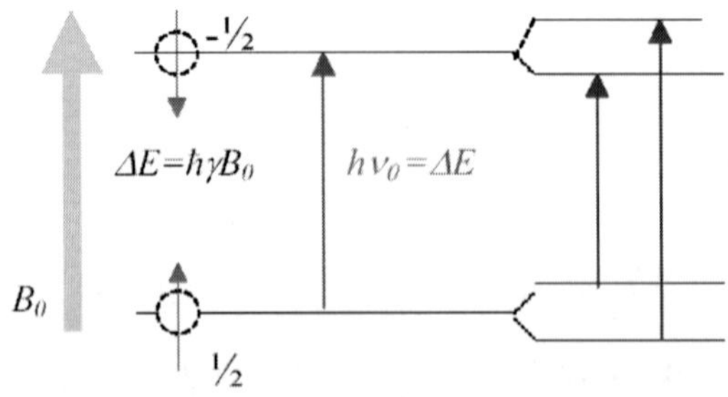

Figure 7.1 Energy levels of a nuclear spin $I = 1/2$ in a magnetic field $\mathbf{B}_0$.

and additional splitting of the energy levels (Fig. 7.1). By measuring these shifts and splittings and, accordingly, the interactions, one derives information on the local crystal and electronic structure of the material under study, the nature of the chemical bonding between the atoms, the processes of charge transfer between the nanoparticle and the molecule or ion it is coupled to, and phase transitions and intrinsic motions in the system on the atomic level. The nuclear spin could be identified with a probe embedded into the crystal lattice of a material, which tracks all the processes occurring in this lattice. An analysis of NMR spectra becomes greatly simplified if one of the interactions plays a dominant part and defines the shape and shift of the line. For instance, in nanodiamond one has to deal mostly with chemical shifts, which will be considered in detail in the next section. Information essential to our understanding of electron-nuclear and internuclear interactions, atomic and molecular motions, and phase transitions can be gained from measurement of the spin–lattice and spin–spin relaxation times T_1 and T_2, which will also be treated below. In summing up, one can say that NMR occupies presently an essential place in the spectroscopy of nanodiamond and plays frequently a fundamental part in unraveling its structure and properties.

7.2 NMR Spectra and the Chemical Shift

In an NMR experiment, the magnetic field at the nucleus is a sum of the strong external magnetic field $\mathbf{B}_0$ and a weak magnetic field

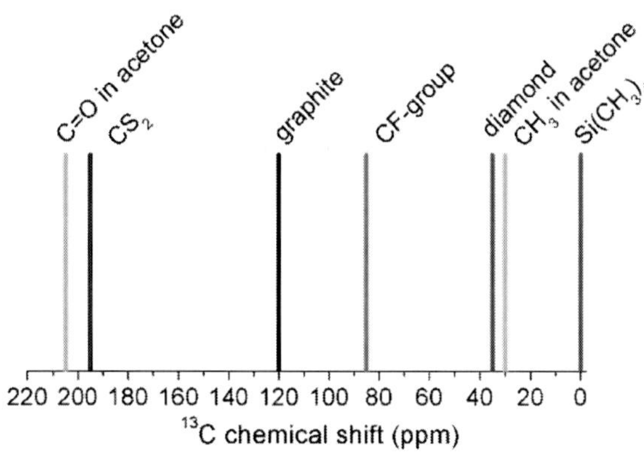

Figure 7.2 Chemical shifts of carbon atoms in various compounds.

induced by electrons rotating around the nucleus. This results in a deviation of the resonance frequency from the Larmor frequency ν_0 called the chemical shift δ [1], which is a measure of the screening action of the electron shell

$$\nu = \nu_0(1 + \delta). \tag{7.1}$$

The chemical shift depends on the immediate chemical environment of the atom and the nature of the chemical bonding. Therefore carbon atoms belonging to different atomic and molecular groups show different chemical shifts and can be identified in an NMR spectrum. Figure 7.2 displays schematically the scale of chemical shifts of carbon atoms. The chemical shift is measured in relative units, parts per million (ppm), which do not depend on the applied magnetic field; $\delta = (\Delta\nu/\nu_0) \times 10^6$ ppm. The stronger the applied magnetic field, the higher the spectral resolution. The chemical shift in NMR is measured relative to the position of the resonance line of a reference compound. In the case of ^{13}C carbon nuclei (only this carbon isotope with the natural abundance of 1.1% has a nuclear spin) tetramethyl silane (TMS) serves as a reference.

Figure 7.2 specifies the positions of the centers of gravity of NMR lines. Chemical shift depends, however, on orientation of the external magnetic field $\mathbf{B}_0$ relative to the directions of chemical bonds of atoms, and represents mathematically a second-rank tensor that describes the anisotropy of the electron cloud surrounding a given

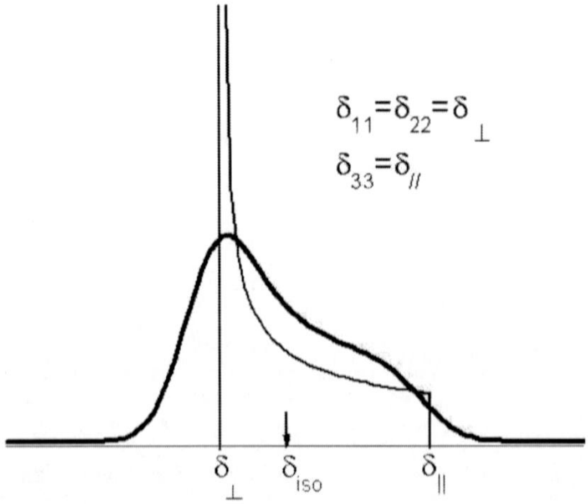

Figure 7.3 NMR line shape for the axially symmetric chemical shift tensor. The thin line is a nonbroadened spectrum, while the thick line is a spectrum broadened by dipole interaction.

nucleus. For an axially symmetric electron cloud we can write

$$\delta = \delta_{\text{iso}} + \alpha(3\cos^2\theta - 1). \tag{7.2}$$

Here $\alpha = (\delta_{||} - \delta_{\perp})/3$, δ_{iso} is the center of gravity of the NMR line, θ is the angle between the applied magnetic field and the principal axis of the tensor, and $\delta_{||}$ and $\delta_{\perp}$ are the parallel and perpendicular components of the chemical shift tensor, respectively.

The spectrum of a polycrystalline sample is a superposition of spectra produced in all the orientations involved. The angular dependence, Eq. (7.2), is responsible for the specific shape of the line shown in Fig. 7.3. Line broadening should be assigned primarily to the dipole–dipole interaction of nuclear spins, which has the same angular dependence as the chemical shift with $\alpha_{\text{dip}} \sim R^{-3}$, where R is the separation between the nuclei.

While chemical shift anisotropy and dipole–dipole interaction contain valuable information on chemical bonding and the electronic and crystal structure of a material, the resultant lines become broadened and overlap to the extent where spectral resolution can become severely curtailed. Fortunately, anisotropic interactions can be effectively averaged out by molecular motion or, in the absence of the latter, by fast rotation of the sample at a "magic angle" $\theta = 54°44'$, which brings ($3\cos^2\theta - 1$) to zero to produce a spectrum of narrow, well-resolved lines. This technique is called magic angle spinning, briefly MAS.

Important additional information can be obtained in cross-polarization (CP) experiments (^{1}H–^{13}C CP MAS), in which polarization of hydrogen nuclei with a high natural abundance of 99.99% is transferred to carbon nuclei having a low natural abundance of 1.1%. These experiments provide a possibility of identifying the atoms of carbon neighboring those of hydrogen (directly or in some cases, through one C–C bond.

7.3 NMR Relaxation

After excitation to a higher level, nuclear spins regain their original equilibrium position at a lower energy level by transferring the excess energy to the lattice vibrations (phonons). This process is called spin–lattice relaxation and proceeds in a characteristic time T_1. This process is, as a rule, fairly slow and takes up a time from a few tens of seconds to a few hours. However, spin–lattice relaxation can be speeded up noticeably through interaction of nuclear spins with paramagnetic centers (Fig. 7.4) due to the strong electron–nuclear and electron–phonon coupling. In solids, the contribution of dipole–dipole interaction of nuclear spins with unpaired electron spins of paramagnetic ions plays a dominant part, with the relaxation rate $R_1 = 1/T_1$ of a nuclear spin I located at a distance r from the electron spin S, given by the expression [1, 2]

$$\frac{1}{T_{1n}} = \frac{2}{5}(\hbar\gamma_I\gamma_S)^2 S(S+1)\left[\frac{\tau_e}{1+\omega_I^2\tau_e^2} + \frac{7\tau_e}{3(1+\omega_e^2\tau_e^2)}\right]\left(\frac{N}{r^6}\right). \tag{7.3}$$

Here N is the number of unpaired electron spins in a nanoparticle, γ_I and γ_S are the nuclear and electronic gyromagnetic ratios, ω_I and ω_e are the nuclear and electronic angular resonance (Larmor) frequencies, and τ_e is the electron spin correlation time, which is actually the lifetime of the electron spin determined by the time of its spin–lattice relaxation T_{1e}. Since τ_e lies usually in the 10^{-7}–10^{-11} second range and $\omega_e\ \tau_e >> 1$, the second term in Eq. (7.3) may be

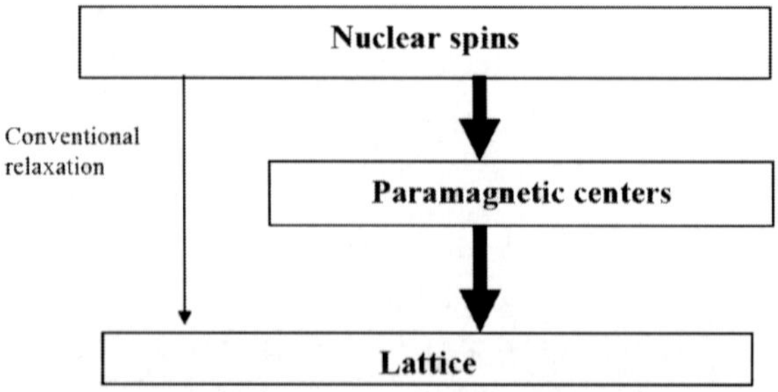

Figure 7.4 Nuclear spin–lattice relaxation via paramagnetic centers.

neglected, which transforms Eq. (7.3) to

$$r = \left[\frac{2}{5}(\hbar\gamma_I\gamma_S)^2 S(S+1) \times \frac{\tau_e}{1+\omega_I^2\tau_e^2} \times T_{1n} \times N\right]^{1/6}. \qquad (7.4)$$

This relation offers a possibility of estimating the distances between the nuclear and electron spins.

Nuclear spin–lattice relaxation in diamagnetic materials with nuclei of one type only results in an exponential recovery of the magnetization of the system, whereas the relaxation involving paramagnetic centers is described by a stretched exponential [3, 4]

$$M(t) = M_\infty \left\{1 - \exp\left[-\left(\frac{t}{T_1}\right)^\alpha\right]\right\}, \qquad (7.5)$$

where M_∞ is the equilibrium magnetization. This relation with the parameter α varying within the $0 < \alpha < 1$ interval is usually employed in describing systems featuring a set of different relaxation parameters [5], which in our particular case appears because of the distribution of relaxation times within a system of nuclear spins located at different distances from the paramagnetic center.

The second type of relaxation is the nuclear spin–spin relaxation (with the characteristic time T_2), which leads to equilibrium within each nuclear spin system (disregarding coupling to the lattice).

7.4 NMR Spectra in Detonation Nanodiamond

In this section we are going to consider NMR spectra of detonation nanodiamond (DND). A "static" ^{13}C NMR spectrum of a purified

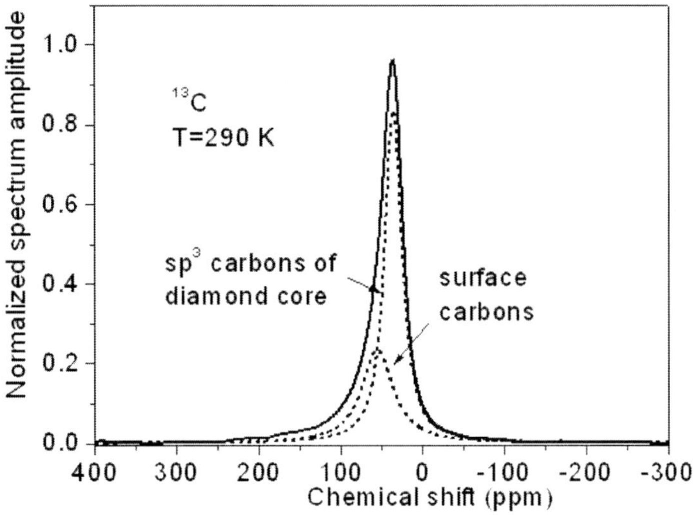

Figure 7.5 "Static" ^{13}C NMR spectrum of a DND sample measured in a magnetic field $\mathbf{B_0} = 8.0196$ T. Deconvolution into two components is shown by dashed lines. Reprinted with permission from Fig. 1 in Panich A. M., Vieth H.-M., Shames A. I., Froumin N., Ōsawa E., and Yao A. (2010) Structure and bonding in fluorinated nanodiamond, *J. Phys. Chem. C*, **114**, pp. 774–782. Copyright 2010, American Chemical Society.

DND sample in a magnetic field $\mathbf{B}_0 = 8.0196$ T (Fig. 7.5) can be deconvoluted into two components, an intense narrow signal with a chemical shift $\delta(^{13}\text{C}) = (35 \pm 1)$ ppm and a weaker one with $\delta(^{13}\text{C}) = 53$ ppm. The chemical shift of 35 ppm is characteristic of bulk diamonds [6–8]; therefore, this line can be reliably identified with the sp^3-hybridized carbon atoms of the diamond core. The line with $\delta = 53$ ppm is assigned to carbon atoms of the DND surface and of a shell that covers the core. More comprehensive information is derived from high-resolution NMR spectra of samples spinned at the magic angle (MAS) (Fig. 7.6) [9–15]. All the DND samples studied reveal an intense line with a chemical shift $\delta(^{13}\text{C}) = 35$ ppm. This line, as a rule, is asymmetric and can be deconvoluted into three components [9, 15] shown in Fig. 7.6 by dashed lines.

The narrow peak with a chemical shift $\delta = 34.4$ ppm and linewidth $\Delta\nu = 3.8$ ppm, which is characteristic of bulk diamond, can be assigned reliably to carbons of the diamond core with perfect

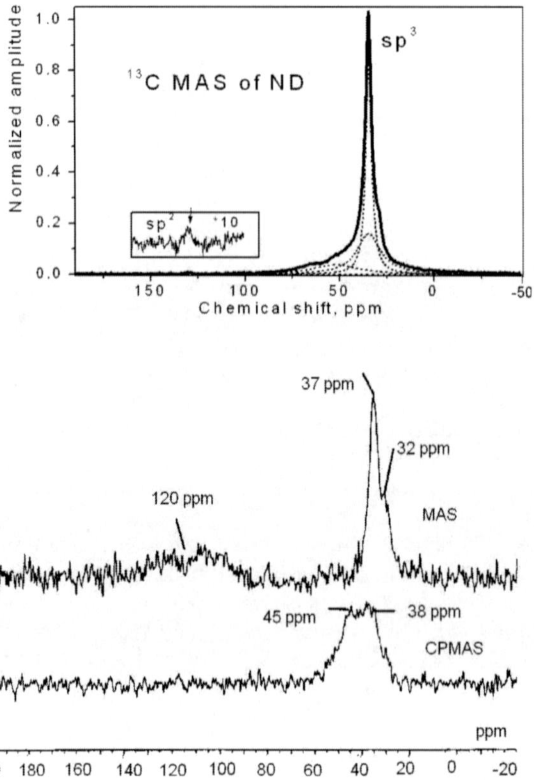

Figure 7.6 (Top) ^{13}C MAS NMR spectrum of nanodiamond deconvoluted into four components (dashed line profiles). The magnified signal of *sp*^{2}-hybridized carbons is shown in the inset (reprinted from Panich A. M., Shames A. I., Zousman B., and Levinson O. (2012) Magnetic resonance study of nanodiamonds prepared by laser-assisted technique, *Diam. Relat. Mater.*, **23**, pp. 150–153. Copyright 2012, with permission from Elsevier). (Bottom) ^{13}C MAS NMR and ^{1}H–^{13}C CP MAS NMR spectra of a DND sample obtained with a 65:35 RDX:TNT ratio (reprinted from Donnet J. B., Fousson E., Delmott L., Samirant M., Baras C., Wang T. K., and Eckhardt A. (2000). ^{13}C NMR characterization of nanodiamonds, *C. R. Acad. Sci. Paris, Serie IIc, Chimie*, **3**, pp. 831–838. Copyright 2000, with permission from Elsevier).

sp^3 configuration. A slightly broader line with a practically the same shift $\delta = 34.6$ ppm and linewidth $\Delta\nu = 15.3$ ppm was attributed to slightly inequivalent *sp*^{3} carbon atoms of the diamond core; this inequivalence may be the result of the slightly distorted tetragonal

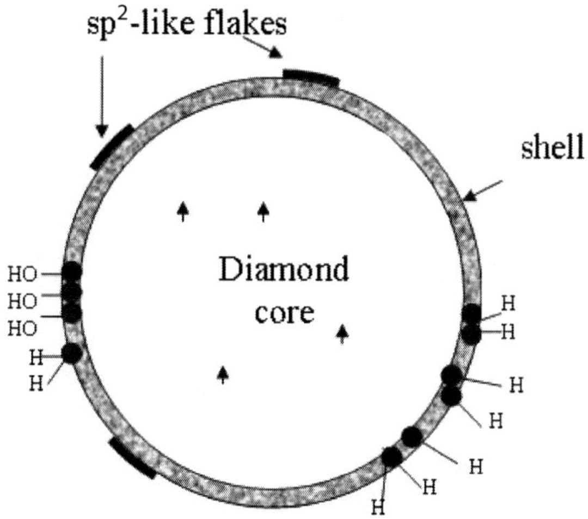

Figure 7.7 Schematic model of the structure of a DND particle. The shell is shown in gray. Filled circles identify carbon atoms forming hydrocarbon and hydroxyl groups. Dangling bonds with unpaired electrons are specified by arrows.

sp^3 configuration. The broad line with $\delta = 52.5$ ppm and $\Delta\nu = 29.2$ ppm is produced apparently by the disordered shell coating the diamond core.

Meanwhile, some samples display also an additional line at about 110–120 ppm (Fig. 7.6) characteristic of aromatic compounds. This line was assigned to sp^2-hybridized carbons of the shell coating the diamond core. This shell can be fullerene-like, as suggested by Raty et al. [16, 17]. The presence (or absence) of this shell depends obviously on the degree of purification and the method of sample treatment employed. The sp^2/sp^3 phase ratio in different DND samples varies over a broad range from 0 to 1.2 [10, 11, 14]. A spherical 4.5 nm diameter nanodiamond particle coated completely by a single graphite layer reveals an sp^2/sp^3 phase ratio of 0.19 [18, 19]. This implies that according to NMR analysis some of the samples studied may not have an sp^2 shell at all, while others are coated completely by such a shell, or embedded only partially in it, as shown in Fig. 7.7 displaying a model of the structure of a nanodiamond particle. Note that the NMR line at 110–120 ppm can

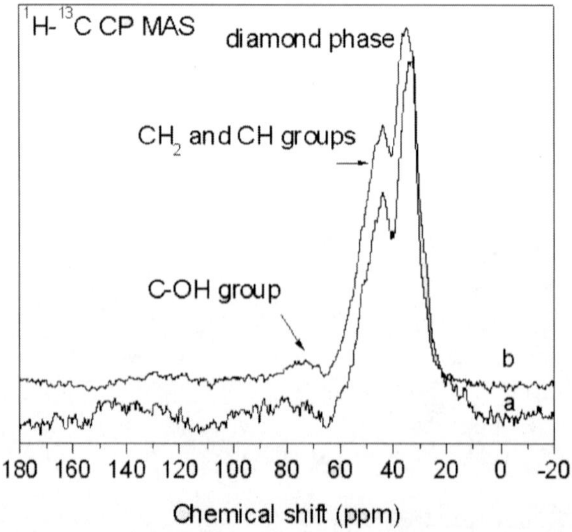

Figure 7.8 ^{1}H-^{13}C CP MAS NMR spectra of a DND sample. Reprinted with permission from Fig. 2 in Panich A. M., Shames A. I., Vieth H.-M., Ōsawa E., Takahashi M., and Vul' A. Ya. (2006). Nuclear magnetic resonance study of ultrananocrystalline diamonds, *Eur. Phys. J. B*, **52**, pp. 397–402. Copyright 2006, Springer Berlin/Heidelberg; Panich A. M. (2007) Solid state nuclear magnetic resonance studies of nanocarbons, *Diam. Relat. Mater.*, **16**, pp. 2044–2049.

also originate from traces of carbon soot. We note that the presence of *sp*^{2} carbon atoms is always detected in Raman spectra of DND (see, e.g., Ref. [20–22]), and the *sp*^{2} shell can be visualized in high-resolution transmission electron microscopy (HRTEM) images (e.g., [23]).

Essential additional information can be obtained in CP experiments (^{1}H–^{13}C CP MAS), the results of which are shown in Figs. 7.6 and 7.8. One clearly identifies here hydrocarbon and hydroxyl groups, such as CH, CH_2, and C–OH [10, 24].

The signatures of hydrocarbon and hydroxyl groups are well manifested in hydrogen NMR spectra [10, 24, 25]. As evident from Fig. 7.9a, a ^{1}H NMR spectrum consists of two lines. The broad component with the linewidth $\Delta\nu$ varying within the 15–36 kHz interval derives from rigidly bound hydrocarbon and hydroxyl groups. Its large width reflects the strong dipole–dipole coupling

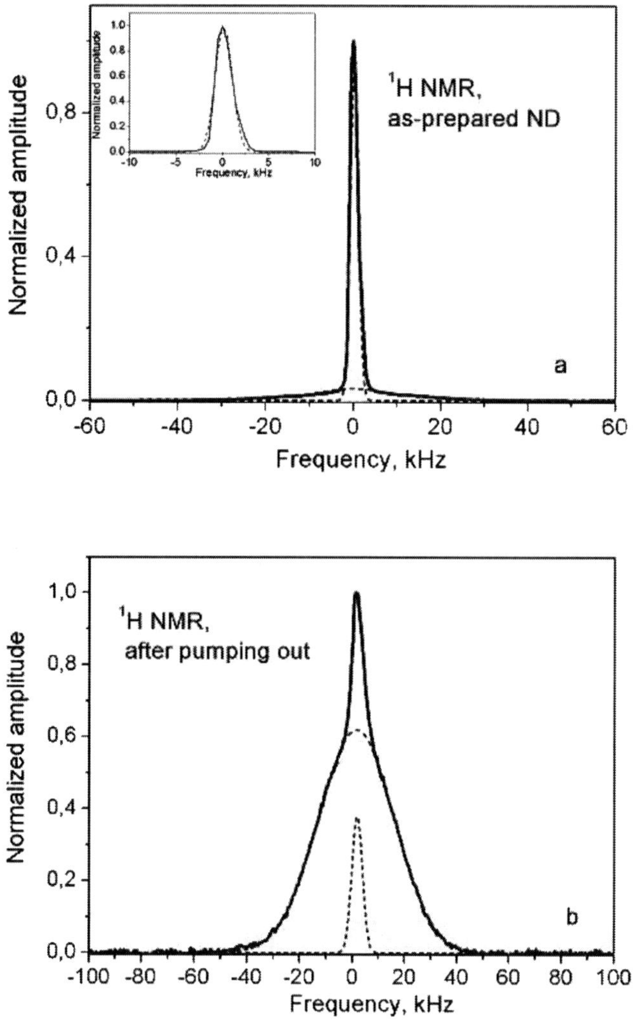

Figure 7.9 (a) ^{1}H NMR spectrum of an as-prepared DND sample. Dashed lines show its deconvolution into two components. The narrow component is shown separately in the inset. (b) ^{1}H NMR spectrum of the same DND sample measured after evacuation to 10^{-4} Torr. Dashed lines show the spectrum deconvoluted into two components. Reprinted with permission from Fig. 6 in Shenderova O., Panich A. M., Moseenkov S., Hens S. C., Kuznetsov V. L., and Vieth H.-M. (2011) Hydroxylated detonation nanodiamond: FTIR, XPS and NMR studies, *J. Phys. Chem. C.*, **115**, pp. 19005–19011. Copyright 2011, American Chemical Society.

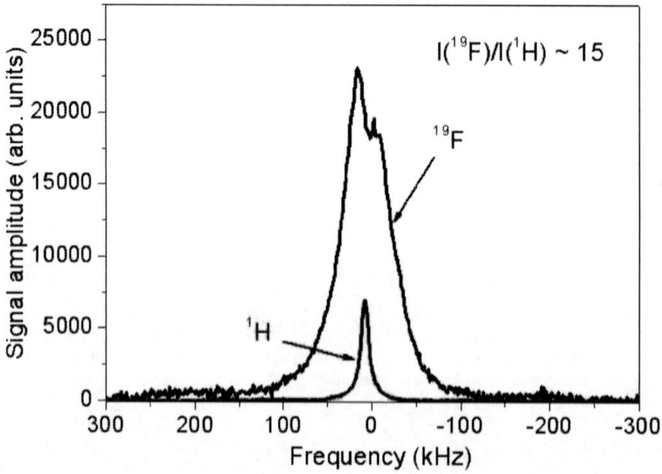

Figure 7.10 Static ^{1}H and ^{19}F NMR spectra of a fluorinated DND sample obtained in the field $\mathbf{B}_0$ = 8.0196 T. Reprinted with permission from Fig. 6 in Panich A. M., Vieth H.-M., Shames A. I., Froumin N., Ōsawa E., and Yao A. (2010) Structure and bonding in fluorinated nanodiamond, *J. Phys. Chem. C*, **114**, pp. 774–782. Copyright 2010, American Chemical Society.

of neighboring spins, thus assuming clusterization of hydrogen atoms on some parts of the nanodiamond surface. The narrow component with the linewidth $\Delta\nu \approx 2$–2.6 kHz originates primarily from moisture adsorbed on the surface of a nanoparticle. This line disappears practically completely upon proper evacuation of the sample (Fig. 7.9b).

Hydrocarbon and hydroxyl groups attached to the DND surface can be replaced with fluorine–carbon groups by fluorination. A ^{19}F NMR spectrum of fluorinated DND (F-DND) obtained in a field $\mathbf{B}_0 = 8.0196$ T (Fig. 7.10) reveals a signal of hydrogen atoms that is very weak compared with that from nonfluorinated DND and is considerably weaker (by approximately 15 times) than that of fluorine atoms. This permits a conclusion that hydrocarbon and hydroxyl groups originally located on the DND surface became effectively substituted by fluorine–carbon groups in the course of fluorination.

The ^{19}F MAS NMR spectrum of an F-DND sample obtained in the field $\mathbf{B}_0 = 7.0492$ T [9] is presented in Fig. 7.11. The dashed

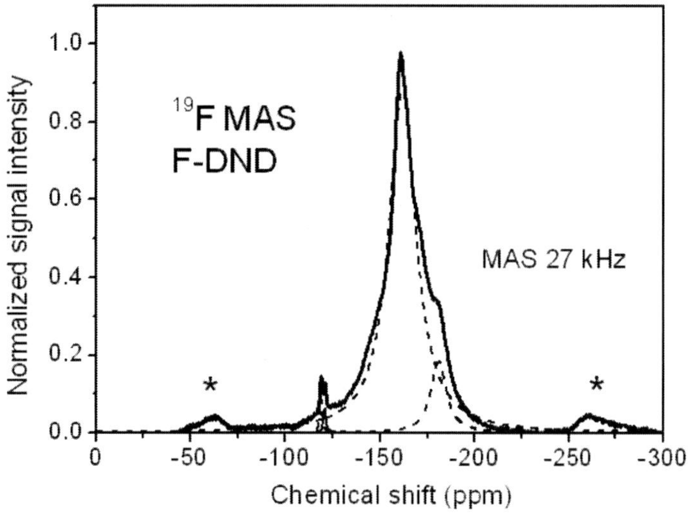

Figure 7.11 ^{19}F MAS NMR spectrum of a fluorinated DND sample obtained in the field $\mathbf{B}_0 = 7.0492$ T. Dashed lines illustrate deconvolution into components. The stars identify the sidebands appearing when the sample is spinned. Reprinted with permission from Fig. 7 in Panich A. M., Vieth H.-M., Shames A. I., Froumin N., Ōsawa E., and Yao A. (2010) Structure and bonding in fluorinated nanodiamond, *J. Phys. Chem. C*, **114**, pp. 774–782. Copyright 2010, American Chemical Society.

lines show the components of the deconvoluted spectrum. The assignment of such lines was verified in the particular cases of fluorinated graphites and fluorine-intercalated graphite compounds [26–28]. The weak signal with $\delta = -120$ ppm relates to CF_2 groups, and the overlapping lines at $\delta = -162$ and $\delta = -181$ ppm should be attributed to two inequivalent C–F groups.

The ^{19}F NMR spectrum of F-DND obtained in a low magnetic field $\mathbf{B}_0 = 0.717$ T, in which chemical shift anisotropy effects manifest themselves very little (Fig. 7.12), reveals a broad line reflecting the strong dipole–dipole coupling of neighboring nuclear spins, which assumes clusterization of fluorine atoms in some areas of the nanodiamond surface, with the remaining areas left practically nonfluorinated.

The ^{13}C MAS NMR spectrum of an F-DND sample obtained in the field $\mathbf{B}_0 = 14.0954$ T [9] (Fig. 7.13) displays at least two well-

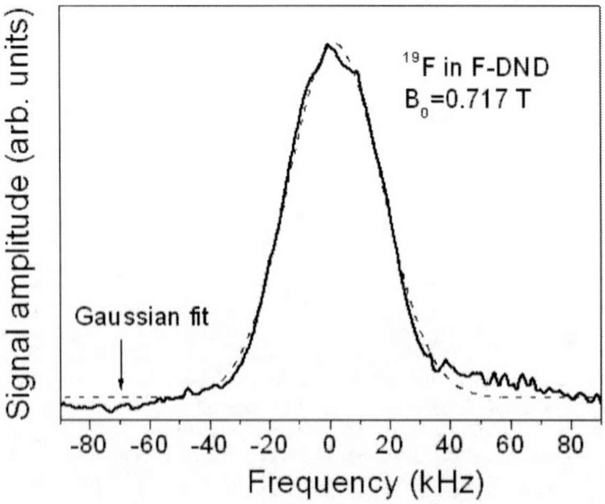

Figure 7.12 ^{19}F NMR spectrum of a fluorinated DND sample obtained in the field $\mathbf{B}_0 = 0.717$ T. Reprinted with permission from Fig. 5 in Panich A. M., Vieth H.-M., Shames A. I., Froumin N., Ōsawa E., and Yao A. (2010) Structure and bonding in fluorinated nanodiamond, *J. Phys. Chem. C*, **114**, pp. 774–782. Copyright 2010, American Chemical Society.

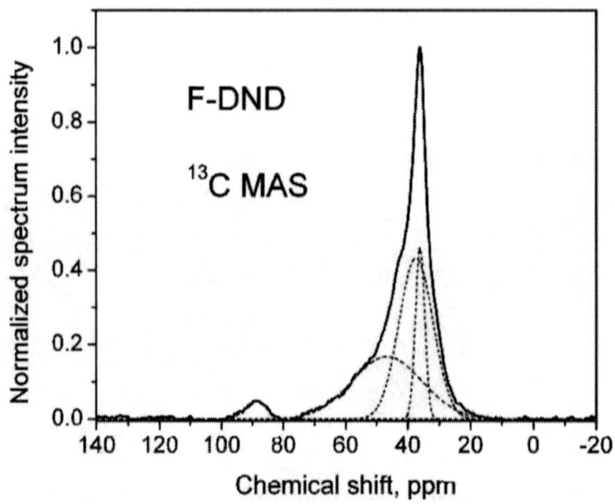

Figure 7.13 ^{13}C MAS NMR spectrum of a fluorinated DND sample obtained in the field $\mathbf{B}_0 = 14.0954$ T. Dashed lines illustrate deconvolution into components. Reprinted with permission from Fig. 2 in Panich A. M., Vieth H.-M., Shames A. I., Froumin N., Ōsawa E., and Yao A. (2010) Structure and bonding in fluorinated nanodiamond, *J. Phys. Chem. C*, **114**, pp. 774–782. Copyright 2010, American Chemical Society.

resolved lines. In accordance with the above line assignment [26–28], the line at $\delta(^{13}C) = 89.0$ ppm is attributed to carbon atoms involved in covalent C–F bonds. The intense signal with $\delta = 36$ ppm is related to sp^3 carbons of the diamond core and can be deconvoluted into three components, with two of them (the narrow ones) belonging to the carbons in the regular and slightly distorted sp^3 configurations and the third one to the disordered shell.

7.5 Nuclear Relaxation in Detonation Nanodiamond

Measurements of the spin–lattice relaxation time T_1 of ^{13}C nuclei of the diamond core reveal an anomalous decrease of T_1 from a few hours in natural diamond [7, 29, 30] to a few hundred milliseconds in nanodiamond. Recovery of nuclear magnetization in the course of relaxation in DND is described by a stretched exponent, Eq. (7.5), which is seen clearly from Fig. 7.14 plotting the variation of magnetization scaling as the 0.57 power of time. Both these features result directly from the interaction of nuclear spins with paramagnetic centers, which in the case in question are dangling bonds with unpaired electrons.

The spin–lattice and spin–spin relaxation of hydrogen atoms in DND is described by a biexponential time dependence of magnetization (t), which originates from the two resonant lines shown in Fig. 7.9. The biexponential magnetization decay stands out clearly in Fig. 7.15. Here the longer spin–spin relaxation time is assigned to the narrow line and the shorter one to the broad one.

In F-DND, spin–lattice relaxation of carbon and fluorine nuclei proceeds also very fast and is also described by a stretched exponential Eq. (7.5). Here $T_1(^{13}C) = 168$ μs, that is, twice shorter than that in nonfluorinated DND, where T_1 varies from 377 μs to 455 μs [9], which correlates, as follows from ESR data, with the twice-higher density of paramagnetic centers in F-DND. The experimental value $\alpha = 0.66$ (Eq. 7.5) suggests that the paramagnetic centers are nonuniformly distributed [4]. Carbon atoms positioned on the surface of a nanoparticle show an order-of-magnitude faster relaxation ($T_1 \sim 17$ ms), thus providing supportive evidence for the model by which noticeable number of paramagnetic centers (dangling bonds) in the diamond core are located not far from the surface [9]. This conclusion is corroborated also by the very short relaxation time of fluorine atoms [9].

Measurements of nuclear spin–lattice relaxation turn out extremely useful in studies of highly purified DND samples whose surface is chemically modified by atoms of transition metal elements such as copper and cobalt. Although the ESR spectrum of these samples does contain a signal with a hyperfine structure due to the Cu^{2+} complexes, it remains unclear whether these complexes are located on the DND surface or exist as a separate phase. The answer to this question comes from measurement of nuclear spin–lattice relaxation [31–33]. The point is that being fixed on the surface and representing actually an integral part of the lattice, paramagnetic ions speed up sharply nuclear paramagnetic relaxation through dipole interaction with nuclear spins.

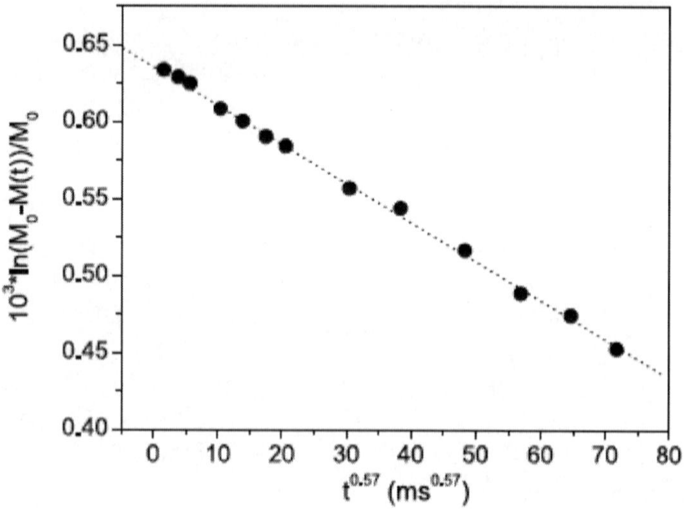

Figure 7.14 ^{13}C magnetization recovery in the process of spin–lattice relaxation in DND plotted on the semilog scale. Reprinted with permission from Fig. 5 in Panich A. M., Shames A. I., Vieth H.-M., Ōsawa E., Takahashi M., and Vul' A. Ya. (2006). Nuclear magnetic resonance study of ultrananocrystalline diamonds, *Eur. Phys. J. B*, **52**, pp. 397–402. Copyright 2006, Springer Berlin/Heidelberg.

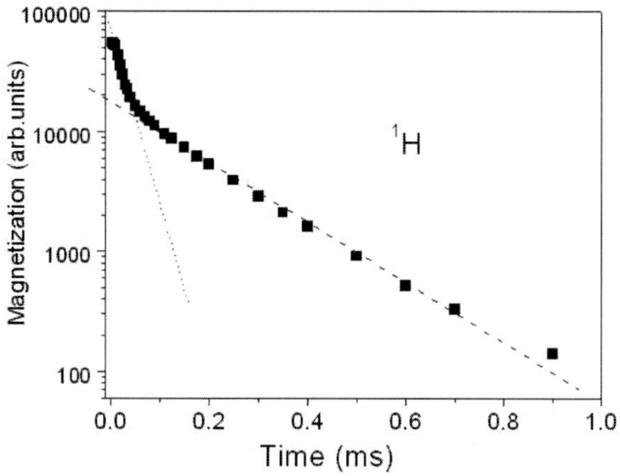

Figure 7.15 ^{1}H magnetization decay (measurement of T_2) in a DND sample plotted on the semilog scale. Reprinted with permission from Fig. 7 in Panich A. M., Shames A. I., Vieth H.-M., Ōsawa E., Takahashi M., and Vul' A. Ya. (2006). Nuclear magnetic resonance study of ultrananocrystalline diamonds, *Eur. Phys. J. B*, **52**, pp. 397–402. Copyright 2006, Springer Berlin/Heidelberg.

By contrast, if these ions exist in the material as a separate phase, the above effect should be extremely small. As seen from Fig. 7.16 and Tables 7.1 and 7.2, copper- and cobalt-decorated samples exhibit increase in the spin–lattice relaxation rate of ^{13}C nuclei with increasing content of paramagnetic ions. This provides compelling evidence for the appearance of the Cu^{2+} and Co^{2+} paramagnetic ions on the surface of DND particles. The mechanism of this relaxation is shown schematically in Fig. 7.17 [31].

Annealing a sample at 550°C and 900°C produces the opposite result, more specifically slowing down of the relaxation rate and recovery of its original level (Fig. 7.16, Tables 7.1 and 7.2), thus demonstrating destruction of the surface paramagnetic complexes of transition metals [31, 32].

The hydrocarbon and hydroxyl groups fixed on the surface are likewise sensitive to the above effects; this follows convincingly from measurements of the spin–lattice relaxation of hydrogen nuclei (Fig. 7.18, Tables 7.3 and 7.4) [33], which demonstrate growth of the relaxation rate induced by chemical modification of the samples

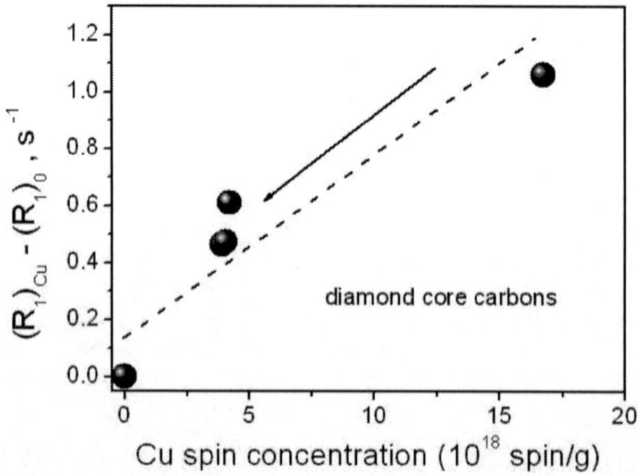

Figure 7.16 ^{13}C spin–lattice relaxation rate R_1 in DND plotted vs. copper ion concentration. Here R_1 is the difference between the relaxation rates in the modified and initial samples. Reprinted with permission from Panich A. M., Shames A. I., Medvedev O., Osipov V. Yu., Alexenskii A. E., and Vul' A. Ya. (2009) Magnetic resonance study of detonation nanodiamonds with surface chemically modified by transition metal ions, *Appl. Magn. Reson.*, **36**, pp. 317–329. Copyright 2009, Springer.; Shames A. I., Panich A. M., Osipov V. Yu., Aleksenskiy A. E., Vul' A. Ya., Enoki T., and Takai K. (2010) Structure and magnetic properties of detonation nanodiamond chemically modified by copper, *J. Appl. Phys.*, **107**, pp. 014318(1–10). The arrow visualizes the decrease of the relaxation rate brought about by sample annealing.

Table 7.1 [31]. ^{13}C spin–lattice relaxation times (T_1) and rates (R_1) in Cu-DND. ΔR_1 is the difference between the relaxation rates in the initial (A1) and modified (A2, A3) samples. The samples annealed in a hydrogen flow at 550°C and 900°C are denoted by A3-1 and A3-2, respectively

Sample	Spin/g Cu, $\times 10^{18}$	T_1 (μs)	$R_1(s^{-1})$	$\Delta R_1(Cu)$ (s^{-1})	α
A1-1	0	598 ± 19	1.672	0	0.58
A2	4.2	438 ± 14	2.283	0.6109	0.61
A3	16.74	366 ± 9	2.732	1.060	0.61
A3-1	4.1	466 ± 12	2.146	0.4737	0.62
A3-2	3.9	468 ± 14	2.137	0.4645	0.61

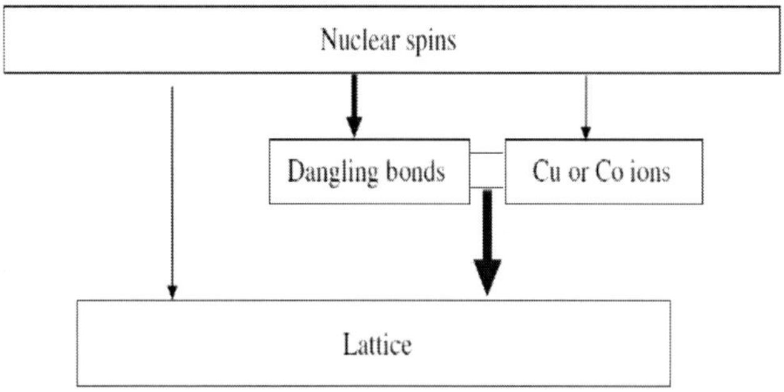

Figure 7.17 Schematic of nuclear spin–lattice relaxation involving paramagnetic ions. Reprinted with permission from Panich A. M., Shames A. I., Medvedev O., Osipov V. Yu., Alexenskii A. E., and Vul' A. Ya. (2009) Magnetic resonance study of detonation nanodiamonds with surface chemically modified by transition metal ions, *Appl. Magn. Reson.*, **36**, pp. 317–329. Copyright 2009, Springer.

Table 7.2 [31]. ^{13}C spin–lattice relaxation times (T_1) in Co-DND. B1, initial sample; B3, cobalt-modified sample; B3-1 and B3-2, samples annealed in a hydrogen flow at 550°C and 900°C, accordingly

Sample	T_1 (μs)
Reference, B1	615 ± 30
Co-B3	412 ± 26
Co-B3-2	575 ± 34
Co-B3-1	545 ± 20

Table 7.3 [33]. ^1H spin–lattice relaxation times (T_1) in Cu-DND for the broad and narrow components of the ^1H spectrum

Sample	T_{1broad} (μs)	T_{1nar} (μs)
A1	70.2 ± 9.1	58.2 ± 0.7
A2-Cu	22.2 ± 1.1	17.2 ± 1.1
A3-Cu	5.15 ± 0.6	14.0 ± 0.8

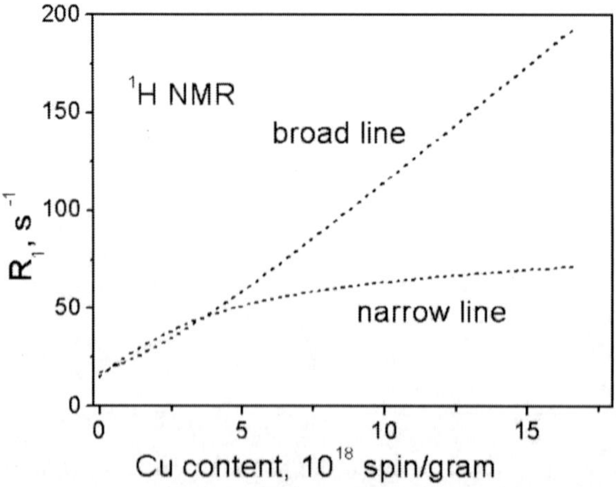

Figure 7.18 ^{1}H spin–lattice relaxation rate R_1 in DND plotted vs. copper ion concentration [33] for the broad and narrow components of the spectrum. Reprinted with permission from Panich A. M., Altman A., Shames A. I., Osipov V. Yu., Alexenskiy A. E., and Vul' A. Ya. (2011) Proton magnetic resonance study of diamond nanoparticles decorated by transition metal ions, *J. Phys. D: Appl. Phys.*, **44**, pp. 125303(1–5).

by copper and cobalt. This may be considered as more evidence for paramagnetic ions being located on the DND surface. A sketch of a chemically modified sample is displayed in Fig. 7.19. Knowing the relaxation times and the number of copper ions per nanoparticle, one can use Eq. (7.4) to estimate the average distances from a hydrogen atom to the copper ions, which turn out to be 0.74 nm and 0.79 nm for samples A3 and A2, respectively.

Table 7.4 [33]. ^{1}H spin–lattice relaxation times (T_1) in Co-DND for the broad and narrow components of the ^{1}H spectrum

Sample	$T_{1\text{broad}}$ (μs)	$T_{1\text{nar}}$ (μs)
B1	70.2 ± 9.1	58.2 ± 0.7
B3-Co	14.5 ± 0.4	6.2 ± 0.3
B3-1-Co	73.8 ± 2.3	86.6 ± 2.0
B3-2-Co	108.7 ± 1.8	108.7 ± 1.8

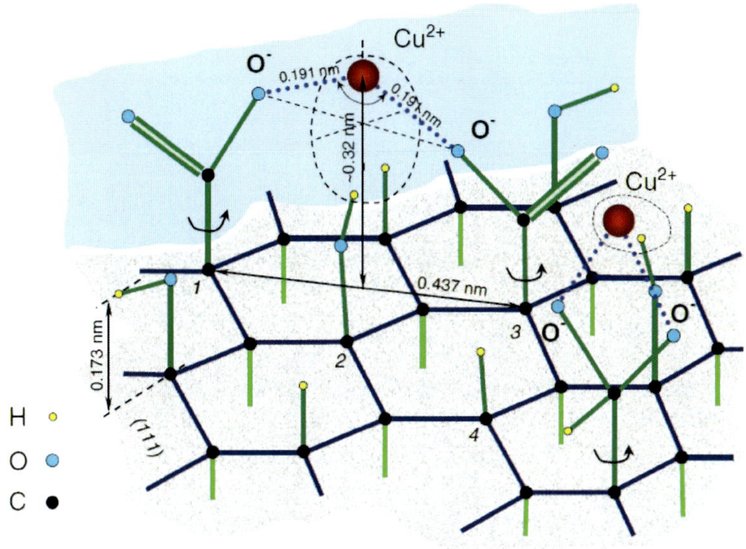

Figure 7.19 Schematic representation of the surface of a chemically modified sample. Reprinted with permission from Panich A. M., Altman A., Shames A. I., Osipov V. Yu., Alexenskiy A. E., and Vul' A. Ya. (2011) Proton magnetic resonance study of diamond nanoparticles decorated by transition metal ions, *J. Phys. D: Appl. Phys.*, **44**, pp. 125303(1–5).

7.6 Conclusion

NMR is an excellent tool to probe the structure, chemical bonds, and electron-nuclear interaction in nanodiamond at the atomic level. This is presently one of the most important spectroscopic methods employed in the investigation of nanodiamonds, which quite frequently serves as a key to understanding their structure and properties.

We also address the readers to the recently published review article on NMR studies of nanodiamonds [34] and recent papers on nanodiamond deaggregation [35] and nanodiamond-to-carbon onion transformation [36] studied by NMR.

References

1. Abragam A. (1961) *The Principles of Nuclear Magnetism* (Clarendon Press, Oxford).
2. Goldman M. (1970) *Spin Temperature and Nuclear Magnetic Resonance in Solids* (Clarendon Press, Oxford).
3. Blumberg W. (1960) Nuclear spin-lattice relaxation caused by paramagnetic impurities, *Phys. Rev.*, **119**, pp. 79–84.
4. Furman G. B., Panich A. M., Yochelis A., Kunoff E. M., and Goren S. D. (1997) Nuclear spin-lattice relaxation of dipolar order caused by paramagnetic impurities, *Phys. Rev. B*, **55**, pp. 439–444.
5. Phillips J. C. (1996) Stretched exponential relaxation in molecular and electronic glasses, *Rep. Prog. Phys.*, **59**, pp. 1133–1207.
6. Duncan T. M. (1987) ^{13}C chemical shieldings in solids, *J. Phys. Chem. Ref. Data*, **16**, pp. 125–151.
7. Duijvestijn M. J., van der Lugt C., Smidt J., Wind R. A., Zilm K. W., and Staplin D. C. (1983) C-13 NMR spectroscopy in diamonds using dynamic nuclear polarization, *Chem. Phys. Lett.*, **102**, pp. 25–28.
8. Retcofsky H. L., and Friedel R. A. (1973). Carbon-13 magnetic resonance in diamonds, coals, and graphite, *J. Phys. Chem.*, **77**, pp. 68–71.
9. Panich A. M., Vieth H.-M., Shames A. I., Froumin N., Ōsawa E., and Yao A. (2010) Structure and bonding in fluorinated nanodiamond, *J. Phys. Chem. C*, **114**, pp. 774–782.
10. Panich A. M., Shames A. I., Vieth H.-M., Ōsawa E., Takahashi M., and Vul' A. Ya. (2006). Nuclear magnetic resonance study of ultrananocrystalline diamonds, *Eur. Phys. J. B*, **52**, pp. 397–402.
11. Donnet J. B., Fousson E., Delmott L., Samirant M., Baras C., Wang T. K., and Eckhardt A. (2000). ^{13}C NMR characterization of nanodiamonds, *C. R. Acad. Sci. Paris, Serie IIc, Chimie*, **3**, pp. 831–838.
12. Alam T. M. (2004) Solid-state ^{13}C magic angle spinning NMR spectroscopy characterization of particle size structural variations in synthetic nanodiamonds, *Mater. Chem. Phys.*, **85**, pp. 310–315.
13. Belobrov P. I., Gordeev S. K., Petrakovskaya E. A., and Falaleev O. V. (2001) Paramagnetic properties of nanodiamond, *Dokl. Phys.*, **46**, p. 459.
14. Dubois M., Guerin K., Petit E., Batisse N., Hamwi A., Komatsu N., Giraudet J., Pirotte P., and Masin F. (2009) Solid-state NMR study of nanodiamonds produced by the detonation technique, *J. Phys. Chem. C*, **113**, pp. 10371–10378.

15. Panich A. M., Shames A. I., Zousman B., and Levinson O. (2012) Magnetic resonance study of nanodiamonds prepared by laser-assisted technique, *Diam. Relat. Mater.*, **23**, pp. 150–153.
16. Raty J. Y., and Galli G. (2003) Ultradispersity of diamond at the nanoscale, *Nat. Mater.*, **2**, pp. 792–795.
17. Raty J. Y., Galli G., Bostedt C., van Buuren T. W., and Terminello L. J. (2003) Quantum confinement and fullerenelike surface reconstructions in nanodiamonds, *Phys. Rev. Lett.*, **90**, pp. 037401(1–4).
18. Bursill L. A., and Fullerton A. L. (2001) Size and surface structure of diamond nano-crystals, *Int. J. Mod. Phys.*, **15**, pp. 4087–4102.
19. Shenderova O. A., Zhirnov V. V., and Brenner D. W. (2002) Carbon nanostructures, *Crit. Rev. Solid State Mater. Sci.*, **27**, pp. 227–356.
20. Aleksenskii A. E., Baidakova M. V., Vul' A. Ya., Davidov V. Yu., and Pevtsova Yu. A. (1997) Diamond-graphite phase transition in ultradisperse-diamond clusters, *Phys. Solid State*, **39**, pp. 1007–1015.
21. Osswald S., Yushin G., Mochalin V., Kucheyev S. O., and Gogotsi Yu. (2006) Control of sp(2)/sp(3) carbon ratio and surface chemistry of nanodiamond powders by selective oxidation in air, *J. Am. Chem. Soc.*, **128**, pp. 11635–11642.
22. Mochalin V., Osswald S., and Gogotsi Yu. (2009) Contribution of functional groups to the Raman spectrum of nanodiamond powders, *Chem. Mater.*, **21**, pp. 273–279.
23. Shames A. I., Panich A. M., Kempiński W., Alexenskii A. E., Baidakova M. V., Dideikin A. T., Osipov V. Yu., Siklitski V. I., Ōsawa E., Ozawa M., and Vul' A. Ya. (2002) Defects and impurities in nanodiamonds: EPR, NMR and TEM study, *J. Phys. Chem. Solids*, **63**, pp. 1993–2001.
24. Panich A. M. (2007) Solid state nuclear magnetic resonance studies of nanocarbons, *Diam. Relat. Mater.*, **16**, pp. 2044–2049.
25. Shenderova O., Panich A. M., Moseenkov S., Hens S. C., Kuznetsov V. L., and Vieth H.-M. (2011) Hydroxylated detonation nanodiamond: FTIR, XPS and NMR studies, *J. Phys. Chem. C.*, **115**, pp. 19005–19011.
26. Panich A. M., Nakajima T., Vieth H.-M., Privalov A., and Goren S. (1998) Nuclear magnetic resonance study of fluorine–graphite intercalation compounds, *J. Phys.: Condens. Matter*, **10**, pp. 7633–7642.
27. Panich A. M., and Nakajima T. (2000) Physical properties and C-F bonding in fluorine-graphite intercalation compounds as seen by NMR, *Mol. Cryst. Liq. Cryst.*, **340**, pp. 77–82.

28. Panich A. M. (1999) Nuclear magnetic resonance study of fluorine-graphite intercalation compounds and graphite fluorides, *Synth. Met.*, **100**, pp. 169–185.

29. Hoch M. J. R., and Reynhardt E. C. (1988) Nuclear spin-lattice relaxation of dilute spins in semiconducting diamond, *Phys. Rev. B*, **37**, pp. 9222–9226.

30. Reynhardt E. C., and Terblanche C. J. (1997) C-13 relaxation in natural diamond, *Chem. Phys. Lett.*, **269**, pp. 464–468.

31. Panich A. M., Shames A. I., Medvedev O., Osipov V. Yu., Alexenskii A. E., and Vul' A. Ya. (2009) Magnetic resonance study of detonation nanodiamonds with surface chemically modified by transition metal ions, *Appl. Magn. Reson.*, **36**, pp. 317–329.

32. Shames A. I., Panich A. M., Osipov V. Yu., Aleksenskiy A. E., Vul' A. Ya., Enoki T., and Takai K. (2010) Structure and magnetic properties of detonation nanodiamond chemically modified by copper, *J. Appl. Phys.*, **107**, pp. 014318(1–10).

33. Panich A. M., Altman A., Shames A. I., Osipov V. Yu., Alexenskiy A. E., and Vul' A. Ya. (2011) Proton magnetic resonance study of diamond nanoparticles decorated by transition metal ions, *J. Phys. D: Appl. Phys.*, **44**, pp. 125303(1–5).

34. Panich A. M. (2012) Nuclear magnetic resonance studies of nanodiamonds, *Crit. Rev. Solid State Mater. Sci.*, **37**, pp. 276–303.

35. Panich A. M., and Aleksenskii A. E. (2012) Deaggregation of diamond nanoparticles studied by NMR, *Diamond Relat. Mater.*, **27–28**, pp. 45–48.

36. Panich A. M., Shames A. I., Sergeev N. A., Olszewski M., McDonough J. K., Mochalin V. N., and Gogotsi Y. (2013) Nanodiamond graphitization: a magnetic resonance study, *J. Phys. Condens. Matter.*, **25**, pp. 245303 (1–8).

Chapter 8

Magnetic and Structural Studies of Multilayered Nanographites Prepared from Detonation Nanodiamond

Vladimir Yu. Osipov
Ioffe Physical-Technical Institute, Solid State Electronics Department, 26 Polytechnicheskaya, St. Petersburg 194021, Russia
Osipov@mail.ioffe.ru

8.1 Introduction

Detonation nanodiamonds (DNDs) are attracting presently considerable interest as an affordable starting material for production of carbon onion, nanographite, and graphene nanoparticles. As mentioned in Chapter 1, transformation of DND particles into carbon onions and nanographite can be readily achieved by high-temperature annealing. The first thorough investigation of this phase transition, DND–carbon onion–nanographite, can be traced to Refs. [1, 2]. A comprehensive study of the mechanism and intrinsic details of the nanodiamond–nanographite structural transformation [3–5] established the possibility of investigating the optical [6, 7] and magnetic [8, 9] effects in the onion-like and nanographite

Detonation Nanodiamonds: Science and Applications
Edited by Alexander Ya. Vul' and Olga A. Shenderova

ISBN 978-981-4411-27-1 (Hardcover), 978-981-4411-28-8 (eBook)
www.panstanford.com

forms of carbon, which derive from the well-developed π electronic system.

Preparation of graphene sheets of monatomic thickness and observation of remarkable quantum effects in this material [10, 11] have stimulated a resurgence of interest in methods employed in the preparation of nanographite and nanographene from DND.

The particular interest focused on the magnetic properties of nanographite finds ready explanation in that while bulk graphite possesses clearly pronounced diamagnetic properties, nanographite reveals paramagnetism generated by the π electronic states forming at the zigzag-shaped edges of graphene sheets [12, 13].

This chapter addresses briefly the issues associated with the technology of nanographite preparation from DND and with methods employed in its identification, with the main emphasis on the magnetic properties of nanographite.

8.2 Magnetic Moment Measurement with a SQUID Magnetometer

Consider first the method used in measurement of magnetic susceptibility.

The magnetic moment **m** of a sample induced in an external magnetic field **H** can be measured by different techniques, the most accurate and sensitive of them being the method based on the superconducting quantum interference device (SQUID). The sensitivity of the SQUID to a magnetic field is two to three orders of magnitude higher than that offered by the best nonsuperconducting magnetometers and is as high as 10^{-14} T·Hz$^{-1/2}$. These instruments are capable even of detecting ultraweak dynamic magnetic fields of bio-objects, which are generated by biocurrents flowing through living tissue.

The SQUID magnetometer consists of two main parts (Fig. 8.1), a high-sensitivity SQUID interferometer proper (or a sensor) and a system of superconducting pickup coils L_1 and L_2 coupled inductively via superconducting circuits and an additional superconducting coil L_S with the SQUID sensor L [14–16]. The equipment is maintained in a cryostat at liquid helium temperature (4.2 K).

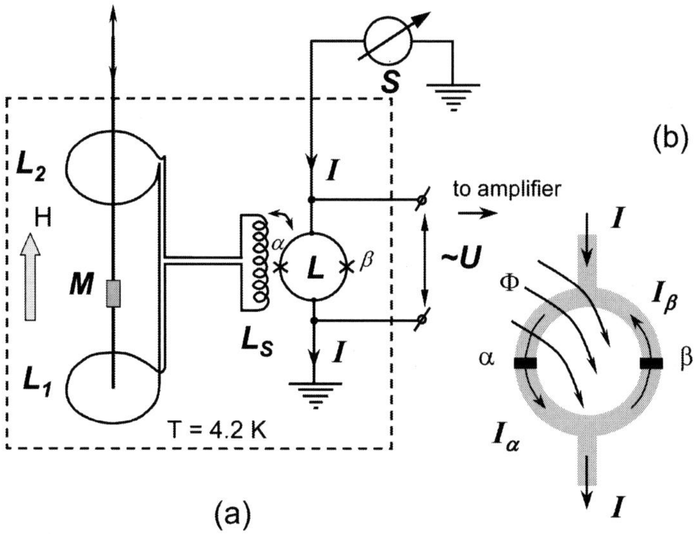

Figure 8.1 Schematic diagram of the SQUID magnetometer. (a) Schematic of the magnetometer; (b) SQUID sensor. *M*, sample reciprocating in measuring chamber; L_1, L_2, superconducting pickup coils of the magnetic flux receiving antenna; *L*, SQUID sensor (superconducting circuit with two Josephson junctions *I*, *II*); L_S output superconducting receiving antenna coil coupled inductively with the SQUID sensor; *S*, DC power supply. Initially, sample *M* is centered between two oppositely connected turns L_1 and L_2. The electric circuit receiving a weak oscillating output voltage signal from the sensor and amplifying it by lock-in detection is not shown. *Abbreviation*: DC, direct current.

The SQUID sensors are of two types, direct current (DC) or alternating current (AC) operated. We are going to consider in what follows only a DC SQUID sensor representing actually a closed superconducting circuit with two Josephson junctions[a] α and β connected in parallel. It can be employed in all cases where a change of the quantity under study can be transformed into a change of magnetic flux. During the measurement, a DC current, larger than twice the critical current of the Josephson

[a]By the Josephson junction one understands the weak contact between two bulk superconductors, which is formed by a tunneling-transparent dielectric layer 1–2 nm thick embedded between these superconductors, a local thin-film interlayer, or a small-area mechanical contact.

junction, flows through the circuit. The system of spaced-apart superconducting pickup coils terminating in an end ring (receiving antenna) is inductively coupled with the SQUID sensor and acts essentially as a superconducting magnetic flux transformer, which transmits the changes in magnetic flux from the pickup coils to the SQUID ring interferometer. A mechanical displacement of a sample possessing a magnetic moment within the superconducting pickup coils translates into a change of the magnetic flux in the transformer, which induces in the closed flux-transformer circuit a superconducting current transmitting the change of the magnetic flux directly to the SQUID sensor. The SQUID sensor generates the primary electric signal proportional to the magnitude of the change of the magnetic flux, followed by its magnification. Significantly, the pickup coils of the receiving antenna within which the magnetized object under study undergoes periodic displacement are maintained in a constant uniform magnetic field of a solenoid.

The operation of a DC SQUID sensor is based essentially on magnetic flux quantization in a superconducting ring with Josephson junctions connected into it. If a magnetic field is present in the interferometer circuit, it induces in the latter a circulating superconducting current. When a small DC current I ($I < 2I_c$, where I_c is the critical current[a] of a single Josephson junction) passes through the circuit, the currents flowing through the two interferometer branches, I_α and I_β, will be different, as will be the phase differences appearing at the two tunneling contacts, δ_α and δ_β. The current I passing through the circuit is the difference between the currents I_α and I_β flowing in the two interferometer branches: $I = I_\alpha - I_\beta = 2I_c \sin\frac{(\delta_\alpha - \delta_\beta)}{2} \cos\frac{\pi\phi}{\phi_o}$. Here ϕ is the magnetic flux crossing the circuit, and $\phi_o = \frac{h}{2e} = 2.07 \times 10^{-7}$ G·cm^2 is a magnetic flux quantum. The phase differences between the two contacts and the flux through the circuit are coupled by the following relation, which reflects flux quantization: $2\pi \cdot n = \delta_\alpha + \delta_\beta + \frac{2\pi}{\phi_o}\phi$,

[a] In a single Josephson junction, current I, if it is smaller than the critical value $I < I_c$, flows with no resistance through the tunneling contact, with $I = I_c \sin\varphi$. Here φ is the phase difference between the phases of the macroscopic wave function of the Cooper pair condensate in the regions before and after the contact. If the current through the Josephson contact $I > I_c$, the voltage across the junction oscillates at microwave frequencies $U \approx \partial\varphi/\partial t$ (the so-called nonstationary Josephson effect).

where n is an integer. Thus we come to the following relation connecting the maximum critical current I_{max} that can pass through the circuit without forming a potential difference across it with the critical current I_c: $I_{max} = 2I_c |\cos(\pi\phi/\phi_o)|$. We readily see that in the $I < 2I_c$ operating mode the critical current of the circuit I_{max} is a periodic function of the ratio ϕ/ϕ_o. In other words, interference[a] manifests itself here as a dependence of the SQUID critical current on the applied external magnetic field. A half-integer number of flux quanta crossing the interferometer gives rise to an unstable superconducting state. For currents $I > 2I_c$, and under variation of the magnetic flux ϕ through the interferometer ring, the output voltage $U = d\phi/dt$ across the circuit oscillates with a period equal to the magnetic flux quantum ϕ_o. The voltage $U(\phi)$ is maximal at the flux $\phi = \phi_o(n + \frac{1}{2})$ and minimal at $\phi = \phi_o \cdot n$, where n is an integer. The maximum peak-to-peak oscillation amplitude $U(\phi)$ is reached at the optimum value of the parameter $LI_c/\phi_o = 1$, where L is the inductance of the interferometer circuit. An analysis of the SQUID sensor output voltage oscillations permits one to estimate with a high accuracy the magnetic moment of the sample. The output signal produced by the electric circuit is proportional to the magnitude of the magnetic moment.

During the time taken by measurement of the magnetic moment, external sources of current are usually disconnected, and the magnetic field of the superconducting solenoid is maintained by the nondissipative current circulating in the solenoid through short-circuited output current leads with the power supply turned off. The sample undergoes a few mechanical displacements within an interval of ± 3 cm relative to the center of the pickup coils already after stabilization of the temperature (to within 0.02 K) and of the magnetic field (the accuracy of adjustment and the field stability are not worse than 0.1 G) in the measurement chamber. The range of magnetic moment measurement by some AC- and DC-operated

[a]We have here actually interference of the Josephson currents passing through the tunneling contacts on different sides of the SQUID circuit, with the phase difference between them being mediated by the magnitude of the magnetic flux crossing the circuit. Said otherwise, the flux of the Bose–Einstein Cooper pair condensate splits in two parts when passing through the circuit, and the two parts of the flux, on acquiring various increments to the phase of the macroscopic wave function, interfere when leaving the circuit.

SQUIDs of a widely used trademarks, for instance, is $\sim \pm 5$ emu for a sensitivity of $10^{-8} - 10^{-7}$ emu (in centimeter-gram-second [CGS] units).

8.3 Determination of Magnetic Susceptibility and the Nature of Diamagnetism in Aromatic Compounds and Graphite

Graphite and other carbon materials belong to the class of diamagnets, whose magnetic susceptibility is negative ($\chi < 0$) and the magnetic permittivity $\mu < 1$. We may recall here that static magnetic susceptibility is defined as the ratio of the magnetization of a substance (or of a medium) **M**[a] to the strength of the magnetic field **H**, the field that generates in the substance under study (or a medium) this magnetization, $\chi = \mathbf{M}/\mathbf{H}$. The magnetic field **H** in the substance, magnetic induction **B**, and the magnetization of the substance **M** are interrelated, in their turn, through the following expression (in SI and CGS units):

$$\begin{aligned} &\mathbf{B} = \mu_o\mathbf{H} + \mu_o\mathbf{M} = \mu_o\mathbf{H} + \mu_o\chi\mathbf{H} = \mu_o(1+\chi)\mathbf{H} = \mu_o\mu\mathbf{H} \quad \text{(SI)} \\ &\mathbf{B} = \mathbf{H} + 4\pi\mathbf{M} = \mathbf{H} + 4\pi\chi\mathbf{H} = (1+4\pi\chi)\mathbf{H} = \mu\mathbf{H} \quad \text{(CGS)} \end{aligned} \tag{8.1}$$

Here $\mu_o = 4\pi \times 10^{-7}$ V·s/A·m is the magnetic constant, and μ is the magnetic permittivity. In the CGS system of units, the magnetic permittivity and magnetic susceptibility are coupled through the relation $\mu = 1 + 4\pi\chi$ and in the SI system through $\mu = 1 + \chi$. Magnetic susceptibility, as follows from the above relations as well, has different values in the SI and CGS systems; indeed, χ (SI) $= 4\pi \cdot \chi$ (CGS).

Thus the magnetic susceptibility of a unit volume of a medium, χ, is a dimensionless quantity characterizing the capability of a medium to become magnetized in a magnetic field. What one

[a]Here magnetization **M** is defined as the ratio of the magnetic moment **m** of a substance (or of a medium) of a small volume to this volume, $\mathbf{M} = \mathbf{m}/\Delta V$, that is, as the magnetic moment of a unit volume of the substance. In the SI system, the magnetic moment **m** is measured in units of A·m^2 and the magnetic field strength **H** in A/m. This makes $\mathbf{m}/\Delta V\mathbf{H}$ a dimensionless quantity because of division by the volume of the magnetized medium ΔV.

measures in an experiment, however, is the total magnetic moment **m** of a sample rather than the magnetization **M**, which should be assigned to the difficulties encountered in measurement of the true volume of small crystallites or of powder samples with a low packed density. For this purpose one usually relates the magnetic moment **m** of a sample to a unit weight or to 1 mole of the substance. Thus for the sake of convenience one specifies the magnetic susceptibility in the CGS system in units of cm^3/g, emu/g (specific m.s.) or cm^3/mol, emu/mol (molar m.s.). In the case of diamagnets, magnetic permittivity shows the factor by which the magnetic field **H** generated by external sources is weakened by the field of the molecular microcurrents of the medium. Bulk graphite is the only one of known substances (with the exception of superconductors in the superconducting state) that possesses the largest-known diamagnetic susceptibility under normal pressure and at a temperature of $\sim 20^\circ C$. This accounts, for instance, for the high strength with which graphite is expelled out of a region with an inhomogeneous magnetic field compared to other substances.

Strong diamagnetism is a rather nontrivial phenomenon. We may recall that isolated atoms or ions with fully occupied core shells[a] are diamagnetic as a result of precession of all core shells of atoms (ions) in an applied magnetic field; this precession generates a circular current around the atomic nucleus, which is effective on the atomic scale and produces a diamagnetic moment **M** in the direction opposite to the applied magnetic field **H**. The internal volume of the atom is thus partially screened by this precession[b] from the influence of the applied magnetic field. In a general case, diamagnetic susceptibility can be anisotropic and specified by a tensor quantity. Planar molecules of aromatic compounds feature the strongest anisotropy of diamagnetic susceptibility. This should be attributed to the presence within each aromatic ring (hexagon consisting of six carbon atoms) of six delocalized π electrons forming a closed collective π electronic shell and to the possibility

[a]The total orbital and spin magnetic moments of such atoms (ions) are zero: $L = S = 0$.

[b]This precession at the circular frequency $\omega_L = eH/2mc$ (CGS) is called the Larmor precession and the corresponding diamagnetism the Larmor diamagnetism. Here e and m are the electron charge and mass, and c is the velocity of light.

of flowing through the latter of circular currents circulating freely clock- or counterclockwise, depending on the orientation of the magnetic field applied perpendicular to the plane of the molecule [17].

A noticeable anisotropy of χ and enhanced orbital diamagnetism become evident already for the molecule of benzene, C_6H_6, consisting of one ring[a] and for the molecule of naphthalene, $C_{10}H_8$, featuring two rings. The origin of the orbital diamagnetism of benzene and of other aromatic molecules of larger sizes was qualitatively explained as far back as 1935 by Lynus Pauling [17] in terms of his concept involving extended diamagnetic ring currents circulating on the periphery of molecules in a perpendicular field. As the number *N* of benzene rings in a compact polycyclic aromatic molecule increases, the specific diamagnetic susceptibility per unit weight increases in absolute magnitude. Indeed, with the number *N* of benzene rings increasing from 3 to 10, the orbital diamagnetic susceptibility χ_{orb} of the molecule changes from -0.72×10^{-6} to -2.0×10^{6} emu/g.[b]

Large aromatic molecules may be considered as molecular analogs of graphite, and therefore, analysis of the specific features revealed by edges in molecules and graphite nanocrystals appears to be essential to understand the nature of diamagnetism in nanometer-size particles. Quantum-chemical methods were employed to a study of superlarge compact aromatic molecules C_{366} H_{54} and C_{438} H_{66} [18].

The concept of aromatic and nonaromatic electronic configurations of a cyclic molecule can be judiciously addressed to analysis of graphene sheets built of a large number of carbon rings, hexagons, as well. Consider a sheet consisting of an even number *N* of atoms, with the number of π electrons in it, likewise equal to *N*, satisfying the condition $N = 4 \times n + 2$, where *n* is an integer. Addition to the system of one delocalized carrier from outside (from a donor[c] or acceptor impurity) culminates in generation at the point of contact, that is, at the point of this charge transfer event, in the vicinity of

---

[a]The lengths of all C–C bonds in the molecule are 0.139 nm, and the benzene ring area is approximately $\sim$0.05 nm^2.

[b]The molecule of ovalene, $C_{32}H_{14}$, is made up of 10 compactly arranged benzene rings.

[c]These donors could be, for instance, atoms of alkali metals, which readily donate electrons in the course of intercalation or adsorption on graphite.

the nearest hexagon, of a local electronic configuration ordered in an arrangement of an odd number of π electrons in the ring (seven in the case of addition of an electron and five for its extraction or an equivalent addition of a π hole). This configuration will have a local spin of 1/2. The excess "conducting" π electron or a hole can move easily over the plane sheet of hexagons and, being actually a mobile 1/2 spin, can be easily detected by electron paramagnetic resonance (EPR). The free migration of such a spin over the sheet can be compared with equivalent displacement of a nonaromatic electron configuration made up of an odd number of π electrons (seven or five).

The effect of orbital diamagnetism reaches a maximum both in a graphite crystal and in a graphene sheet of a limited size with the number of rings $N > 6 \times 10^3$. For a perfect graphite crystal, $\chi_\perp = \chi_{\text{orb}} \approx -22.5 \times 10^{-6}$ emu/g and $\chi_{\|} \approx -0.5\times 10^{-6}$ emu/g for the directions perpendicular and parallel to the aromatic ring plane, respectively. The large diamagnetic moment is induced in a graphite crystal by the circular orbital currents flowing over the aromatic planes of graphite sheets and transported by delocalized π electrons. Incidentally, the equivalent single π carrier contributing to the creation of the diamagnetic moment traces in the magnetic field a contour of a few tens of aromatic rings in area.

The average space-averaged magnetic susceptibility of micro- and submicrocrystalline graphite powder is $\chi = (\chi_{\text{orb}} + 2\chi_{\|})/3 \approx -7.8\times 10^{-6}$ emu/g and tends to decrease sharply with the crystal grain decreasing sharply in size below a certain critical value (for <18–20 nm) and an increasing content of defects and other crystalline imperfections [19]. The latter becomes realized, for instance, as mechanical milling of microcrystalline graphite reduces it to fine- and ultradisperse states. Mechanical milling produces, as a rule, powders with particles exhibiting a large scatter in size (from ~0.05 to 1 micron), which makes investigation of the magnetic properties of a nanographite fraction of particles less than 50 nm in size and with a narrow size distribution practically impossible. By contrast, nanographite particles obtained from DND exhibit a very narrow size distribution (not more than ~1–2 nm wide), a point of a certain academic interest for researchers studying their magnetic properties as a function of size. The magnetic properties of multilayer nanographites will be addressed in some detail in Section 8.5.

8.4 Edge π Electronic States in Nanographenes and Nanographites

In 1996, Fujita et al. were the first to predict the existence of unusual π electronic states evolving at zigzag edges of graphene sheets [12, 13]. Edge states produce a narrow peak in the density of electronic states, DOS(E), of nanographene or a graphene sheet with a large fraction of edges, through holes and point-like vacancies. The DOS(E) function[a] of graphene with a large fraction of zigzag edges is exemplified graphically in Fig. 8.2. The main singularities in the density of π electronic states spaced on the energy scale by $2\gamma_0$, as well as the specific forms of the π and π^* bands, do not differ practically here from those expected for an ideal infinite graphite sheet. Here $\gamma_0 \approx 3$ eV is the band parameter of a two-dimensional graphite sheet. To cite an example, single-electron interband transitions between singularities spaced by $2\gamma_0$ initiate the appearance of a maximum in optical absorption spectra of practically all nanocarbon materials [6, 20], a feature making possible their specific identification.

The narrow peak in the DOS(E) function produced by the zigzag edges lies close to the $E = 0$ point, where the π and π^* bands, occupied by π electrons and empty, respectively, cross, and the Fermi level is located. Localized states at zigzag boundaries are of the acceptor type and usually trap electrons and donate an equivalent amount of delocalized carriers (π holes) to the graphene sheets. While this mechanism, we shall call it "self-doping," makes the nanographene sheets more conducting, and the concentration of the excess delocalized carriers does not exceed a few holes per $\sim$10,000 carbon atoms, which is, thus, much less than the total concentration of valence electrons in the π band of a graphene sheet. In the case of graphite-like nanosized particles of arbitrary topology

[a]The density of electronic states DOS(E) $= \mathrm{d}N/\mathrm{d}E$ is defined as the number of discrete electronic states $\mathrm{d}N$ contained within a narrow interval from E to $E+\mathrm{d}E$ on the energy scale.

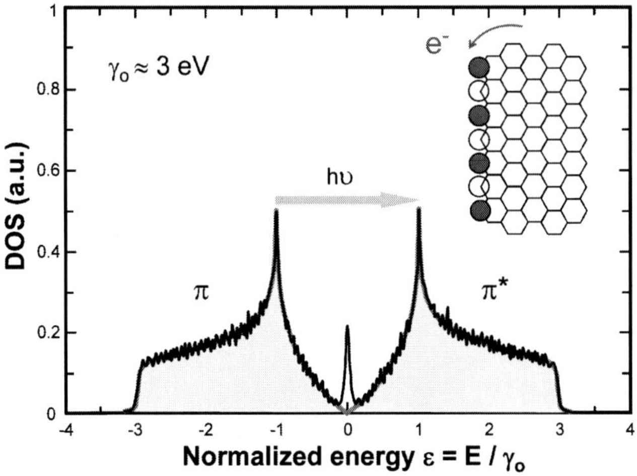

Figure 8.2 DOS(E) of a graphene sheet with a large content of edges or holes. The arrow in the energy diagram specifies optical transitions between singularities in the π and π^* bands, which account for maximum optical absorption in the UV range of the spectrum. Inset: Zigzag-type edge of a graphene sheet and schematic of "self-doping" of material. *Abbreviation*: UV, ultraviolet.

the shape of the singularities in the density of π electron states may turn out extremely sensitive to the concentration of vacancies and holes in graphene sheets and local curvature, that is, to defects in the layers making up the nanoparticle.

It is the edge π electronic states of nanographites and other nanostructured carbon materials (e.g., nanohorns) account for their remarkable electronic properties, including the anomalous paramagnetism and unusual sensitivity of the EPR signal to molecular oxygen in the triplet state [21–23], and it is the paramagnetism revealing a characteristic temperature dependence of magnetic susceptibility obeying the Curie–Weis law ($\sim$1/T) that is one of the fundamental properties of edge π electronic states, which had been successfully predicted by theory in the very beginning of the investigation of these systems [24, 25].

The concepts underlying the model of Fujita assume the outer dangling bonds of edge carbon atoms at a zigzag boundary to be uniformly saturated by atomic hydrogen as in C(sp^2)-H, although in actual practice they can be saturated by oxygen-containing (hydroxyl, carboxyl, etc.) groups as well.

The realistic boundary of an arbitrary shape in a graphene sheet (the outer uneven edge or the edge of an inner hole) can be visualized as a sequence of short chains of the zigzag or armchair type. At such an uneven boundary, zigzag chains consisting of three to about seven aromatic rings may play a significant role. Figure 8.3 displays an image of the zigzag edge of a graphene sheet about seven aromatic rings long, obtained by scanning tunneling microscopy (STM). The bright spots in the STM image reflect the regions of the edge-state macroscopic wave function with a maximum amplitude

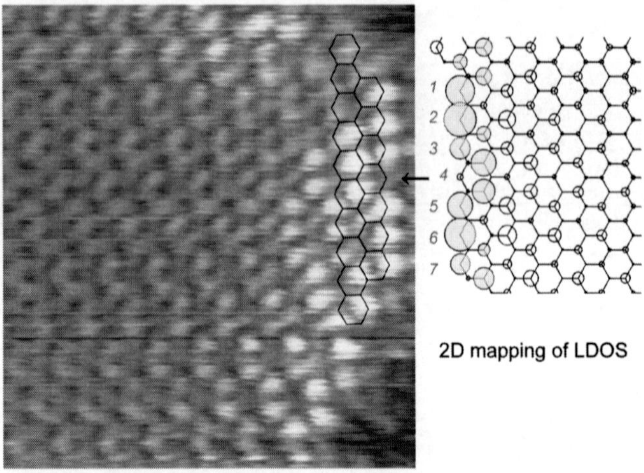

Figure 8.3 STM image of a graphene sheet zigzag edge consisting of seven outer carbon atoms and zigzag fragments. The zero of electron density amplitude is located at the center of this section (site 4). The graphene sheet rests on a substrate of highly oriented pyrolythic graphite. The area of the displayed spot of the STM image is 3.8 × 3.8 nm^2. Reprinted with permission from Kobayashi Y., Fukui K., Enoki T., and Kusakabe K. (2006) Edge state on hydrogen-terminated graphite edges investigated by scanning tunneling microscopy, *Phys. Rev. B*, **73**, pp. 125415(1–8). Copyright 2006, by the American Physical Society; Kobayashi Y., Fukui K., Enoki T., Kusakabe K., and Kaburagi Y. (2005) Observation of zigzag and armchair edges of graphite using scanning tunneling microscopy and spectroscopy, *Phys. Rev. B*, **71**(19), pp. 193406 (1–4). Copyright 2005, by the American Physical Society.

[26, 27]. The amplitude of an edge-state wave function reaches a maximum, as a rule, at edge sites (carbon atoms) of the graphene sheet, to decay smoothly in the direction perpendicular to the boundary at a distance equivalent to two to three aromatic rings. For the edge about seven aromatic rings long shown in Fig. 8.3, the wave function of the edge state drops to zero at the fourth site, in the middle of the chain, while at sites 1, 2, and 3 and 5, 6, and 7 it is at a maximum. The zeros of the wave function can, as a rule, be detected in STM images of the edge states of zigzag chains consisting of an odd number of aromatic rings. If the wave function passes through zero in the first layer, in the second and third layers of aromatic rings beyond the edge the wave function reveals a mosaic structure.

Scanning tunneling spectroscopy (STS) permits experimental determination of the local density of electronic states LDOS(E) at zigzag- and armchair-type boundaries. This is done by fixing the tip of the scanning tunneling microscope over the area of the graphene sheet under study and measuring the first derivative of tunneling current with respect to voltage ($\mathrm{d}I_t/\mathrm{d}U$) at different bias voltages (U) between the tip and the substrate (within the interval of –0.5 V to +0.5 V). In STS, the experimentally measured function ($\mathrm{d}I_t/\mathrm{d}U$) of bias ($U$) is directly proportional to the LDOS(E) function. Here $E = eU$, and e is the electron charge. For a zigzag boundary, the ($\mathrm{d}I_t/\mathrm{d}U$) function of U has a characteristic peak in the region of the zero bias voltages with a full width at half maximum of $\sim$50 mV [8, 27]. This peak can be identified with the LDOS(E) peak close to the Fermi level $E_\mathrm{F} \approx 0$. For an armchair-type edge, the LDOS(E) graph does not have such a peak at $E_\mathrm{F} \approx 0$. Thus, both STM and STS studies yield compelling experimental evidence[a] for the existence of π electronic edge states at graphene zigzag boundaries [28].

8.5 Preparation of Nanographite from Nanodiamond and Methods of its Characterization

In the second half of the 1990s it was proposed that DND be used for the preparation of multilayered nanographite [2, 29].

[a]The corresponding STM images and convincing STS data were first obtained by the Japanese scientists Y. Kobayashi, K. Fukui, T. Enoki et al. in 2005–2006.

The uniqueness of using DND for these purposes (see Chapter 1) follows from the specific characteristics of this material: indeed, the crystalline diamond core of DND powder grains is typically 4–5 nm in size, with an extremely narrow size distribution of the particles, and nanodiamond itself transforms readily to graphite already at annealing temperatures above 1,500°C.

Annealing of DND powder in an inert atmosphere or in vacuum at temperatures not lower than 1,600°C produces multilayered nanographenes/nanographites, or, accepting the terminology used from the beginning of the 1990s, carbon onions with layers of quasi-spherical or polyhedral types [30].

The annealing temperature of 1,600°C and higher for preparation of multilayered nanographites was not chosen at random. The temperature of 1,600°C is close to the Debye temperature of crystalline diamond, at which all phonon modes of the diamond lattice are strongly excited, so a diamond nanoparticle with a tetrahedrally coordinated covalent lattice converts easily into a particle of multilayered nanographite in time on the order of 2–10 minutes. The process of graphitization, propagating continuously from the surface into the diamond core, realizes as sequential transformation of a series of eight (111) planes of the diamond octahedron/cuboctahedron, one after another, into curved graphene planes, which make up a quasi-closed carbon framework and are spaced by ~0.335 nm. In this model, the front of graphitization proceeds from the surface of a nanodiamond particle, which contains a host of point-like defects and, hence, is easy to become graphitized, into the diamond core [4]. At $T = 1,250°C$, the velocity of propagation of the graphitization front is small, ~0.05 nm/hour, to increase by more than a factor of 250 at ~1,600°C. Increasing the time of the high-temperature treatment of the material (at $T = 1,600°C$) from ~10 minutes to ~120 minutes results in local straightening of the curved nanographene planes through the concurrent hopping and self-diffusion of carbon atoms over the interstices and sites of a defective nanographene sheet, and formation of multilayered nanographite particles of polyhedral type with clear-cut faces. Incidentally, the number of point defects, both in the inner regions of a nanographite particle and at zigzag edges, decreases with increasing annealing time.

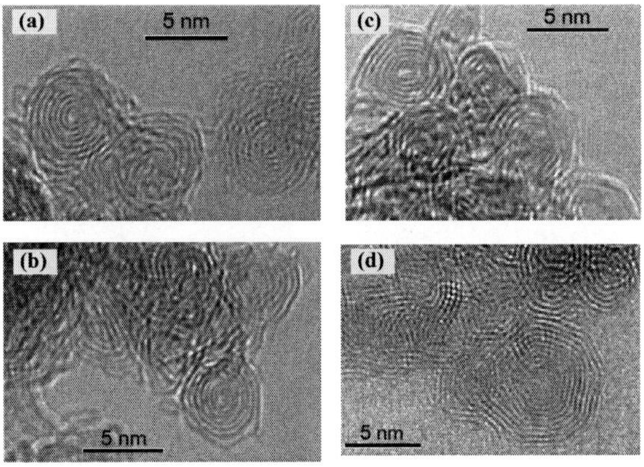

Figure 8.4 Images of nanographite particles obtained by high-resolution transmission electron microscopy. Electron energy in the probing beam 200 keV. Samples: (a) ng2, (b) ng7, (c) ng35, and (d) ng120. The number in sample designation specifies time of material annealing at 1,600°C in minutes.

Figure 8.4 displays images of nanographite particles produced from nanodiamond at different annealing times (from 2 to 120 minutes) and obtained by high-resolution transmission electron microscopy. Annealing at $T = 1,600°\text{C}$ converts nanodiamond particles into quasi-spherical carbon onion (particle size 5–7 nm) already after two minutes of the thermal treatment (Fig. 8.4a). When the annealing time is increased to 30–120 minutes, part of the quasi-spherical particles transforms into polyhedral ones with a hollow core (Fig. 8.4c,d). Typical defects of graphene sheets and zigzag edges are five- and seven-member carbon rings, which, if present in noticeable amounts (above 5%), result in amorphization[a] of a perfect aromatic sheet. A decrease of the number of five- and seven-member carbon rings with increasing annealing time brings about further development of the π electron subsystem of aromatic sheets and formation of edge π electronic states along the zigzag boundaries formed by aromatic hexagons. The average separation d_{002} between graphite layers in multilayered nanographite is now

[a]The presence of five- and seven-member carbon rings in nanographite obtained at short annealing times (<15 minutes) at $T = 1,600°\text{C}$ or at lower temperatures of 1,500–1,550°C is supported convincingly by observation of an additional band at $\sim$1,530 cm^{-1} in the Raman scattering spectra of nanographites.

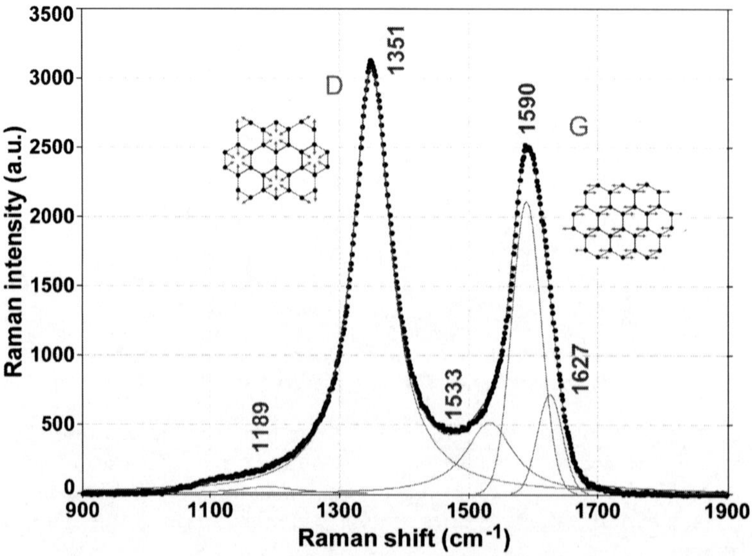

Figure 8.5 First-order Raman spectrum of multilayered nanographite in the region from 900 cm^{-1} to 1,900 cm^{-1}. Wavelength of incident radiation is 514.5 nm. Sample ng7. The figure specifies Lorentzian or Gaussian contours centered at ~1,190 cm^{-1}, ~1,350 cm^{-1}, ~1,530 cm^{-1}, 1,590 cm^{-1}, and 1,627 cm^{-1}.

0.345 nm, which is somewhat larger than the corresponding value for bulk graphite (0.335 nm). The larger interlayer spacing suggests that nanographite particles are of turbostratic structure with no mutual three-dimensional ordering of layers.

Raman spectroscopy is a powerful tool for probing the structure of nanographite particles [7]. A Raman spectrum obtained at $\lambda_0 =$ 514.5 nm consists primarily of the so-called D and G bands (Fig. 8.5). The G band (~1,590 cm^{-1}) derives, in both bulk and pyrolythic graphite, from scattering from the E_{2g} phonon modes of the hexagonal lattice, which is allowed by selection rules, and the D band (~1,350 cm^{-1}) from scattering from vibrations of lattice defects located close to the armchair and zigzag edges. Incidentally, the efficiency of scattering from defects at the armchair-type boundaries is a few times larger than that at zigzag edges.

Significantly, the appearance of the D band is due to light scattering from the A_{1g} breathing modes of aromatic ring vibrations, which in the case of a perfect defect-free graphene sheet is forbidden by the fundamental wave vector selection rules for incident and scattered light. At crystallite edges, however, as well as close to defect regions containing five- and seven-member carbon rings, the selection rules are relaxed to allow photon scattering from the hexagon breathing vibration modes. It is the presence of defects that activates this process of Raman scattering and formation of the D band, which is not present in the case of a perfect graphite crystal or a single graphene sheet.

Graphite crystallite edges may be considered as intrinsic structural defects. There is a high probability of existence close to them of isolated five- and seven-member carbon rings and of their various complexes.

In a first approximation, it may be assumed that if the intensity of the G band is proportional to the total number of carbon atoms within a laser beam–illuminated area, that of the D band will be proportional to the number of carbon atoms located near the edges or to the overall length of these edges within the illuminated spot. The smaller the size of defect-free crystallite, L_a, in the graphene sheet plane, the larger the specific fraction of edge atoms ($\sim 1/L_a$) and, hence, the relative intensity of the D band normalized against the G band intensity. For nanographite particles larger than 2 nm the size L_a can be calculated from the well-known empirical relation [31]: $L_a\,[nm] = 4.4\frac{I_G}{I_D}$. Here I_G and I_D are the peak intensities of the G and D bands and 4.4 is a coefficient derived empirically from an analysis of Raman spectra of nanographite samples varying in size from 3 nm to 50 nm, which were measured at the incident wavelength of 514.5 nm. In our particular case of multilayered nanographites obtained from nanodiamond, the average size of a crystallite (a perfect defect-free region of nanographene) in the graphene sheet plane is $\sim$3.5 nm and is practically independent of the annealing time at 1,600°C [7]. Thus, data amassed from Raman spectroscopy, as well as from X-ray diffractometry measurements, suggests that in a fully graphitized nanodiamond particle one can isolate at the very least several nanographite blocks (crystallites) with a size in the graphene sheet plane about two times smaller than the diameter of the multilayered particle formed.

On the whole, the structure of nanographite particles prepared from DND may be considered to have been studied in good enough detail by various methods, including X-ray diffractometry, EPR, electron energy loss spectroscopy, etc. [32–36].

8.6 Magnetic Properties of Nanographites: Paramagnetism and Diamagnetism

To prepare nanographite powders for measurement of their magnetic characteristics, they are usually degassed by prolonged (many hours) heating at a temperature of 450°C in high vacuum and sealed in a special quartz tube in a vacuum not worse than 3×10^{-6} Torr. Nanographite particles are subjected to degassing to remove the paramagnetic molecular oxygen and other gaseous impurities sorbed on their surface, a procedure necessary to ensure correct magnetic measurements at low temperatures ($T < 100$ K).

The behavior with temperature of the magnetic susceptibility of various samples of multilayered nanographites is shown graphically in Fig. 8.6.

The increase of the susceptibility observed to occur with decreasing temperature at $T < 50$ K finds ready explanation in the paramagnetism of edge π-electronic states of nanographite [9].

The total magnetic susceptibility of nanographite powder can be presented as a sum of four terms: $\chi = \chi_{\mathrm{core}} + \chi_{\mathrm{Pauli}} + 1/3\chi_{\mathrm{orb}} + \chi_{\mathrm{CW}}$, where χ_{core} is the temperature-independent diamagnetic susceptibility of carbon atoms ($\approx -0.5 \times 10^{-6}$ emu/g), χ_{Pauli} is the Pauli paramagnetism, χ_{orb} is the orbital diamagnetism in the graphene sheet plane, and χ_{CW} is the Curie–Weiss paramagnetism.

The coefficient 1/3 in the relation accounts for spatial averaging of all possible random orientations of individual graphite crystallites in the powder. The component $\chi_{\mathrm{Pauli}} > 0$ and is practically temperature independent, while $\chi_{\mathrm{orb}} < 0$ and depends weakly on temperature. χ_{CW} exhibits the strongest dependence on temperature and is described by the Curie–Weiss law $\chi_{\mathrm{CW}} = \frac{C}{T-\theta}$, where C is the Curie constant, which is proportional to the concentration of

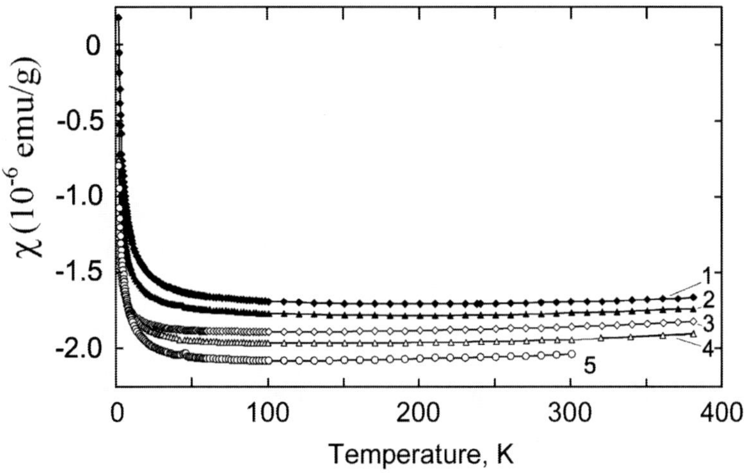

Figure 8.6 Temperature dependence of experimentally measured magnetic susceptibility (χ) of multilayered nanographites for samples annealed for different times (at 1,600°C). Curves 1–5 correspond to samples with annealing times of 2, 7, 15, 35, and 120 minutes. Data was obtained at a fixed magnetic field of $\sim$1 T (10^4 G).

localized spins or paramagnetic centers N_s in a paramagnetic spin subsystem, and θ is the Weiss temperature characterizing the type and strength of magnetic interaction among localized spins in the system. $C = \frac{N_S S(S+1) g^2 \mu_B^2}{3k_B}$, where μ_B is the Bohr magneton, k_B is the Boltzmann constant, S is the total spin of the paramagnetic center dominant in the system and responsible for the paramagnetism in it, g is the Landé or gyromagnetic factor[a] (i.e., the ratio of the magnetic moment of a spin center to its angular momentum), and N_s is the concentration of paramagnetic centers in the system expressed in units of g^{-1}.

In an analysis of the contribution of χ_{CW} to magnetic susceptibility at $T < 50$ K one may neglect the weak temperature dependence $\chi_{orb}(T)$ and accept the approximate form: $\chi = \chi_o + \frac{C}{T-\theta}$. Here $\chi_o = \chi_{core} + \chi_{Pauli} + 1/3\chi_{orb} \approx$ const.

Figure 8.7 visualizes the dependences of $(\chi - \chi_o)^{-1}$ on T constructed for two samples, ng7 and ng120, which were obtained

[a]The relative gyromagnetic ratio, expressed in units of $e/2mc$, is a dimensionless quantity.

under short and long temperature treatments at 1,600°C. The $\chi(T)$ relations can be readily straightened in the $(\chi - \chi_o)^{-1}$ versus T coordinates in the 0–80 K interval if an appropriate numerical value of χ_o is chosen, and the value of the θ cutoff on the temperature scale at which $(\chi - \chi_o)^{-1}$ vanishes is $|\theta| < 1$ K. This strongly suggests that the Curie paramagnetism derives from the subsystem of quasi-localized spins of edge states that are coupled by weak antiferromagnetic interaction ($\theta \approx -0.7$ K). The experimentally determined Curie constants (C) permit calculation of the concentration of localized edge spins with the relation $N_s = \frac{3k_B C}{S(S+1)g^2\mu_B^2}$. For the values of S and g of the edge-localized spins in nanographite, which are to be substituted into the relation for N_s, we set $S = 1/2$, $g = 2$. The latter choice finds support in the data obtained by EPR obtained at a frequency of ~9.5 GHz. Indeed, the g factor of a Lorentzian EPR signal from an edge spin is 2.0014. The edge spin concentration in a series of nanographite samples (ng2–ng120) obtained from DND falls off from $\sim 7.5 \times 10^{18}$ g^{-1} to $\sim 3.6 \times 10^{18}$ g^{-1}. Reducing N_s to the weight of one nanographite particle[a] ($\sim 2.30 \times 10^{-19}$ g), we find that this corresponds to a decrease of the number of spins from ~2 to ~1 for one nanoparticle.

The dependence of the magnetization (M) of nanographite on the field (H) at $T = 2$ K can be fitted well by two components: $M \approx \chi_o \cdot H + M_{loc}(H)$. Here the first component is, on the whole, diamagnetic (because $\chi_o < 0$), linear in the magnetic field **H**, and practically temperature independent as a result of χ_{orb} and χ_{Pauli} depending only weakly on temperature. The second component, deriving from the paramagnetism of localized spins, depends nonlinearly with saturation on the magnetic field, the saturation pattern being strongly dependent on temperature.

Consider the nonlinear-in-field component of magnetization in more detail. The low-temperature orientational magnetization of the "gas" of localized spins, which is determined by the average projection of magnetic moments of individual spins $\langle M_Z \rangle$ on the

[a] We are assuming here that the weight of one particle of nanographite is equal (to within ~5% and disregarding its internal structure and topography) to that of one particle of nanodiamond from which the former was formed. For the density of crystalline diamond, the weight of a nanodiamond particle is estimated to be $\sim 2.30 \times 10^{-19}$ g.

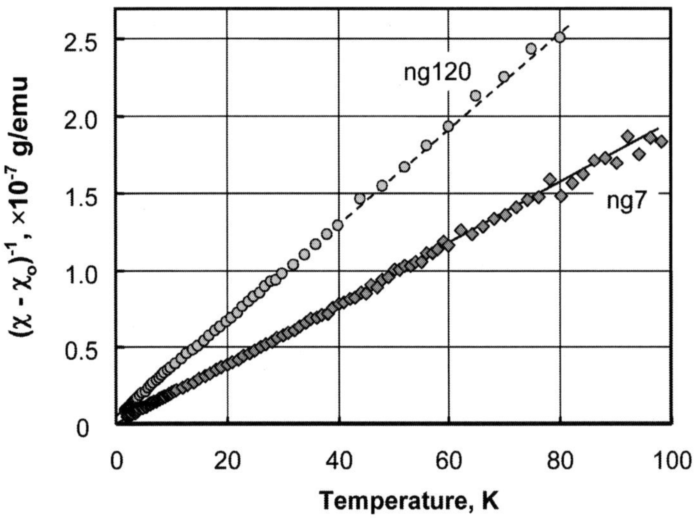

Figure 8.7 Temperature dependence of inverse magnetic susceptibility measured at $T < 80$ K on two nanographite samples with a short (7 min) and a long (120 min) annealing time.

direction of the magnetic field **H**, is described by the Curie–Brillouin law

$$M_{\text{loc}} = \langle M_z \rangle = \widetilde{N}_S g S\, \mu_B\, B_S(x), \text{ where } x \equiv gS\, \mu_B H / k_B T \tag{8.2}$$

Here g, S, μ_B, and k_B have already been specified above, $\widetilde{N}_S$ is the concentration of magnetic moments of isolated spins or of paramagnetic centers, and $B_S(x)$ is the Brillouin function:

$$B_S(x) = \frac{2S+1}{2S} \coth \frac{(2S+1)x}{2S} - \frac{1}{2S} \coth \frac{x}{2S} \tag{8.3}$$

Equations (8.2) and (8.3) assume Boltzmann statistics for an ensemble of magnetic moments of spins, regardless of their nature, a discrete energy spectrum E_n of the spin magnetic moment in a magnetic field, and the ensemble itself consisting of noninteracting identical magnetic moments.

In weak magnetic fields or at high temperatures, the conditions in which $x << 1$, Eqs. (8.2) and (8.3) become simplified noticeably to yield for $\langle M_Z \rangle$

$$M_{\text{loc}} = \langle M_Z \rangle = \frac{\widetilde{N}_S g^2 \mu_B^2 S(S+1)}{3k_B} \left(\frac{H}{T} \right), \tag{8.4}$$

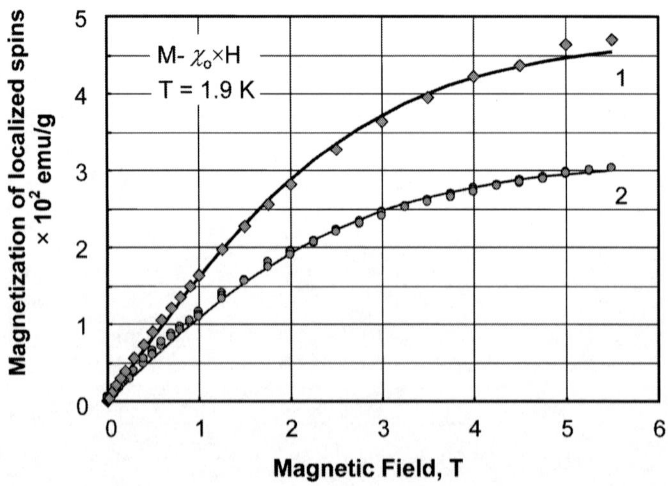

Figure 8.8 Magnetization curves $M_{\text{loc}}(H)$ for a subsystem of nanographite edge-localized spins obtained at $T = 1.9$ K. The magnetization component related to linear magnetization through the term with **H** was subtracted from M: $M_{\text{loc}}(H) = M - \chi_o \cdot H$. Curves 1 and 2 identify nanographite samples with a short (7 min) and a long (120 min) annealing time.

that is, a linear dependence of M_{loc} on **H** with a coefficient of proportionality χ_{CW} depending hyperbolically on temperature (the Curie law)—see above. In the simplest case of $S = 1/2$, we obtain for the Brillouin function $B_S(x) = \tanh(x)$.

Figure 8.8 plots the magnetization of localized spins $(M - \chi_o H)$ versus the magnetic field **H** for two nanographite samples, ng7 and ng120 [37]. These graphs are approximated satisfactorily by Brillouin curves with two fitting parameters, for S and $\widetilde{N}_S$. In the approximations, the spin was assumed everywhere to be $S = 1/2$, and the value of $\widetilde{N}_S$ was determined by selection from the set of allowed values for the edge spin concentration; it turned out that for the ng7 and ng120 nanographites $\widetilde{N}_S$ is 6×10^{18} g^{-1} and 3.6×10^{18} g^{-1}, respectively. These values are approximately equal to the values of N_S derived above from the Curie constants.

It is remarkable that it is the spins of the edge states localized at zigzag boundaries that the paramagnetic properties of nanographite observed experimentally are related to. Indeed, EPR measurements demonstrate the absence of paramagnetic ions of 3d transition metals (iron, cobalt, titanium) in the system, with the concentration of intrinsic paramagnetic defects associated with unpaired σ orbitals of dangling C–C sp^3 bonds between neighboring graphene sheets being lower by one to two orders of magnitude than the concentration N_S determined above.

At the same time, EPR spectroscopy yields practically the same value for the spin concentration of edge states (EPR signal with the g factor ≈ 2.0014) as the data obtained with a SQUID magnetometer. A distinctive feature of this EPR signal is the extremely high sensitivity of signal width to the concentration of paramagnetic molecular oxygen ($S = 1$), which is present in the air and broadens the EPR signal of edge spins $S = 1/2$ through the dipole–dipole interaction (Fig. 8.9). This broadening is proportional to $\sim 1/r^3$, where r is the average distance between an edge spin and an oxygen molecule adsorbed on the surface of a nanographene sheet. EPR signal broadening is also proportional to the partial pressure of oxygen in the ampoule with the sample. At $T = 300$ K and a partial oxygen pressure of 0.2 bar, the additional signal broadening $\Delta H_{\rm pp}^{\rm air} - \Delta H_{\rm pp}^{\rm vac}$ produced by external atmosphere is $\sim$3.5 mT. In vacuum, the EPR signal narrows immediately down to its minimal value $\Delta H_{\rm pp}^{\rm vac}$ (from $\sim$10 mT to 1.2 mT for particles obtained with 2–120-minute annealing time). The broadening-narrowing signature of the EPR signal of edge spins followed in numerous cycles of evacuation-admission of air into the chamber with the sample was observed to be fully reversible.

We note that EPR signals ($g = 2.0022$) of point defects, which originate from dangling sp^3 bonds present in the system in residual concentrations, are insensitive to molecular oxygen. This should be attributed to the sp^3 defects being located at some depth under the nanoparticle surface and to their being topologically inaccessible for oxygen molecules. Thus the unusual intrinsic paramagnetism of nanographite originates from extended π electronic states at zigzag boundaries.

Figure 8.10a illustrates the result of subtraction of the constant $\chi_{\rm core}$ and of the temperature-dependent Curie–Weiss component from the experimentally measured magnetic susceptibility of the same ng2–ng120 nanographite samples $\chi - \chi_{\rm CW} - \chi_{\rm core} = \chi_{\rm Pauli} + 1/3\chi_{\rm orb}$.

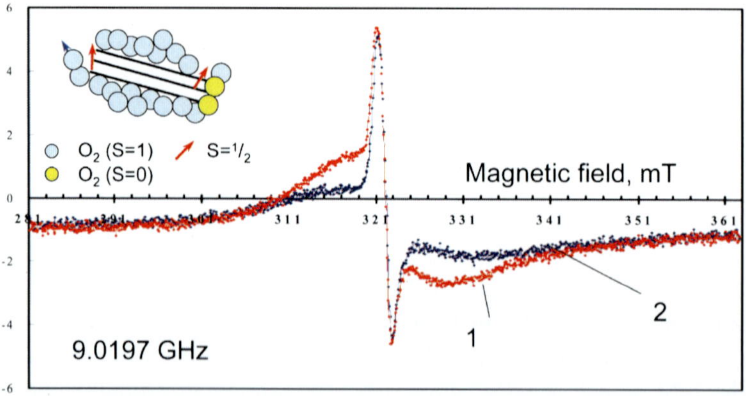

Figure 8.9 EPR spectra of nanographite with a short annealing time (2 min) recorded in technical-grade vacuum (curve 1) and in air (curve 2). The narrow line ($g = 2.0022$) at the center of the spectrum is related to point defects with unpaired σ orbitals (vacancies or dangling sp^3 bonds between graphene sheets) and the broad line to spins of edge π electron states, which is highly sensitive to paramagnetic molecular oxygen. Microwave radiation frequency 9.0197 GHz ($T = 300$K). Inset top: Schematic arrangement of molecules of physi- and chemisorbed oxygen with respect to nanographite edges and fragments of basal planes.

The reason for the weak variation of the $\chi_{\text{Pauli}}+1/3\chi_{\text{orb}}$ term with temperature should be looked for in the temperature dependence of the orbital diamagnetism $\chi_{\text{orb}}(T)$. The behavior with temperature of $\chi_{\text{orb}}(T)$ can be analyzed in the $\chi_{\text{Pauli}}(T) \approx$const approximation in the context of the phenomenological approach proposed in Refs. [38, 39].

Accepting the band model of two-dimensional graphite [38, 39], the modified relation for orbital susceptibility $\chi_{\text{orb}}(T)$ can be written as[a]

$$\chi_{\text{orb}}\,[\text{emu/g}] = -1.3625 \times 10^{-3}\gamma_0^2 \frac{\sec\text{h}^2\,(E_{\text{F}}/2k_{\text{B}}\,(T + \Delta T))}{(T + \Delta T)} \tag{8.5}$$

[a]The numerical coefficient in the relation is specified by the following fundamental constants and parameters: $-\frac{\sqrt{3}}{32\pi}\left(e/\hbar c\right)^2 \frac{N_a a^4}{Mk_{\text{B}}}$, where $a = 0.246$ nm is the lattice constant in the graphite layer, N_a is the Avogadro number, M is the molar weight of carbon, and $\hbar$ is the Planck constant divided by 2π.

Here $\gamma_o \approx 3$ eV is the two-dimensional band parameter of a graphite sheet in eV, E_F is the Fermi energy, and ΔT is an additional temperature term taking into account electron scattering of nonthermal origin from the edges and structural defects of a graphite nanoparticle, or, in other words, spreading of the DOS(E) function for an ensemble of particles in the vicinity of the Fermi level.

Equation (8.5) proposed in Refs. [38, 39] is a result of straightforward modification of the relation[a] for the diamagnetic susceptibility of an ideal two-dimensional graphite sheet [40]:

$$\chi_{\text{orb}}(T) = -\text{const} \cdot \frac{1}{k_B T} \cdot \text{sec h}^2\left(E_F/2k_B T\right), \quad (8.6)$$

which would fit satisfactorily the experimental data obtained for disordered graphite-like nanomaterials with weak temperature dependence $\chi_{\text{orb}}(T)$.

The behavior with temperature of the orbital diamagnetic susceptibility of nanographite samples derived from experimental data by the conversion $\chi_{\text{orb}}(T) = 3 \times (\chi - \chi_{\text{CW}} - \chi_{\text{core}} - \chi_{\text{Pauli}})$ and approximated with the use of Eq. (8.5) is displayed in graphical form in Fig. 8.10b [9]. The quantities ΔT and E_F in Eq. (8.5) and the a priori unknown χ_{Pauli} were found as parameters from the best fit of the theoretical curve to the experimental data. The parameter ΔT, which is a measure of disordering or spreading of the density of states (DOS) function for an ensemble of nanographite particles in the vicinity of the conical point of contact of the π and π^* bands by $\Delta E = k_B \Delta T$, varies from 1,100 K to 1,280 K, with the Fermi level remaining within the interval from ~0.16 eV to ~0.17 eV for samples with the annealing time of 2–120 minutes. In these conditions, the ratio $k_B \Delta T / E_F$ is about ~0.5. χ_{Pauli} decreases in magnitude from 0.70×10^{-6} emu/g to 0.35×10^{-6} emu/g within the annealing time interval of 2–120 minutes.

All nanographites prepared at 1,600°C and with different annealing times demonstrate fairly close values of orbital susceptibility (from -5.7×10^{-6} to -5.5×10^{-6} emu/g within the temperature interval of 200–350 K) and, as shown by Raman spectroscopy, the same crystallite size (~3.5 nm) in the graphene

[a]This relation was derived in Ref. [40] for the case of small Fermi energies E_F (<0.3 eV).

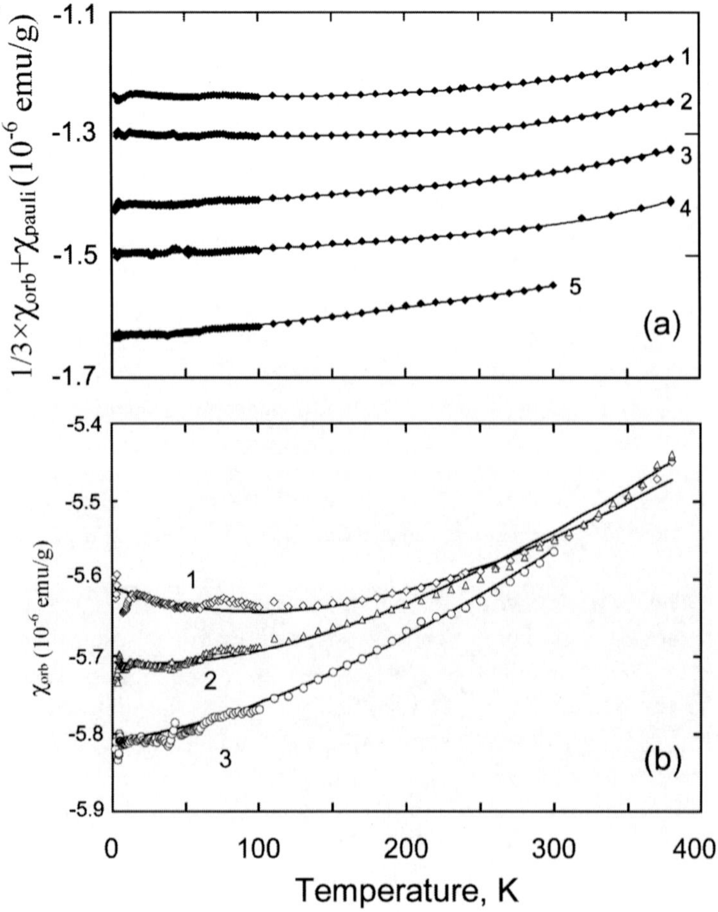

Figure 8.10 (a) Temperature dependence of the total contribution to magnetic susceptibility of nanographites from two components, orbital diamagnetism and Pauli paramagnetism. Curves 1–5 relate to samples prepared with annealing times of 2, 7, 15, 35, and 120 minutes. (b) Temperature dependence of orbital diamagnetic susceptibility of nanographites for samples prepared at different annealing times. Curves: 1, ng2; 2, ng35; 3, ng120. The solid curves drawn through experimental points were obtained with relation (6), where E_F, ΔT, and χ_P are parameters derived in the curve-fitting procedure.

sheet plane. A graphene sheet of such a diameter accommodates about ~200 hexagons. One comes thus to the conclusion that an aromatic system constructed of ~200 hexagons should have $\chi_{orb} \approx -(5.6 \pm 0.1) \times 10^{-6}$ emu/g at $T = 300$ K. This is about

$\sim$4 times smaller than χ_{orb} for a perfect graphite crystal. This figure should not come as a surprise considering the small size of the graphene sheet (crystallite) in which the magnetic field-induced circulating π electron currents evolve.

Incidentally, both the orbital diamagnetism and the intrinsic paramagnetism of nanographite with a large contribution of edges can be readily understood in terms of the texture of local circular currents circulating within graphene sheets and near their edges. The corresponding calculations of the magnetic susceptibility of nanographene strips can be found in Ref. [24].

The weak variation of orbital nanographite susceptibility with temperature should apparently be attributed to the strongly pronounced spreading of the integral DOS(E) function of the ensemble of nanoparticles as a whole. Considered from a phenomenological viewpoint, the actual causes accounting for this spreading are not really important in the case of a macroscopic ensemble of nanoparticles. These may be specific structural defects and geometric irregularities of the layers, which define the individual unique shape of the DOS$^{(i)}$(E) function for each of the N nanoparticles of an ensemble or processes involved in charge transport that may occur even in the case of identical, indistinguishable particles of the ensemble (Fig. 8.11).

What this actually means is that while electric charge transport itself from one particle to another leaves the DOS functions of individual but identical nanoparticles unchanged, they become displaced along the energy scale E. This displacement along the energy scale can be toward both positive and negative values of E, depending on the actual sign and magnitude of the charge transferred ($q = \pm e, \pm 2e, \pm 3e, \ldots$). It appears appropriate to note that while this process of charge transport leaves the ensemble of nanoparticles, on the whole, electrically neutral, individual nanoparticles will carry a discrete charge distributed in a random way over the ensemble, and this may initiate displacement of individual energy band diagrams of nanoparticles relative to the overall Fermi level of the system as a whole (Fig. 8.11) [41].

Viewed in the context of this approach, the weak temperature dependence of the function $\chi_{\text{orb}}(T)$ can be attributed to a strong

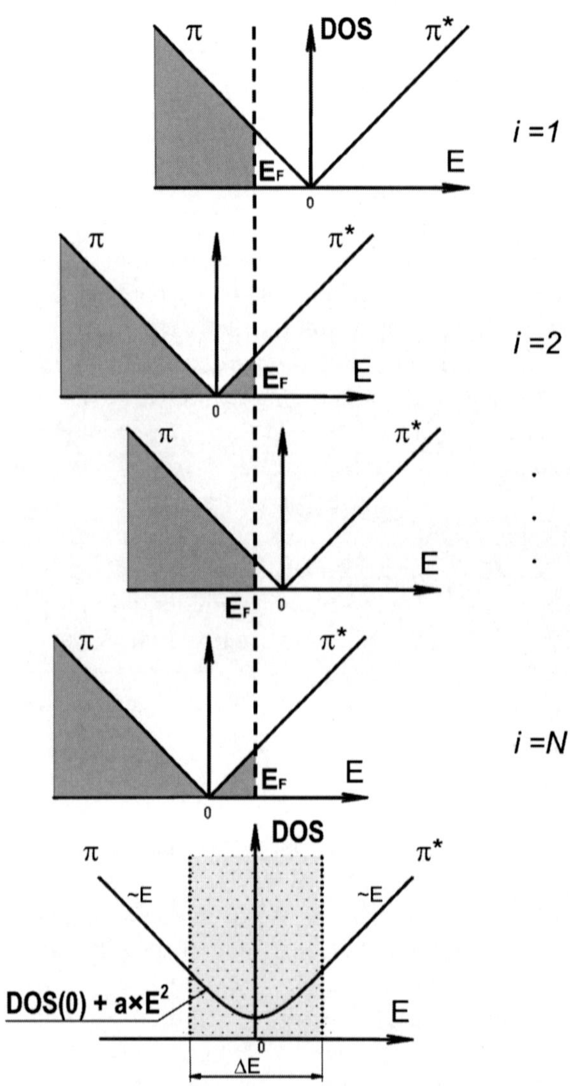

Figure 8.11 Schematic illustrating spreading of the DOS function for an ensemble of nanographite particles, DOS(E), in the vicinity of the point of contact of the π and π^* bands, which originates from charge exchange and transfer of a discrete charge from one particle to another in an ensemble. The statistically averaged DOS function $\text{DOS}(E) = 1/N \sum_{i=1}^{N} \text{DOS}^{(i)}(E)$ is plotted at the bottom of the figure.

scatter of the values of $E_F^{(i)}$[a] over the ensemble of nanoparticles as a whole. But then Eq. (8.6) can be generalized to the form applicable to the ensemble of nanoparticles

$$\chi_{\text{orb}}\,[\text{emu / g}] = -1.3625 \times 10^{-3}\gamma_0^2 \cdot \frac{1}{N}\sum_{i=1}^{N}\frac{\operatorname{sech}^2\left(E_F^{(i)}/2k_B T\right)}{T}. \tag{8.7}$$

Here $E_F^{(i)}$ is the Fermi level for the i-th particle of the ensemble, and $N > 10^9$ is the total number of particles in the ensemble. The distribution of the magnitude of $E_F^{(i)}$ over the ensemble of particles $(i = 1, \ldots, N)$ extends from -0.2 eV to $+0.2$ eV, and the distribution itself is bell shaped with a maximum at $E_F \approx 0$.

Significantly, Eq. (8.6) was derived in the approximation assuming small values of the Fermi energy ($E_F/\gamma_0 << 1$), that is, it yields acceptable results for $|E_F| < 0.3$ eV. A more general expression for orbital susceptibility was derived [42] in terms of the theory of magnetic susceptibility of graphite-based intercalation compounds, which was developed for values of $|E_F|$ lying in the interval from 0 eV to ~3 eV.[b] For small values of the parameter $E_F/\gamma_0 << 1$, the predictions of the general theory [42] coincide fully with the conclusions drawn for specific cases [40]. In this connection, the model of the temperature dependence $\chi_{\text{orb}}(T)$ developed using Eq. (8.7) for an ensemble of nanoparticles with individual values of $E_F^{(i)}$ spread over the particles will function well enough for the range of $\left|E_F^{(i)}\right| < 0.3$eV, thus supporting the experimentally observed low values of $|\chi_{\text{orb}}(T)|$ and the weak temperature dependence $\chi_{\text{orb}}(T)$ for nanographites compared with that revealed in a crystal of bulk graphite.

To sum up, it is the edge π electron spin states that account for the unusual low-temperature Curie–Weiss paramagnetism of nanographites, whereas at high temperatures the material behaves

[a]For each nanoparticle, $E_F^{(i)}$ is reckoned from the point of contact of the π and π^* bands.

[b]The Fermi energy for bulk graphite is ~−0.024 eV and is determined by the parameters of the intrinsic acceptor-like defects of the material. By intercalating graphite with alkali metals or other atomic groups one can shift the Fermi level noticeably into the π^*- band, with the material acquiring a certain number of π-electrons (conduction carriers).

as a conventional diamagnet with a specific temperature dependence of orbital diamagnetism.

References

1. Kuznetsov V. L., Malkov I. Yu., Chuvilin A. L., Moroz E. M., Kolomiichuk V. N., Shaichutdinov Sh. K., and Butenko Yu. V. (1994) Effect of explosion conditions on the structure of detonation soots: ultradisperse diamonds and onion carbon, *Carbon,* **32**, pp. 873–882.
2. Kuznetsov V. L., Chuvilin A. L., Butenko Y. V., Mal'kov I. Y., and Titov V. M. (1994) Onion-like carbon from ultra-disperse diamond, *Chem. Phys. Lett.*, **222**(4), pp. 343–348.
3. Alexenskii A. E., Baidakova M. V., Vul' A. Ya., Davydov V. Yu., and Pevtsova Yu. A. (1997) Diamond-graphite phase transition in ultradisperse-diamond clusters, *Phys. Solid. State*, **39**, pp. 1007–1015.
4. Butenko Yu. V., Kuznetsov V. L., Chuvilin A. L., Kolomiichuk V. N., Stankus S. V., Khairulin R. A., and Segall B. (2000) Kinetics of the graphitization of dispersed diamonds at "low" temperatures, *J. Appl. Phys.*, **88**, pp. 4380–4388.
5. Okotrub A. V., Bulusheva L. G., Kuznetsov V. L., Butenko Yu. V., Chuvilin A. L., and Heggie M. I. (2001) X-ray emission studies of the valence band of nanodiamonds annealed at different temperatures, *J. Phys. Chem. A*, **105**, pp. 9781–9787.
6. Tomita S., Fujii M., and Hayashi S. (2002) Optical extinction properties of carbon onions prepared from diamond nanoparticles, *Phys. Rev. B*, **66**, 245424(1–7).
7. Osipov V. Yu., Baranov A. V., Ermakov V. A., Makarova T. L., Chungong L. F., Shames A. I., Takai K., Enoki T., Kaburagi Y., Endo M., and Vul' A. Ya (2011) Raman characterization and UV optical absorption studies of surface plasmon resonance in multishell nanographite, *Diam. Relat. Mater.*, **20**, pp. 205–209.
8. Enoki T., Kobayashi Y., Katsuyama C., Osipov V. Yu., Baidakova M. V., Takai K., Fukui K., and Vul' A. Ya. (2007) Structures and electronic properties of surface/edges of nanodiamond and nanographite, *Diam. Relat. Mater.*, **16**(12), pp. 2029–2034.
9. Osipov V. Yu., Enoki T., Takai K., Takahara K., Endo M., Hayashi T., Hishiyama Y., Kaburagi Y., and Vul' A. Ya. (2006) Magnetic and high

resolution TEM studies of nanographite derived from nanodiamond, *Carbon*, **44**, pp. 1225–1234.

10. Novoselov K. S., Geim A. K., Morozov S. V., Jiang D., Zhang Y., Dubonos S. V., Grigorieva I. V., and Firsov A. A. (2004) Electric field effect in atomically thin carbon films, *Science*, **306**, pp. 666–669.
11. Novoselov K. S., Geim A. K., Morozov S. V., Jiang D., Katsnelson M. I., Grigorieva I. V., Dubonos S. V., and Firsov A. A. (2005) Two-dimensional gas of massless Dirac fermions in grapheme, *Nature*, **438**, pp. 197–200.
12. Fujita M., Wakabayashi K., Nakada K., and Kusakabe K. (1996) Peculiar localized state at zigzag graphite edge, *J. Phys. Soc. Jpn.*, **65**, pp. 1920–1923.
13. Nakada K., Fujita M., Dresselhaus G., and Dresselhaus M. S. (1996) Edge state in graphene ribbons: nanometer size effect and edge shape dependence, *Phys. Rev. B*, **54**(24), pp. 17954–17961.
14. Clarke J. (1976) *SQUID—Current Performance of Superconducting Quantum Interference Devices* (Walter de Gruyter, Berlin).
15. Gallop J. C. (1991) *SQUIDs, the Josephson Effect and Superconducting Electronics* (Adam Hilger, Bristol, England, Philadelphia).
16. Mercereau J. E. (1970) Superconducting magnetometers with sensitivities approaching 10^{-10} G, *Rev. Phys. Appl.*, **5**, pp. 21–24.
17. Pauling L. (1936) The diamagnetic anisotropy of aromatic molecules, *J. Chem. Phys.*, **4**, pp. 673–677.
18. Steiner E., Fowler P. W., Soncini A., and Jenneskens L. W. (2007) Current-density maps as probes of aromaticity: global and Clar π ring currents in totally resonant polycyclic aromatic hydrocarbons, *Faraday Discuss.*, **135**, pp. 309–323.
19. Pinnick H. T. (1954) Magnetic susceptibility of carbons and polycrystalline graphites I, *Phys. Rev.*, **94**(2), pp. 319–326.
20. Duley W. W., and Seahra S. (1998) Graphite, polycyclic aromatic hydrocarbons, and the 2175 A extinction feature, *Astrophys. J.*, **507**, pp. 874–888.
21. Bandow S., Kokai F., Takahashi K., Yudasaka M., and Iijima S. (2001) Unique magnetism observed in single-wall carbon nanohorns, *Appl. Phys. A*, **73**, pp. 281–285.
22. Enoki T., and Takai K. (2009) The edge state of nanographene and the magnetism of the edge-state spins, *Solid State Commun.*, **149**, pp. 1144–1150.

23. Osipov V. Yu., Shames A. I., Enoki T., Takai K., Endo M., Kaburagi Y., and Vul' A. Ya. (2010) Interaction between edge-localized spins and molecular oxygen in multishell nanographites derived from nanodiamonds, *Diam. Relat. Mater.*, **19**(5–6), pp. 492–495.
24. Wakabayashi K., Fujita M., Ajiki H., and Sigrist M. (1999) Electronic and magnetic properties of nanographite ribbons, *Phys. Rev. B.*, **59**(12), pp. 8271–8282.
25. Wakabayashi K., Fujita M., Ajiki H., and Sigrist M. (2000) Magnetic properties of nano-graphites at low temperature, *Physica B*, **280**, pp. 388–389.
26. Kobayashi Y., Fukui K., Enoki T., and Kusakabe K. (2006) Edge state on hydrogen-terminated graphite edges investigated by scanning tunneling microscopy, *Phys. Rev. B*, **73**, pp. 125415(1–8).
27. Kobayashi Y., Fukui K., Enoki T., Kusakabe K., and Kaburagi Y. (2005) Observation of zigzag and armchair edges of graphite using scanning tunneling microscopy and spectroscopy, *Phys. Rev. B*, **71**(19), pp. 193406 (1–4).
28. Tao C., Jiao L., Yazyev O. V., Chen Y.-C., Feng J., Zhang X., Capaz R. B., Tour J. M., Zettl A., Louie S. G., Dai H., and Crommie M. F. (2011). Spatially resolving edge states of chiral graphene nanoribbons, *Nat. Phys.*, **7**, pp. 616–620.
29. Qin L., and Iijima S. (1996) Onion-like graphitic particles produced from diamond, *Chem. Phys. Lett.*, **262**(3–4), pp. 252–258.
30. Andersson O. E., Prasad B. L. V., Sato H., Enoki T., Kaburagi H., Hishiyama Y., Yoshikawa M., and Bandow S. (1998) Structure and electronic properties of graphite nanoparticles, *Phys. Rev. B*, **58**, pp. 16387–16395.
31. Tuinstra F., and Koenig J. (1970) Raman spectrum of graphite, *J. Chem. Phys.*, **53**, pp. 1126–1130.
32. Tomita S., Burian A., Dore J. C., LeBolloch D., Fujii M., and Hayashi S. (2002) Diamond nanoparticles to carbon onions transformation: X-ray diffraction studies, *Carbon*, **40**, pp. 1469–1474.
33. Tomita S., Sakurai T., Ohta H., Fujii M., and Hayashi S. (2001) Structure and electronic properties of carbon onions, *J. Chem. Phys.*, **114**, pp. 7477–7482.
34. Tomita S., Fujii M., Hayashi S., and Yamamoto K. (1999) Electron energy-loss spectroscopy of carbon onions, *Chem. Phys. Lett.*, **305**, pp. 225–229.
35. Obraztsova E. D., Fujii M., Hayashi S., Kuznetsov V. L., Butenko Yu. V., and Chuvilin A. L. (1998) Raman identification of onion-like carbon, *Carbon*, **36**(5–6), pp. 821–826.

36. Kuznetsov V. L., and Butenko Yu. V. (2005) *Properties and Applications of Ultrananocrystalline Diamond*, NATO Science Series III: Physics and Astronomy Synthesis, Vol. 192, "Nanodiamond graphitization and properties of onion-like carbon," pp. 199–216.

37. Osipov V. Yu., Shames A. I., Enoki T., Takai K., Endo M., Hayashi T., Kaburagi Y., and Vul' A. Ya. (2009) Magnetic and EPR studies of edge-localized spin paramagnetism in multishell nanographites derived from nanodiamonds, *Diam. Relat. Mater.*, **18**(2–3), pp. 220–223.

38. Kotosonov A. S. (1986) Diamagnetism of quasi-two-dimensional graphites, *JETF Lett.*, **43** (1), pp. 37–40.

39. Kotosonov A. S. (1991) Diamagnetism of carbon fibers, *Sov. Phys.: Sol. State* **33**(9), pp.1477–1480.

40. McClure J. W. (1956) Diamagnetism of graphite, *Phys. Rev.*, **104**(3), pp. 666–671.

41. Takai K., Oga M., Sato H., Enoki T., Ohki Y., Taomoto A., Suenaga K., and Iijima S. (2003) Structure and electronic properties of a nongraphitic disordered carbon system and its heat-treatment effects, *Phys. Rev. B*, **67**, pp. 214202(1–11).

42. Safran S. A., and DiSalvo F. J. (1979) Theory of magnetic susceptibility of graphite intercalation compounds, *Phys. Rev. B*, **20**(12), pp. 4889–4895.

Chapter 9

Applications of Detonation Nanodiamonds

Artur T. Dideikin
Ioffe Physical-Technical Institute, Solid State Electronics Department, 26 Polytechnicheskaya, St. Petersburg 194021, Russia
dideikin@mail.ioffe.ru

This chapter addresses the issues associated with the broad array of emerging applications of detonation nanodiamonds (DND). We are purposefully not going to dwell on the use of DND in biology and medicine, because this subject is treated in Chapter 10 of the present book, and Chapter 5 discusses application of nanodiamonds as biomarkers.

9.1 Applications of DND in Present-Day Technology

In view of the unique combination of detonation nanodiamonds (DND) properties, it appeared only reasonable to assume that this material would find broad application in various areas of technology immediately after its discovery. As it happens, however, quite frequently in the development of new technologies, the areas where DND attracted broad interest evolved gradually as new information appeared on the properties and structure of this material.

Detonation Nanodiamonds: Science and Applications
Edited by Alexander Ya. Vul' and Olga A. Shenderova

ISBN 978-981-4411-27-1 (Hardcover), 978-981-4411-28-8 (eBook)
www.panstanford.com

9.1.1 *Abrasives*

DND found its first application naturally in abrasive compositions intended for ultrafine mechanical polishing of hard surfaces of materials that did not contain iron, among them optical components, semiconductor plates, hard magnetic discs for information storage, parts of high-precision mechanical devices, etc.

Present-day polishing compositions based on DND offer a possibility of obtaining surfaces with an average roughness not in excess of 0.3 nm, which is apparently a limit attainable by the techniques employed in mechanical polishing. There is, however, a factor still remaining unclear, namely, that such outstanding results of processing were demonstrated before the problems involved in DND deaggregation have been solved (see Chapter 2). Indeed, the presence of solid aggregates of DND particles about 100 nm in size should act as a serious deterrent to attaining such a low roughness index. This either suggests that DND deaggregates in the course of preparation of the abrasive composition or evidences a relatively low strength of such aggregates, including those in which 4 nm DND particles become coupled through impurities of ferromagnetic metals. The latter can be assigned to the fact that DND of industrial manufacture used in preparation of polishing compositions has not been until recently subjected to purification from traces of metals.

As stressed in relevant publications, to prevent saturation of surfaces subjected to final polishing by DND particles present in the corresponding compositions, special provisions should be made [1], a practice accepted usually when employing diamond powder-based abrasives.

9.1.2 *Binder for Tools and Abrasive Ceramic*

Besides the use as an abrasive component of polishing compositions, DND also found application in abrasive diamond-based tools as a component of the binder to bond abrasive grains together. Application of DND permits increasing the strength and elasticity of a diamond-based abrasive tool up to 150%. Materials of moderate hardness can be processed also with an abrasive tool of diamond ceramic prepared by sintering DND with glass-forming oxides at high pressure and temperature. A side effect of considerable significance associated with the use of DND in abrasive compositions is improved removal of heat from the cutting region [2].

9.1.3 *Lubricants*

Using DND as an additive to industrial lubricating oils has also emerged as a promising application already during the first few years following its discovery. Adding DND to oils permitted in some cases reducing noticeably (by up to 8–12%) fuel consumption and increasing the service life of internal combustion engines. The efficiency of DND additions can be raised by combining them with small amounts of nanographite. Positive results were obtained also by adding to oils detonation soot with the minimal degree of its purification.

The mechanism by which DND operates in lubricating oils remains largely unclear. At first it was believed that diamond nanoparticles located on surfaces in friction play the part of balls in ball bearings that transform the sliding to rolling friction [3]. This effect can naturally be realized only if the surfaces in friction are very hard and the DND concentration in the lubricating oil is high enough. In actual fact, the operation of a friction pair reached by adding DND to a lubricant is observed to improve over a considerably broader range of conditions. There is a more plausible mechanism by which the DND particles added to the lubricant first favor running-in of the surfaces in friction by playing the part of a polishing abrasive, followed by penetration into these surfaces, a process improving their microhardness and durability. This suggestion is borne out by the observation that the effect of these additives persists for a fairly long time after the oil had been replaced by a new one, which does not contain DND [2]. The nanographite component plays an auxiliary part, and manufacturers are presently looking for ways to discontinue its use as additive to liquid motor oils, because it makes oil opaque and not marketable. Intense search is presently under way to find optimal conditions for the use of lubricants containing isolated 4 nm DND particles [4].

9.1.4 *Metal-Nanodiamond Coatings*

One of the most promising areas of DND application in present-day mechanical engineering is development of Cu-, Zn-, Sn-, Au-, Ag-, Cr-, and Ni-based composite metal coatings prepared by electroplating. A particularly effective approach involves introduction of DND into the compositions of nickel- and chromium-based plating. The wear resistance of such plating increases four to nine times, a feature employed to advantage in various machines and mechanisms with friction surfaces, which are designed to operate in very hard conditions.

Among such engines and devices are, in particular, aircraft engines, immersion pumps for use in oil wells, valves, stoppers and pumps for the chemical industry, etc. Besides the remarkable wear resistance, galvanic plating containing DND exhibits high elasticity, a feature accounting for its ability to withstand high strains. The friction coefficient of such plating paired with various metallic and nonmetallic materials decreases to one-fourth the value characteristic of similar coatings that do not contain DND. Application of composite galvanic plating with DND makes it possible in some cases to reduce the thickness of metal-nanodiamond films by two to three times compared to traditional protective films and, in this way, to increase the efficiency of the production equipment employed in the manufacture of such plating as well as to reach saving of material.

The technology of depositing metal-based diamond plating is actually very simple: dry DND powder is poured directly into the water-based electrolyte of the electroplating bath. The quality of plating improves with increasing DND concentration in the electrolyte up to the level specifying the beginning of precipitation (20 g/L) [5].

The mechanism underlying incorporation of DND particles into the metallic base layer involves presumably electrophoresis of the dielectric particles of nanodiamond, paralleled by galvanic deposition of the metal [6]. In these conditions, the relative content of diamond in the deposited film is fairly low. For nickel-based plating, for instance, it does not exceed 0.6 wt.% of the maximum possible DND concentration in the electrolyte. The part played by DND in metal-based composite coatings consists apparently of reducing the size of the grains making up the structure of the metal film and increasing accordingly its microhardness. The DND particles located at grain boundaries strengthen presumably their coupling and block crumbling out of the plating with wear. The lowering of friction is attributed to DND aggregates with sizes of about 100 nm emerging onto the surface of the plating. This process resembles in main features the way in which DND acts when added to a lubricant. The question of whether DND does deaggregate as a metal–nanodiamond film wears out remains open. None of the available publications has reported thus far the observation of single 4 nm diamond particles in a lubricant that is employed with friction pairs coated by metal-based nanodiamond plating.

9.1.5 *Bulk Metal–Nanodiamond Composites*

The next obvious step in the area of metal–nanodiamond composites is development of corresponding bulk materials. Of particular interest as bases of such materials are copper, aluminum, and its alloys. Incorporating nanodiamond powder into an aluminum matrix opens the way of obtaining materials with high thermal conductivity [7] and specific strength for use in aircraft engineering and atomic and electronic industries. The volume fraction of diamond in an aluminum- or copper-based composite may be as high as 50%. Use of DND in such composites appears particularly promising due to the small size of the diamond particles and the possibility of modifying within a broad range their surface. This offers a possibility of reaching the maximal possible adhesion to the metal and, accordingly, the best attainable strength of the composite. The technology of production of metal–nanodiamond composites is based on the methodology employed in powder metallurgy, which includes mixing of disperse diamond with metallic powders, followed by sintering at high pressure and temperature.

9.1.6 *Diamond-Based Ceramic*

As already mentioned, the first hard ceramic composites with DND were employed as binders for diamond-based cutting tools and, in some cases, as an independent cutting material for processing of nonferrous metals and nonmetallic materials of moderate hardness.

DND is used to manufacture nanoporous ceramic filters for biochemical engineering. Silicon carbide (SiC)- and nanodiamond-based ceramic has recently attracted considerable interest as a promising material for this area. This material exhibits high hardness, resistance, and elasticity [8, 9]. The ceramic is sintered at a temperature of 1,700–1,800°C and a pressure of up to 8 GPa. The volume content of SiC in the material is as high as 28%, with its hardness reaching 80 GPa.

A certain advantage of this ceramic is the comparatively weak dependence of the parameters of the material obtained on the conditions of synthesis, which should be assigned to the ease with which the covalent Si–C and C–C bonds form on the surface of nanodiamond particles in the course of sintering with silicon carbide.

9.1.7 *Bulk Composites with Organic Polymers*

Of particular interest for the present-day industry are composites based on organic polymers. They are used in overwhelming amounts as construction and sealing materials. Application of DND as a component of such composites is presently revealing steady growth.

In contrast to other carbon components, for instance, soot, which is used as a bulk filler in rubber-based elastic polymer materials in amounts of up to 45–60%, DND is added in tens of times smaller amounts and plays here the part of an efficient modifier of the mechanical and chemical properties of a polymer material. For instance, incorporation of DND in resins based on polyisoprene rubbers increases its elasticity (the stress at 300% elongation increases by a factor of 1.6), strength, and rupture resistance (by a factor of 1.6–1.8).

A similar improvement of properties is exhibited also by resins based on other rubbers, including isoprene, butadiene styrene, butadiene nitrile, and natural rubber.

A number of patents for use of DND as an additive to resin for car tires with improved strength and service life have been issued. The content of DND in a resin is 0.1–0.5%. It was found that besides improved strength and wear resistance, addition of DND to the resin of car tires improves their force of friction on a smooth surface, even if the latter is coated by a water film [10].

The mechanism underlying improvement of the characteristics of rubber-based elastic polymer materials involves the formation of additional bonds between polymer molecules and is in a large measure mediated by the surface activity of DND particles. Considering that the above results were obtained without deaggregation and purposeful surface modification of DND particles introduced into the resins, one can expect a noticeable increase in the efficiency of addition of isolated deaggregated DND particles whose surface is specially modified for formation of chemical bonds with molecules of a specific polymer.

A positive effect of DND introduction is observed in thermoplastic elastomers. It was demonstrated that introducing DND in amounts of up to 3–5% into a thermoplastic elastomer based on polypropylene and ethylene-propylene-diene rubber, the friction coefficient of the polymer surface decreases and its resistance to scratches improves as does the thermal stability of the material [11]. Adding DND in an amount of as little as 0.6% improves significantly the elastic properties and strength of composite plastic materials based on polyvinyl alcohol [12].

Progress has recently been reached in application of polymer–nanodiamond composites based on polyimide and polyurethane with DND contents of up to 1–3% in microelectronics as material for insulating films possessing low dielectric permittivity and high thermal conductivity [13]. Another possible application of such composites includes insulation for magnet wires in transformers and heavy-power small electric motors for use in transport facilities. The high strength and thermal conductivity of such insulation will make it possible in due time to reduce the overall dimensions and increase the specific power of these mass production devices.

9.1.8 *Polymer Film Coatings*

Polymer film coatings based on fluoroelastomers and polysiloxanes are widely employed in machine building and chemical technology to reduce friction and protect metals and other materials against deleterious effects of the atmosphere and aggressive media, as might be encountered, in particular, in land and sea transport facilities and aircraft and in chemical and mining industries. DND embedded in such coatings reduces their permeability to polar solvents by a factor of 50. The highest chemical resistance to salts, acidic and alkaline media is demonstrated by coatings based on ethylene with perfluoroalkylvynil esters. Adding nanodiamond to these coatings reduces the coefficient of dry friction over metals, increases by more than a factor of 10 the coating rupture strength, and improves by three to five times the strength of the adhesion contact of a polymer film to a metal. Incorporation of DND into the composition of a coating increases its resistance to abrasive wear by 1.5–2 times.

Another class of the plastic–nanodiamond composites that has already gained broad popularity encompasses various cosmetics and skin care products. A number of patents have even been issued in this area [14].

As a rule, DND particles are introduced into polymer coatings before the onset of polymerization. DND is added in the form of dry powder or a suspension in a solvent, which reacts with the polymer used as the base of the composite.

Thin (1–100 μm) polymer–nanodiamond films and coatings can be produced also by combined electrophoresis from the suspension of nanodiamond in a solvent containing the polymer in the desired concentration, or a mixture of components to form the base of the coating in the course of polymerization.

A comparable efficiency is reached in the powder approach to coating production, in which a layer of a mixture of thermoplastic polymer and DND grains is formed on a metallic surface by electrostatic transport, which is subsequently heated to the polymer melting temperature.

DND particles embedded in polymer coatings play a dual part. First, due to the presence on the DND surface of carboxyl functional groups (COOH), they, just as in the case of bulk composites, become chemically bonded to the polymer, a process that increases the number of its intermolecular bonds and, accordingly, improves its elastic characteristics and strength [15]. It is these additional chemical bonds coupling the surface to the nanodiamond particles embedded in the polymer coating that underlie the increase in the strength of the adhesion contact connecting coatings with the surfaces to be protected. Second, being located on the polymer surface, DND particles are in continuous contact with the surface in friction and act as a hyperfine abrasive, thus favoring its smoothing down. In due time, the properties of this surface stabilize as a result of its saturation by DND particles, and the friction coefficient decreases, similar to the way this happens when using lubricants with DND.

9.1.9 *Adhesives*

Recent progress in adhesive joint technology has modified markedly the approach to design and engineering in many areas of machine building and construction. Adding DND to epoxy-based bonding agents improves significantly the strength of both the bond itself and, through adhesion, the bonded materials as well. Successful application of DND in epoxy-based bonding agents rests on appropriately chosen functionalization of the surface, which should favor chemical incorporation of nanodiamond particles into the polymer forming as the bonding agent solidifies.

To cite an example, adding to the epoxy-based adhesive 7 wt.% DND with the surface modified by ethylenediamine, which becomes bonded to it through NH_2 amino groups, increases the hardness of the solidified adhesive by up to three times, increases Young's modulus by 50%, and reduces the plastic shear strain by a factor of 5. Because the ethylenediamine attached to DND particles is capable of initiating polymerization of epoxy resin, this opens up a possibility of discontinuing completely the application of a specific hardener in the bonding agent [16].

9.1.10 *Ferromagnetic Composites*

Interest in polymer-based ferromagnetic composites was originally aroused by a search for ways to improve the characteristics of flexible-base magnetic information carriers (discs and tapes). Introduction of DND into the composition of flexible polymer–ferromagnetic information carriers (tapes) for backup systems permitted one to reduce the friction coefficient in the tape drive by up to a factor of 1.4 and to prolong the carrier service life up to 4.5 times [17].

Besides, DND embedded in ferromagnetic coatings made it also possible to reduce significantly the size of magnetic domains and to increase the recording density [2]. The content of DND in ferromagnetic coatings of magnetic carriers was 1–2%. Interesting results were obtained in deposition of ferromagnetic metals (Ni, Co, Cr) directly on the surface of diamond nanoparticles [18], but DND aggregation, the problem that was not yet solved at the time, made their practical application impossible. The demand for flexible magnetic information carriers became ever weaker with time as electronic memory devices based on silicon field-effect transistors (FETs) were gaining in popularity. The information amassed in the course of development of ferromagnetic composites is presently used successfully in designing very promising DND-incorporated magnetic abrasive lapping materials (MALMs) intended for selective final smoothing and polishing of high-precision devices [19]. These materials are based on abrasive particles that are coated, just as in ferromagnetic composites, by a layer of magnetic metals.

9.1.11 *Crystallization Centers for CVD Technology of Diamond Films*

The technology of deposition of diamond and diamond-like films on various surfaces intended for their protection and reduction of friction has become an issue of widespread recognition in the area of DND application, which is presently demonstrating remarkable progress.

Layers of polycrystalline diamond produced by chemical deposition from a gas phase (chemical vapor deposition [CVD]) [20] are enjoying presently wide use as protective and antifriction coatings in engineering, high-precision instrumentation, optics, and medicine in order to confer biological inertness to elements of the equipment used, surgical tools, and metallic and ceramic implants. To improve the abrasive stability against dust, diamond CVD films are deposited on glass panels installed in buildings, motor transport, and aircraft. Diamond and diamond-like CVD films enjoy an ever-increasing use in optical instrumentation as antireflection and protective coatings.

Present-day possibilities of the CVD technique permit manufacture of solid diamond polycrystalline plates up to 5 mm thick and 100–200 mm in diameter, in considerable demand for scientific equipment employed in nuclear and space research. Some producers deposit on the inner surface of the plastic wrapping a diamond-like film to prolong the shelf life of food and biological products.

For a crystalline DVD film to start growing on an arbitrary solid surface, it must have centers of crystallization (precursors). Particles of natural or artificial diamond produce the most favorable conditions for crystallization of CVD diamond. The quality of diamond CVD films is determined primarily by the degree of their uniformity, which, in its turn, directly depends on the conditions governing the initial stage of growth. Thus, for a uniform CVD film to form, on the surface of the substrate there should be as many as possible uniformly distributed diamond particles. The limiting rate of growth of a diamond CVD film is also governed by the density and uniformity of distribution of crystallization centers on the substrate.

By using DND as a precursor in the preparation of diamond CVD films, one can produce crystallization center densities on the substrate surface of 10^{12} cm^2, a figure higher than that reached with any other diamond material [21].

To form precursors on a glass or a metal surface, one ordinarily makes use of the properties of DND particles to saturate this surface under mechanical friction. This is done in the course of final polish of the surface intended for CVD deposition by a DND-containing abrasive composition.

To deposit diamond particles on a rough surface, it is subjected to ultrasonic treatment in suspensions with DND content of 3–5%, a procedure that also contributes to incorporation of DND particles into the surface layer of the material until the desired concentration of uniformly distributed crystallization centers is reached. Use of DND for formation of precursors improves also substantially adhesion of diamond and diamond-like CVD films to various surfaces, the mechanism at work here being the same as the one involved in preparation of the above-mentioned polymer-DND composite coatings.

The successful solution of the problem associated with deaggregation and preparation of suspensions of single 4 nm diamond particles has made possible not only a significant improvement of the quality and uniformity of diamond CVD films produced on various surfaces but also obtaining of radically novel results through development of spatial micro- and nanostructures.

9.1.12 *Photonic Crystals Produced by CVD*

Photonic crystals have been spurring recently enormous research interest [22]. These crystals are actually periodic two- or three-dimensional structures transparent for optical radiation, in which the refraction coefficient varies with a period close to or exactly equal to the wavelength of light passing through them. Diffraction accompanying propagation of photons in such crystals brings to life the phenomena similar to those governing the motion of charge carriers in semiconductor crystals. A bandgap forms for photons of specific energies, just as it does for carriers in semiconductors.

Photonic crystal structures open up the way to designing adaptive optical waveguides, high-efficiency converters of solar energy, sources of illumination, and devices for direct processing and display of optical information without its preliminary conversion into electrical signals. This can be readily reached by creating inside such periodic structures regions with different average refraction indices, regions with a nonlinear medium, and active luminescent regions playing the part of radiation emitters [23].

Until very recently, synthetic opals, that is, structures composed of SiO_2 spheres of submicron size distributed uniformly over polystyrene matrices, were the only real illustration of photonic crystals. [24]

Progress in the technology of formation of crystallization centers based on the use of suspensions of 4 nm DND particles permits one presently to prepare bulk photonic crystals composed of regularly arranged SiO_2 spheres and polycrystalline CVD diamond filling up the space between them. The first to form is a film made up of closely packed SiO_2 spheres, about 500 nm in diameter, after which the space between the spheres is filled by the DND water suspension by the capillary effect. Subsequent evaporation of water leaves the surface of the spheres making up the film coated by DND particles. Next the film is placed into a reactor, where CVD diamond forms in microwave plasma discharge. The CVD regime is set such that the diamond that has crystallized on the DND particles filled uniformly the space separating the SiO_2 spheres. The final product is a bulk structure composed of polycrystalline diamond with regularly distributed SiO_2 spheres in contact with one another, which reveals the properties of a photonic crystal. That the spheres are in contact is a favorable factor providing the possibility of removing them completely by straightforward chemical etching. The three-dimensional structure with regularly arranged spherical voids about 500 nm in diameter left after this treatment represents the so-called inverted photonic crystal consisting fully of polycrystalline diamond [25]. The new technology involving DND opens up new vistas in the area of research and application of photonic crystals, by permitting, for instance, incorporation into their composition of luminescent centers created preliminarily in the particles acting as centers of crystallization and of nonlinear optical media by filling with them the voids, and employment of the methods developed in lithography for polycrystalline diamond films.

9.1.13 *Adsorbents and Chromatography*

As stressed in some of the earlier chapters, the most significant feature of DND, as actually of any other nanodisperse material, is that a noticeable part of the carbon atoms making up its particles is located on the surface of these particles.

The specific surface of dry DND powder is 370–450 m^2/g, depending on the extent of aggregation. While this figure is several times smaller than the extreme characteristics exhibited by other nanocarbon materials, including single-wall nanotubes, nanofibers, disordered graphene, and nanoporous carbon, the combination of mechanical strength with chemical inertness of particles with the possibilities evolving from surface functionalization opens up broad vistas for its application as a high-efficiency absorbent, including binding of toxins in living organisms [26].

As a material with a high sorption capacity, liquid chromatography occupies by right a separate place in DND application.

Hot-pressed granulated DND is used as a stationary phase in ion chromatography, which is indispensable in separation of ions of alkaline-earth and transition metals [27]. When separating polyaromatic hydrocarbons, monoalkyl benzenes, polymethyl benzenes, etc., a stable phase of sintered DND permits one to reach separation efficiencies of up to 45,000 theoretical plates per meter, a figure substantially larger than the corresponding characteristic for the materials employed traditionally in chromatography as a stationary phase, including silica gel, aluminum oxide, and various porous carbon materials [28]. Still higher efficiencies with DND used as a stable phase can be attained in separation of biologically active substances, including proteins [29]. The nontoxic nature of DND is a basic factor underlying this possibility.

Application of DND in biology and medicine is an area exhibiting presently the fastest growth. It covers not only sorbents and chromatography but also systems of targeted drug delivery through cell membranes, various biological markers, systems intended for determination of the concentration of bioactive agents in living organisms, and a host of others. All of them are discussed in considerable detail in the last chapter of the monograph.

9.1.14 *Catalysts*

Recent years have been highlighted by a growth of interest in the use of DND as a carrier of metallic catalysts, in particular of platinum group metals. Application of particles of a neutral, chemically stable phase, with particles of a catalyst, for which one typically uses metals of the platinum group, deposited on their surface, opens the way to increasing noticeably the efficiency of catalysis, while at the same time reducing the expenditure of precious and rare metals.

DND may be considered as a practically ideal carrier for a catalyst due to its chemical stability, mechanical strength, and small size of its particles.

Metallic catalysts with DND particles used as carriers can be designed, depending on the particular application intended for it, by a number of techniques, including chemical reduction from salts added preliminarily to the DND water suspension, electrophoresis of metallic nanoparticles prepared for the purpose, or direct galvanic reduction of metal ions on the surface of DND grains [30]. In this way high-efficiency catalysts for oxidation of methanol in fuel cells can be readily produced.

Platinum film-coated DND grains manufactured for use as high-sensitivity sensors of carbon oxide in the atmosphere operate by oxidizing carbon oxide to carbon dioxide, the principle underlying operation of fuel elements, and, thus, do not require additional power supply [31].

DND–palladium complexes exhibit catalytic activity in the hydrogenation of unsaturated hydrocarbons high enough to conduct the reaction even at room temperature [32]. The area where DND-based catalysts can be employed to advantage is not limited to oxidation and hydrogenation of hydrocarbons. To cite an example, DND–copper complexes produced by reduction from copper sulfate accelerate noticeably decomposition of ammonium perchlorate and trimethylene trinitro amine, substances used as rocket fuel components [33].

It was demonstrated that, besides the capability of functioning as a catalyst carrier, DND particles themselves reveal under certain conditions catalytic activity. It was found that DND particles are capable of increasing the rate of oxidative dehydration of ethyl benzene, a reaction used to produce styrene. In this case, the catalytic activity of DND derives from the presence of areas on the surface of nanodiamond particles where sp^2-hybridized carbon persists.

Such an activity is characteristic also of spherical carbon onion particles produced from DND by thermal annealing [34]. There is no doubt whatsoever that DND-based catalysts are facing broad vistas in various areas of chemistry, biology, and medicine. The next essential step in DND application in this direction should be expected development of selective catalytic systems.

9.1.15 *Coolant Liquids*

A new, potentially promising area where DND may become of great value is development of high-efficiency heat transport media for cooling systems based on the so-called nanoliquids, that is, liquid suspensions of nanoparticles possessing high thermal conductivity.

It was found that adding 0.3 mass.% DND can increase the thermal conductivity of a liquid by more than 40% [35]. Such a large increase of thermal conductivity for a comparatively small addition of DND cannot be accounted for in the context of generally accepted theory. The mechanism underlying this phenomenon is based apparently on resonant propagation of phonons, which sets in at a certain ratio of the concentration and dimensions of the nanoparticles suspended in the liquid [36]. DND was used in practice for the first time as an addition to oil filling high-voltage transformers to increase its thermal conductivity. This opened up a way for eliminating internal breakdowns initiated by local oil overheating and lowering the overall operating temperature [35].

9.2 Unsolved Problems in DND Applications

Considerable progress has become evident in recent years in important areas associated closely with the technology of DND production; it initiated, in particular, the beginning of intense research aimed at finding a possibility of employing this material in various domains of biology and medicine.

First, although the commercial DND supplied by most manufacturers still contains a considerable amount of impurities, some producers have started production of limited amounts of DND of a higher degree of purification, which, after a certain preparation, can be accepted for research in biology and medicine.

The residual impurities found in DND are, as a rule, chemically stable metal oxides originating from the walls of the equipment employed for DND synthesis and isolation; they are difficult to remove, and reducing their concentration requires application of high-cost purification technologies involving nanomembranes, ion exchange, gravity-assisted separation and so on.

Although DND of a high degree of purification is presently a high-cost material supplied in small amounts, it is available for laboratories engaged in biological and medical research. The content of residual impurities in highly purified DND is measured by a method based on analysis of electron paramagnetic resonance spectra, which has a sensitivity of up to 10^{-7} mass [37] but is capable of detecting ferromagnetic impurities only. For this reason, preparation of DND of a high degree of purification must be conducted in conditions precluding formation of impurities to which the electron paramagnetic resonance (EPR) method is insensitive.

The second, not less essential factor is the need for solving the problem of DND deaggregation without resorting to mechanical milling, which, while having been in the past the only technique of preparation of suspensions of isolated 4 nm diamond particles, brought about surface graphitization of DND, as well as its contamination by fragments of the zirconium oxide balls disintegrating in the course of milling. A separate chapter of the present book deals with description of the new methods of DND deaggregation that are based on thermal annealing of dry powder in various atmospheric environments, which preclude contamination and graphitization of the surface of DND particles. This provided a basis for realization of the rich potential of DND as a nanomaterial. This will, in particular, open the way to reaching substantially higher technical characteristics of composite materials with DND, of coatings, crystallization centers for further development of CVD technology, catalysts, etc., in the areas where this material was earlier employed in the form of aggregates only.

Despite the impressive results demonstrated in the recent years, further progress in DND technology is curtailed by problems whose solution could favor successful application of DND in other fields of science and technology, which thus far have not been approached with this purpose.

To begin with, one has not yet succeeded in obtaining suspensions of 4 nm DND particles with a relative content of aggregated particles less than 10^{-4}, so it should be added that not more than 20% of original DND become transformed to single 4 nm particles. The reason for this lies in the fact that the bond structure inside the strong DND aggregates remains largely unknown. Neither has one yet deciphered the mechanism involved in deaggregation of DND particles occurring under thermal annealing in various atmospheres. Finding the answers to these issues would be all the more significant given that large aggregates present even in comparatively small amounts among 4 nm DND particles place severe constraints on the possibility of using it as a basis for development of promising electronic devices, for instance, planar field electron emitters; it has to be added that the fairly low yield of 4 nm particles requires the use of centrifugal fractionation, a process that increases substantially the cost of the final product.

Another serious problem lies in the uncertainty with which we know the factors responsible for formation of the electrokinetic potential of single 4 nm DND particles in liquids, whose magnitude governs the suspension stability. One does not presently know efficient ways to charge DND particles to a specified electrokinetic potential in different liquid media, and this accounts for the fact that stable suspensions of 4 nm DND particles can now be obtained in deionized water only. DND suspensions in organic solvents advertised actively by a number of manufacturers consist actually of aggregates with a minimum size of 80 nm. This problem becomes particularly acute in biological and medical areas of DND application because of the decrease of the electrokinetic potential the 4 nm DND particles experience when emerging from water, even into the physiological solution they coagulate and precipitate.

Production of nanodiamond particles below 4 nm in size is a problem of not lower potential significance. Many issues still remaining unresolved in molecular biology and medical applications of nanodiamond would possibly find solution if particles 2–3 nm and smaller in size were available. Such particles could probably be produced by ozone treatment or plasma etching, but the question of the stability of DND particles of such dimensions under normal conditions remains open.

There can be no doubt that a solution of the above problems will provide a fresh impetus to the penetration of DND into new areas of technology.

9.3 Future Promising DND Applications in Engineering and Technology

9.3.1 *Field Emitters*

A promising, while still not realized, possibility of employing electronic properties of DND in electronic devices with radically new parameters is its use in designing uniform two-dimensional field emitters for light sources and high-resolution visual information displays.

Intense studies directed at development of large-area field electron emitters with diamond coatings to reduce the work function, which have been conducted for more than 15 years, still have not been crowned with results of practical significance [35]. This can be traced, first of all, to the problem of uniformity of emission currents from such structures still remaining unsolved, which, in its turn, results from the absence of an adequate model of the mechanism responsible for generation of emission currents in carbon nanostructures. Because emission currents emerging from carbon nanomaterials concentrate in point regions (emission centers), development of structures in the form of isolated, uniformly distributed DND particles on a smooth conducting (metallic, graphite, or semiconductor) surface would solve the problem of a uniform distribution of emission centers over an area, an arrangement that would finally evolve into radically new planar electroluminescent displays.

It is probable that development of a uniform planar field emitter will find its solution through formation of combined structures of nanocarbon particles of different types, for instance, of DND-decorated graphene layers [38].

9.3.2 *Nanothermoelements*

Similar structures made up of uniform layers of DND particles arranged on smooth conducting surfaces will provide a basis for development of thermoelectric energy converters making use of the anomalous thermoelectric effect, which is driven by electron drag by ballistic phonons in nanoparticles [39]. Development of such structures will become possible after the abovementioned problem of obtaining really monodisperse suspensions of DND particles finds its solution. It is such a DND suspension that would permit preparation of uniform layers of diamond particles 4 nm thick, with no large particles that would project above their surface.

The possibility of growing continuous conducting carbon onion layers on the surface of 4–5 nm nanodiamond particles in the course of thermal or electron beam annealing opens up new prospects of designing quantum dot electronic structures that could be employed for switching of electric signals and detection and generation of electromagnetic radiation.

9.3.3 *Nano- and Micromechanical Devices*

It would be a technologically attractive idea to combine isolated dielectric DND particles and the onion structures formed of them with carbon nanotubes. Point electron-beam machining can, under certain conditions, initiate formation of chemical bonds between closely spaced carbon atoms, provided excess heat is adequately removed. Such an approach could possibly open the way to designing novel elements of nanomechanisms whose operation would be based on the elastic properties of nanotubes and displacement of nanodiamond and carbon onion grains attached to them and on the use of properly modulated electric fields as a source of driving force.

9.3.4 *Luminescence and Magnetometry*

An independent promising application of DND is based on using its luminescent properties deriving from nitrogen and nitrogen-vacancy (NV) defects in the crystal structure of its particles. A long time passed before one succeeded in experimentally detecting the DND luminescence characteristic of nitrogen defects in the presence of a sizable amount of nitrogen in its composition, which was observed by other techniques [40]. Recent studies revealed in DND particles subjected to electron irradiation with subsequent anneal a luminescence characteristic of neutral and negatively charged (NV) centers [41]. High-pressure sintering of DND also reveals the presence of nitrogen-luminescent centers N^0 $(NV)^-$ in concentrations as high as 10^{20} cm^{-3} [42]. Besides the light markers designed for use in living biological systems mentioned in Chapter 5, the luminescent nitrogen defects in DND particles provide a basis for their use in single-photon sources for future quantum computers, quantum cryptography systems, and data transmission lines whose operation would rely on entangled photon states [43].

The energy level diagram of the $(NV)^-$ centers in diamond offers a remarkable possibility of detection and measurement of magnetic fields on the scale of individual molecules. The first pioneering models of instruments making use of the principle of electron spin resonance on $(NV)^-$ centers in artificial diamonds have already been developed, which permit measurement of magnetic fields generated by spins of single electrons and even by nuclear spins at room temperature [44].

A nanomagnetometer providing spatial resolution and optical detection of the signal is actually an atomic force microscope with a probe representing a diamond nanocrystal containing $(NV)^-$ centers. The probe is moved to scan the surface of the sample to which 2.87-GHz microwave radiation is supplied through a waveguide. The 637 nm luminescence of the $(NV)^-$ centers is excited by a 532 nm laser beam. The luminescence is directed through a confocal objective onto the detector. With no external magnetic field present, the ground state of the two unpaired electrons of a $(NV)^-$ center represents actually a three-level system (a triplet) m_0 and $m_{\pm 1}$ separated in energy between m_0 and $m_{+1} = m_{-1}$ by 2.87 GHz.

Excitation by 532 nm radiation drives electrons through the level system with a radiative transition at 637 nm into the m_0 ground state, thus resulting in its predominant filling. Resonant absorption of the 2.87 GHz microwave radiation transfers the electrons through interaction of the free spins with the field of the electromagnetic wave primarily into the $m_{\pm 1}$ state. As a result, the electron excited optically from this ground state level has a slightly higher energy than if excited from the m_0 level. In this case, recombination from the excited state occurs primarily by the alternate channel having no radiative transitions, with the eventual substantial falloff in luminescence intensity. When an external magnetic field is applied, the resonance condition $m_{+1} = m_{-1}$ holds no longer, the occupancy of the m_0 level grows, and the luminescence intensity increases. Thus, the luminescence intensity turns out to be dependent on external magnetic field. By moving a diamond particle containing $(N\text{-}V)^-$ centers above the surface of the sample, one can scan the two-dimensional distribution of the magnetic field with a sensitivity of up to 3 nT, with a possibility of observing the distribution of functional groups with single spins and even of C^{13} atoms. Application of DND particles as sensitive elements in such instruments will improve substantially the spatial resolution due to their small size.

9.3.5 *Selective Chemical Sensors*

DND particles with an appropriately processed surface can act as selective elements of electronic sensors responding to the concentration of a substance under study in liquid and gaseous media. For this purpose particles with the surface modified by the corresponding chemical agents are immobilized on the surface of the electrodes connected into the electric signal detection circuit. In another method of detection, DND particles with the surface modified for efficient selective adsorption of specified molecules are deposited onto the surface of a submicron-size silicon rod in which mechanical vibrations are excited. The resonance frequency of such a device is so sensitive to an attached mass as to permit not only detection of individual molecules adsorbed by DND particles on its surface but even, in some cases, their identification [45].

Summing up, we cannot but come to the conclusion that the scope of the fields where the use of DND has already gained broad recognition, as well as the remarkable potential of its future applications, leaves no doubt in the possibility of this unique material to form a basis of novel technologies (see Fig. 9.1).

The review presented here of the already realized and promising applications of DND, a truly wonderful material, which, while already having found its niche is still hiding its secrets, can in no case claim to be exhaustive. The reason for this lies in the very fast expansion of the domains of DND application in science, technology, medicine, and other spheres of human creative activities. We shall consider the goal set in this chapter accomplished if the reader has derived from it valuable information he or she did not have before, all the more so if this contributes to further progress in research and broadening of the areas where DNDs can find new domains of application.

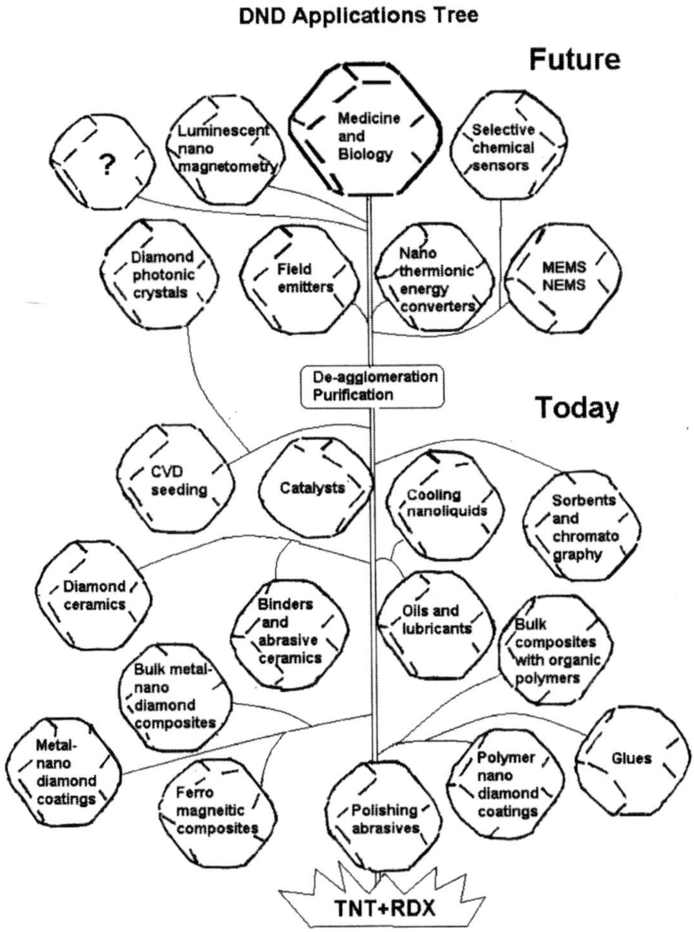

Figure 9.1 Applications of DND: today's and future.

References

1. Zhu Y., Feng Z., Wang B., and Xu X. (2004) Dispersion of nanodiamond and ultra-fine polishing of quartz wafer, *China Particuol.*, **2**(4), pp. 153–156.
2. Dolmatov V. Y. (2001) Detonation synthesis ultradispersed diamonds: properties and applications, *Russ. Chem. Rev.*, **70**, p. 607.

3. Tao X., Jiazheng Z., and Kang X. (1996) The ball-bearing effect of diamond nanoparticles as an oil additive, *J. Phys. D: Appl. Phys.*, **29**, pp. 2932–2937.

4. Chou C. C., and Lee S. H. (2008) Rheological behavior and tribological performance of a nanodiamond-dispersed lubricant, *J. Mater. Process. Technol.*, **201**, pp. 542–547.

5. Petrov I., Detkov P., Drovosekov A., Ivanov M. V., Tyler T., Shenderova O., Voznecova N. P., Toporov Y. P., and Schulz D. (2006) Nickel galvanic coatings co-deposited with fractions of detonation nanodiamond, *Diam. Relat. Mater.*, **15**, pp. 2035–2038.

6. Mandich N. V., and Dennis J. K. (2001) Codeposition of nanodiamonds with chromium, *Met. Finish.*, **99**(6), pp. 117–119.

7. Mizuuchi K., Inoue K., Agari Y., Morisada Y., Sugioka M., Tanaka M., Takeuchi T., Tani J.-i., Kawahara M., and Makino Y. (2011) Processing of diamond particle dispersed aluminum matrix composites in continuous solid–liquid co-existent state by SPS and their thermal properties, *Composites: Part B*, **42**, pp. 825–831.

8. Wachtman J. B., Ownby P. D., Jr., and Liu J. (2008) Nano-diamond enhanced silicon carbide matrix composites, *Proc 15th Ann. Conf. Composites Adv. Ceramic Mater., Part 1 of 2: Ceramic Eng. Sci. Proc.*, **12**(7–8).

9. Ekimov E. A., Gromnitskaya E. L., Gierlotka S., Lojkowski W., Palosz B., Swiderska-Sroda A., Kozubowski J. A., and Naletov A. M. (2002) Mechanical behavior and microstructure of nanodiamond-based composite materials, *J. Mater. Sci. Lett.*, **21**, pp. 1699–1702.

10. Wanga Y.-X., Wua Y.-P., Li W.-J., and Zhanga L.-Q. (2011) Influence of filler type on wet skid resistance of SSBR/BR composites: effects from roughness and micro-hardness of rubber surface, *Appl. Surf. Sci.*, **257**, pp. 2058–2065.

11. Hejazi Jahromi F., and Katbab A. A. (2012) Nanodiamond-based PP/EPDM thermoplastic elastomer composites: microstructure, tribodynamic, and thermal properties, *J. Appl. Polym. Sci.*, **125**, 1942–1950.

12. Maitra U., Prasad K. E., Ramamurty U., and Rao C. N. R. (2009) Mechanical properties of nanodiamond-reinforced polymer-matrix composites, *Sol. State Comm.*, **149**, pp. 1693–1697.

13. Shenderova O., Tyler T., Cunningham G., Ray M., Walsh J., Casulli M., Hens S., McGuire G., Kuznetsov V., Lipa S. (2007) Nanodiamond and onion-like carbon polymer nanocomposites, *Diam. Relat. Mater.*, **16**, pp. 1213–1217.

14. Nanodiamonds for cosmetics November 13, 2007, US patent (2007) 7,294,340 issues.
15. Zhang Q., Naito K., Tanaka Y., and Kagawa Y. (2008) Grafting polyimides from nanodiamonds, *Macromol.*, **41**(3), pp. 536–538.
16. Mochalin V. N., Neitzel I., Etzold B. J. M., Peterson A., Palmese G., and Gogotsi Y. (2011) Covalent incorporation of aminated nanodiamond into an epoxy polymer network, *ACS Nano*, **5**(9), pp. 7494–7502.
17. Kurmachev V. I., Timoshkov Y. V., Orehovskaja T. I., and Timoshkov V. Y. (2004) Nanodiamonds in magnetic recording system technologies, *Phys. Sol. State*, **46**(4), pp. 696–702.
18. Gubin S. P., Popkov O. V., Yurkov G. Yu., Nikiforov V. N., Koksharov Yu. A., and Eremenko N. K. (2007) Magnetic nanoparticles fixed on the surface of detonation nanodiamond microgranules, *Diam. Relat. Mater.*, **16**, pp. 1924–1928.
19. Zheng Z. X., Wang R., and Wang C. M. (2011) Electroless plating of Co-Zn-P thin film onto nanodiamond cores, *Curr. Appl. Phys.*, **11**, pp. 227–230.
20. Gruen D. M., and Buckley-Golder I. (eds.) (1998) Diamond films: recent developments, *MRS Bull.*, **23**(9), p. 16.
21. Baidakova M., and Vul' A. Ya. (2007) New prospects and frontiers of nanodiamond clusters, *J. Phys. D: Appl. Phys.*, **40**, pp. 6300–6311.
22. Galisteo-López J. F., Ibisate M., Sapienza R., Froufe-Pérez L. S., Blanco A., and López C. (2011) Self-assembled photonic structures, *Adv. Mater.*, **23**(1), pp. 30–69.
23. Colvin V. L. (2001) From opals to optics: colloidal photonic crystals, *MRS Bull.*, **26**(8), pp. 637–642.
24. Krauss T. F., and De La Rue R. M. (1999) Photonic crystals in the optical regime—past, present and future, *Prog. Quantum Electron.*, **23**, pp. 51–96.
25. Kurdyukov D. A., Feoktistov N. A., Nashchekin A. V., Zadiranov Yu. M., Aleksenskii A. E., Vul' A. Ya., and Golubev V. G. (2012) Ordered porous diamond films fabricated by colloidal crystal templating, *Nanotechnology*, **23**, pp. 01560(1–8).
26. Gibson N., Fitzgerald Z., Luo T.-J., Shenderova O., Grichko V., Bondar V., Puzur A., and Brenner D. (2007) *NSTI Nanotech: Technical Proceedings*, Vol. 2, "Bio materials and tissues" (CRC Press), p. 838.
27. Nesterenko P. N., Fedyanina O. N., Volgin Y. V., and Jones P. (2007) Ion chromatographic investigation of the ion-exchange properties of microdisperse sintered nanodiamonds, *J. Chromatogr. A*, **1155**, pp. 2–7.

28. Nesterenko P. N., and Fedyanina O. N. (2010) Properties of microdispersed sintered nanodiamonds as a stationary phase for normal-phase high performance liquid chromatography, *J. Chromatogr. A*, **1217**, pp. 498–505.
29. Purtov K. V., Puzyr' A. P., and Bondar' V. S. Nanodiamond sorbents: new carriers for column chromatography of proteins, *Dokl. Biochem. Biophys.*, **419**, pp. 72–74.
30. Bian L. Y., Wang Y. H., Zang J. B., Yu J. K., and Huang H. (2010) Electrodeposition of Pt nanoparticles on undoped nanodiamond powder for methanol oxidation electrocatalysts, *J. Electroanal. Chem.*, **644**, pp. 85–88.
31. Vershinin N. N., Efimov O. N., Bakaev V. A., Aleksenskii A. E., Baidakova M. V., Sitnikova A. A., and Vul' A. Ya. (2011) Detonation nanodiamonds as catalyst supports. Fullerenes, nanotubes and carbon nanostructures, *Fullerenes Nanotubes Carbon Nanostruct.*, **19**, pp. 63–68.
32. Turova O. V., Starodubtseva E. V., Vinogradov M. G., Sokolov V. I., Abramova N. V., Vul' A. Ya., and Alexenskiy A. E. (2011) Palladium supported on detonation nanodiamonds a highly effective catalyst of the C=C and C≡C bond hydrogenation, *Catal. Commun.*, **12**, pp. 577–579.
33. Shi X., Jiang X., Lu L., Yang X., and Wang X. (2008) Structure and catalytic activity of nanodiamond/Cu nanocomposites, *Mater. Lett.*, **62**, pp. 1238–1241.
34. Su D., Maksimova N. I., Mestl G., Kuznetsov V. L., Keller V., Schlo R., Keller N. (2007) Oxidative dehydrogenation of ethylbenzene to styrene over ultra-dispersed diamond and onion-like carbon, *Carbon*, **45**, pp. 2145–2151.
35. Davidson J. L., and Kang W. P. Applying CVD diamond and particulate nanodiamond, NATO Science Series, Series II: Mathematics, Physics and Chemistry, Vol. **122**, pp. 357–372.
36. Wang X. W., and Xu X. (1999) Thermal conductivity of nanoparticle-fluid mixtures, *J. Thermophys. Heat Transfer*, **13**(4), pp. 474–480.
37. Panich A. M., Altman A., Shames A. I., Osipov V. Yu., Aleksenskiy A. E., and Vul' A. Ya. (2011) Proton magnetic resonance study of diamond nanoparticles decorated by transition metal ions, *J. Phys. D: Appl. Phys.*, **44**, pp. 125303–125307.
38. Wang Y., Jaiswal M., Lin M., Saha S., Ozyilmaz B., and Loh K. P. (2012) Electronic properties of nanodiamond decorated grapheme, *ACS Nano*, **6**(2), pp. 1018–1025.

39. Eidelman E. D., and Vul' A. Ya. (2007) The strong thermoelectric effect in nanocarbon generated by the ballistic phonon drag of electrons, *J. Phys.: Condens. Matter*, **19**, pp. 266210–266221.

40. Mochalin V. N., Shenderova O., Ho D., and Gogotsi Y. (2012) The properties and applications of nanodiamonds, *Nat. Nanotechnol.*, **7**, pp. 11–23.

41. Vlasov I. I., Shenderova O., Turner S., Lebedev O. I., Basov A. A., Sildos I., Rahn M., Shiryaev A. A., and Tendeloo G. V. (2010) Nitrogen and luminescent nitrogen-vacancy defects in detonation nanodiamond, *Small*, **6**(5), pp. 687–694.

42. Baranov P. G., Soltamova A. A., Tolmachev D. O., Romanov N. G., Babunts R. A., Shakhov F. M., Kidalov S. V., Vul' A. Ya., Mamin G. V., Orlinskii S. B., and Silkin N. I. (2011) Enormously high concentrations of fluorescent nitrogen-vacancy centers fabricated by sintering of detonation nanodiamonds, *Small*, **7**(11), pp. 533–1537.

43. Aharonovich I., Castelletto S., Simpson D. A., Stacey A., McCallum J., Greentree A. D., and Prawer S. (2009) Two-level ultrabright single photon emission from diamond nanocrystals, *Nano Lett.*, **9**(9), pp. 3191–3195.

44. Maze R., Stanwix P. L., Hodges J. S., Hong S., Taylor J. M., Cappellaro P., Jiang L., Gurudev Dutt M. V., Togan E., Zibrov A. S., Yacoby A., Walsworth R. L., Lukin M. D. (2008) Nanoscale magnetic sensing with an individual electronic spin in diamond, *Nature*, **455**, pp. 644–647.

45. Kulha P., Kromka A., Babchenko O., Vanecek M., Husak M., Williams O. A., and Haenen K. (2010) Nanocrystalline diamond piezoresistive sensor, *Vacuum*, **84**, pp. 53–56.

Chapter 10

Biomedical Application of Nanodiamonds: Reality and Perspectives

Levon B. Piotrovskiy,[a] Dmitrii N. Nikolaev,[a] and Olga A. Shenderova[b]

[a] *Institute of Experimental Medicine of the North-West Branch of the Russian Academy of Medical Sciences, 12 Akad. Pavlov str., St. Petersburg 197376, Russia*
[b] *International Technology Center, 8100 Brownleigh Dr., Raleigh, NC 27617, USA*
levon-piotrovsky@yandex.ru

10.1 Introduction

Diamonds have always been arousing considerable interest in people, and not just as precious gems. Legends about the intriguing therapeutic properties of diamonds can be traced back throughout the history of mankind. In ancient India, diamond was believed to reduce fewer; to help in healing inflammations, infections, vomiting, jaundice, and other liver diseases; to alleviate sclerosis, etc.; to produce a favorable impact on the course of treatment of various chronic diseases; to relieve fatigue and irritation; and to improve metabolism, while the owner of a diamond would not suffer from stomach pains and deterioration of memory until old age. Some people traditionally believe that even the water into which this stone had been dipped acquires and retains medicinal properties.

Detonation Nanodiamonds: Science and Applications
Edited by Alexander Ya. Vul' and Olga A. Shenderova

ISBN 978-981-4411-27-1 (Hardcover), 978-981-4411-28-8 (eBook)
www.panstanford.com

Presently, we have been witnessing renewed interest in diamond-like molecules from chemists and biologists. For confirmation, we refer to the publication entitled "Diamonds Are a Chemist's Best Friend" in such a respectable journal as *Angewandte Chemie* [1]. Lower diamondoids, particularly the adamantan itself, have won a notable place in biology and medicine [2–5]. The surface of diamondoids is regular and terminated by covalently bonded hydrogen atoms, which confers on their molecules clearly pronounced lipophilic (hydrophobic) properties. Therefore, the lower and higher diamondoids can be classified as organic substances. At the same time, due to close similarities in the core structure of the diamondoids (especially high diamondoids) and nanodiamonds, chemical properties of these two materials should be similar [6].

The emergence in the 1980s of methods based on detonation synthesis and processing of ultradisperse diamonds (UDDs) had initiated work on the application of this material in many areas of science and technology, including biology and, eventually, medicine. After the prefix "nano" had established its rightful place in the scientific language, the UDDs were renamed to become detonation nanodiamonds (DNDs)[a] to reach a still higher respect among researchers [7]. Anyway, the flood of publications dealing with their possible applications in biology and medicine, including nanoscale medicine [8], is growing from year to year. It appears pertinent to cite here a few of the recently published reviews [9, 10–14] and, particularly, the recent extensive review [15] and, especially, a monograph [16].

The potential of application of nanodiamonds in biology and medicine derives fully from their unique properties, more specifically:

- highly developed surface area: up to 15% of all atoms in DND particles 4 nm in diameter reside on the surface [15], and their total specific surface area may reach as high as 300–400 m^2/g and more (see, e.g., Ref. [17]). This accounts for the high DND sorption capacity;

---

[a]The abbreviation DND only applies to nanodiamonds obtained by detonation synthesis; in all other cases the term "nanodiamonds" will be used, including a general term for diamond nanoparticles, without emphasizing a method of their synthesis.

- hydrophilic surface of DNDs:[a] this feature sets them apart from all the other carbon nanostructures (fullerenes, nanotubes, graphene, etc.), which are hydrophobic by nature. DND surface accommodates a variety of oxygen-containing functional groups (carboxyl, ether, ester, hydroxyl and keto groups, anhydrides, and lactone cycles) [7, 17–19];
- the possibility of carrying out redox reactions on DND-coated electrodes [12], a feature potentially attractive for DND biosensor applications;
- chemical inertness of the diamond core itself, which accounts for its metabolic stability as well [12, 14, 20–23];
- the possibility to produce fluorescent nanodiamonds [10, 24];
- the DND ability of internalizing into cells [25–27]; and
- last but not the least attractive feature of DNDs, their availability on a commercial scale [12].

10.2 Surface Properties of Detonation Nanodiamonds

The diamond surface may be considered as a revealing illustration of an interface between inorganic solid crystals and organic molecules [28]. While the DND surface, just as that of diamondoids, is primarily formed of sp^3-hybridized carbon atoms, the DND carbon atoms, in contrast to those of diamondoids in which all valence-free carbon bonds are terminated by hydrogen only, are bound to hydrogen atoms or oxygen-containing groups. In nanodiamond particles one demonstrated also the presence of unpaired electrons ($\sim$40 electrons per particle about 4.8 nm in size). It was suggested that rather than representing dangling bonds on the surface, they are located inside a particle, 0.4 nm to 1 nm distant from the surface [29].

Nanoscaled diamond particulates obtained by grinding of conventional diamonds in ball and bead mills differ strongly in properties from those produced by detonation synthesis [30, 31]. These differences stem mostly, if not predominantly, from the isolation

[a]Some authors report that the surface of natural diamonds is hydrophobic [17].

and purification treatment of the DNDs involving acid oxidation of impurities. This is what gives rise to the formation on the surface of oxygen-containing functional groups [6]. This is why the DND surface is hydrophilic and carries, as already mentioned, various polar oxygen-containing functional groups (carboxyl, lactone, keto, and carbonyl), with alkyl groups having been observed in some cases [6, 7, 14, 32, 33].

10.2.1 *Surface Modification*

The highly developed surface and the presence on it of various chemical functional groups, as well as the relative ease with which they can be modified, open up new vistas on the way both to realizing direct interaction of DND particles with biological molecules and structures and to devising purposeful modifications that would be capable of enhancing the interaction of DND particles with various objects and improving their selectivity [34, 35].

It was found possible to introduce on the surface of DND particles nitrogroups [36]; esters; $-(CH_2)_nCOOH$-type groups [37]; fluorine atoms; alkyl and amino groups; and amino acid residues [38]. Attachment of aromatic residues and fullerenes is also possible [39]. To learn of other possible chemical modifications of the DND surface, the reader is referred to Refs. [9, 40–43].

As in the case of other carbon nanostructures, in particular nanotubes [44], there are two major approaches to surface modification, namely, noncovalent bonding and covalent bonding.

The first approach can be best exemplified by physisorption of poly-L-lysine polypeptide (via ionic interaction of acidic groups on the surface with basic poly-L-lysine amino groups), which culminates in the formation of the so-called "aminated" nanodiamonds [45, 46].

The second approach can be illustrated by covalent linking of the same poly-L-lysine to carboxyl groups on the nanodiamond surface [26, 47].

One can envisage also a combination of these two approaches, in which polymers having hydroxyl or amino groups are adsorbed onto nanodiamond particles, after which various molecules, including peptides, can be covalently bonded to these precursors [48].

10.2.2 *Surface Modification by Proteins*

A process playing a major role in biological applications involves specific bonding of DNDs to proteins. Both covalent [49] and noncovalent [45] bonding strategies are presently enjoying broad use. For example, covalent attachment of the growth hormone to DNDs about 100 nm in size was demonstrated [49]. Biotin bonded covalently to a fluorescent marker (rhodamine, $\sim$1 molecule/nm^2) was attached to amino-functionalized DND. It was found that such compounds are capable of specifically bonding streptavidine [50]. Covalent bonding of bovine serum albumin via a comparatively long linker to the DND surface was described recently [51]. The advantages inherent in this approach to surface modification include also reduced tendency to DND aggregation, an aspect that is a problem in itself [41, 52]. It can be overcome, however, by forming a covalently bonded polyethylene glycol coating on nanodiamonds. This modification offers a possibility of producing particulates with a persistent dispersion in physiological solutions, with particles capable of internalizing into cells; consider, in particular, the HeLa line[a] [53].

Covalent attachment of biological macromolecules to the DND surface is a method revealing fairly obvious advantages; indeed, the product thus obtained exhibits higher stability mediated by strong covalent bonding rather than by electrostatic interaction [41]. Following this way peptide-functionalized DND was synthesized [41]. Noncovalent modification, that is, sorption of specific molecules on the surface, may possess in some cases benefits, particularly in the area of transportation of biomolecules where desorption after delivery is required [54].

These two methods can be readily employed in attaching to the DND surface both small molecules [54–57] and biological polymers, such as proteins and nucleic acids [58–60].

There are reports of producing DND functionalized by a variety of proteins, covalent ferment [61], cytochrome *c* [62], lysozyme [63, 64], and a noncovalent poly-L-lysine adduct [56]. Functionalization of DND with various antigens [65], growth hormone [49], a variety

[a] This is the cell line derived from the cancer tumor of a patient named Henrietta Lacks who died of cancer in 1951.

of neurotoxins [66], and deoxyribonucleic acid (DNA) [31, 67–69] was also described.

Preparation of peptide-modified DND [41] is a most promising research area from the standpoint of pure chemistry as well; indeed, one can use DNDs as platforms in solid-phase chemical synthesis. This method, developed as far back as 1963 by Merrifield for synthesis of peptides, remains a widespread technique in present-day chemistry. The method consists essentially of immobilizing the first link on an insoluble carrier (e.g., the first monomer of the future oligomer, which may be a peptide, oligosaccharide, or oligonucleotide), with subsequent bonding to it of the next links [70]. As an insoluble platform one most frequently uses as a resin chloromethylated copolymer of styrene with divynylbenzene, an organic resin (the Merrifield resin). Solid-phase synthesis is employed to advantage also in the preparation of low-molecular compounds [71], including high-throughput combinatorial techniques [72]. On the other hand, employing DNDs as a solid-phase support is capable of broadening substantially the range of compounds synthesized by solid-phase synthesis, because the large surface area and, hence, the comparatively high concentration of functional groups (up to 1 molecule/nm^2 [50]), the high chemical stability, high resistance to swelling, and the possibility of using a broad variety of solvents for synthesis and isolation of the final product characteristic of DNDs make the latter preferable to the organic polymers employed presently.

Surface modification by proteins opens up a possibility of preparation of DND-based biochips, because it is immobilization of proteins on a substrate that is truly a critical step in the development of biochips [73–75], while the chemical and physical stability of DNDs confers high strength to the support. One of the ways of modification consists essentially in attaching amino groups to the DND surface [76, 77]. This can be achieved by a two-stage method [62] by which DNDs are coated by poly-L-lysine to form a precursor, to which another protein, for instance, cytochrome *c*, is bonded (see also Ref. [50]). On the DND surface can also be immobilized antibodies to *Salmonella* and *Staphylococcus aureus*, which retain affinity toward the corresponding bacteria [78]. Attachment of complex protein molecules to modified nanodiamond offers also a possibility to employ electrophoresis for purification of proteins [50]. It should be pointed out, however, that due to their surface being charged, the DNDs themselves are mobile in electrophoresis, which permits one to use this method to attach them to a surface [79]. One can also immobilize DNDs on a monolithic platform [80]. The DNDs used in this study were particulates 1–2 μm in size. They were fluorinated by the technique described earlier [38], which imparted to them solubility in polar solvents and suppressed strongly the tendency towards agglomeration. Moreover, fluorine atoms attached to the DND surface can react with nucleophiles, thus making it possible to immobilize DND particles onto a 3-aminopropylsilane-modified glass platform. In the resultant structure intended for a sensor application, DNDs were attached to the substrate via a linker, rather than a diamond film grown, which makes the fabrication process simpler. Importantly, the fluorine atoms remaining on the DND surface can be exchanged for other biomolecules, a feature that may have application potential for fabrication of various sensors [80].

In summary, the following methods of DND surface modification should be emphasized [6, 15, 41]:

(1) Indirect methods:
- Noncovalent binding to polymers (including peptides). This modification is capable of changing radically the nature of the particle surface up to imparting to it hydrophobic properties.
- Using functional groups of the new coverage for subsequent covalent attachment to it of various ligands.

(2) Direct methods:
- Modification of the type of surface functional groups (reduction, substitution, etc.).
- Covalent attachment to functional groups on the surface of various molecules (both as linkers for subsequent binding of biologically significant groups and for direct immobilization of biologically active molecules [BAMs]).
- Use of intermediate functional groups for further functionalization. In some cases this may evolve into a multistage process. Indeed, preparation of biotin-modified

nanodiamond particles includes (a) homofunctionalization of the surface by reduction of all acidic functional groups to the hydroxyl functionalities; (b) silylation with (3-aminopropyl)trimethoxysilane, which gives rise to the formation of a linker with an end amino group; and (c) covalent attachment to the linker of a biologically significant ligand (biotin molecule) [81].

These methods combined with the inertness of the crystalline core itself and the highly developed surface open up new avenues for application of DNDs in technology.

10.3 Possible Areas of DND Applications

In full analogy with the application of any other material in biology and medicine, one can envisage at least three major ways in which DNDs can be used in practice:

1. As a material used without a direct contact with biological objects, that is, in the form of external sensors, chips, etc., in biological and medical analysis
2. In direct contact with biological objects, such as:
   (i) in vitro (as intracellular probes, markers, etc.); and
   (ii) in vivo (as therapeutic agents, vehicles for drug delivery, diagnostic probes, etc.).

Each of the above ways offers advantages and, more significantly, reveals inherent limitations, which we are going to consider now separately.

10.3.1 *Bioanalytical Applications of Nanodiamonds*

10.3.1.1 Chromatography

Progress in modern methods of purification requires development of less time-consuming and more efficient techniques for separation of samples having complex composition. In chromatography, this can be reached by properly increasing the pressure and temperature, as well as by using novel eluents, which quite frequently are chemically aggressive in these conditions. The main problem emerging in connection with these requirements consists essentially of a lack of appropriate sorbents. At the same time, a variety of carbon materials enjoy wide use in different variants of chromatographic separation [82]; they were shown to operate well with nanodiamond materials as well [83], both pure and coated [84].

The surface of "conventional" synthetic diamonds is hydrophobic [85] and carries only a limited number of functional groups. This suggests DNDs as a viable alternative to the former because their surface, as already mentioned, is hydrophilic and nonuniform. It is these features that underlie the possibility of using them as sorbents, including the areas associated with separation of molecules of biological significance [86–88]. Study of the surface of DND in terms of their potential use as sorbents for various types of chromatography was carried out in [89].

The surface of nanodiamonds is highly developed, with an area as large as 3 m^2/g to 22 m^2/g even for particles 0.5 μm in size, while for smaller particles it may reach 250–450 m^2/g [90]. This makes them applicable as sorbents practically in all kinds of chromatography, namely, liquid adsorption (both at normal and high pressure), ion exchange, gas, and affinity [59, 87, 91–94]. One can find readily also the description of the use of nanodiamonds in high-performance liquid chromatography (HPLC) and ion exchange chromatography [83, 95] and as a stationary phase in gas chromatography [96].

Literature contains also information on the application of DNDs to separation of proteins, in particular recombinant apoobelin and recombinant luciferase from *Escherichia coli* (*E. coli*) cells, and to isolation of small molecules (aflatoxin B1) [86, 97, 98]. Regrettably, the above papers, as some others as well, do not specify the characteristics of the DNDs used, in particular the particle size. Considering, however, that column chromatography was performed by the authors under standard conditions, under normal pressure, the sorbent grain size should be not less than 40 mesh (i.e., not less than 0.3 mm). The same authors point out, however, that the DNDs used by them were poor sorbents altogether because they destruct in the course of chromatographic measurements [99]. By contrast, there are reports suggesting not only a good adsorption capacity of a DND-based sorbent but its high stability as well [88].

One could guess, however, that the extra stability of the sorbent accrues from the surface modification in which the nanodiamond sorbent (particles 38–68 μm in size) is coated by polyallylamine cross-linked to confer to it additional strength.

The high efficiency of covalently modified DNDs in separation and analysis of protein mixtures was demonstrated, in particular in selective isolation of glycoproteins [59]. Thus, properly modified DNDs can be employed to advantage not only in specific sorption of biological macromolecules but also in separation of complex mixtures [see also Ref. 15].

10.3.1.2 Solid-phase sorbents

Apart from chromatography, DND can be employed in straightforward solid-phase extraction of a variety of substances as well. Indeed, a study of the ability of nanodiamond gel to adsorb various anions of the type of arsenide, selenite, bichromate, etc., revealed its high selectivity with respect to the tungstate [94]. Significantly, this study demonstrated also a high stability of DNDs to oxidizers; indeed, interaction with the sorbent did not reduce the bichromate ion $Cr_2O_7^{2-}$. DNDs about 100 nm in size were successfully employed to isolate from dilute solutions such proteins as cytochrome *c*, bovine serum albumin, myoglobin, angiotensin I, bradykinin, and some others, thus attesting to a high affinity of the sorbent to proteins and peptides [87, 91, 100].

A comparison of sorption of lysozyme onto DND and silica gel powders 100 nm in size showed the binding capacity of DNDs to be 10 times that of silica gel. This should be most probably assigned to the substantially larger surface area of porous nanodiamond aggregates because while each protein molecule occupies on silica gel 2 nm^2, on nanodiamond the sorbent area per molecule is 10 nm^2 [101]. Demonstration of noncovalent attachment of lysozyme to nanodiamonds [28] was complemented by showing that particles about 100 nm in size are capable of immobilizing on their surface about 3,700 protein molecules [45].

A study was performed of the sorption capacity of DNDs toward natural BAMs, in particular the BAM extracted from the above-ground parts of the medicinal plants *Chamerion angustifolium (L)*

Holub and *Agrimonia pilosa Ledeb* [102]. As model compounds for adsorption, rutin, quercetin, dihydroquercetin, and tanin were used. It was demonstrated that all the compounds contained in the model solutions in their basic form (flavonoids: aglycones quercetin and dihydroquercetin; glycoside rutin; and polyphenol tannin) are adsorbed completely by DND. Calculation of the DND sorption capacity for individual compounds yielded the following figures: 0.1 g DND powder sorbs 0.0013 g of rutin, 0.0013 g of quercetin, 0.00136 g of dihydroquercetin, and 0.00138 g of tanin. Investigation of DND sorption of a BAM extracted from the above-ground part of medicinal plants started with identifying the qualitative composition and quantitative content of the main BAM groups contained in the extracted material. It was revealed that DNDs sorb not only flavonoids and polyphenols but other BAM groups as well, more specifically hydroxycinnamic acids, coumarins, and chlorophylls. Nanodiamonds absorb flavonoids from the extracts treated by nanoscale diamonds to a larger extent (up to 96%) than other BAM groups.

10.3.1.3 Biochips and sensors

DNDs appear to be appropriate templates for various biosensors, both in the form of the nanoparticles themselves and of the films and coatings assembled from them. However, one can properly modify the surface not only of particles but also of nanocrystalline or ultrananocrystalline diamond (UNCD) films grown by chemical vapor deposition (CVD). For example, an interdigitated gold-nanodiamond electrode fabricated on the nanodiamond surface was used as a basis for development of a sensor for the antigen of the C-reactive protein; it could, in principle, be employed in the detection of risk marker proteins, a factor conducive to early diagnostics of various cardiovascular diseases [103, 104]. A plane electrode formed of a nanodiamond film coated onto a silicon substrate was capable of detecting dopamine [105]. The capacitive effect of the field in the electrolyte-diamond insulator-semiconductor architecture was used properly to develop a biosensor for penicillin detection [106]. A catalase ferment-functionalized nanodiamond film was demonstrated to be promising as a basis for a sensor for hydrogen peroxide [61]. Nanodiamond particles were shown to have inherent advantages when used in various biosensor systems [107, 108]. A detailed description of electrochemical sensors based on UNCD films can be found in the literature [109].

One can bind to a diamond film a single-chain DNA, followed by noncovalent adsorption of fluorescently labeled DNA [110]. DNA covalently immobilized on an amino-modified UNCD film is believed to form an ideal substrate for microelectronics-integrated biosensors [69]. Another illustration of UNCD-modified films is described in Ref. [111]. Development and use of DND-modified electrodes for electrochemical identification of a variety of molecules are discussed in Refs. [5, 112, 113].

DNDs are suggested for use in the field of mass spectrometry [46, 91], in particular in studies of noncovalent interactions [114].

UNCD films are a good substrate for growth cells of various lines (HeLa, PC12[a], and MC3T3,[b]) with growth on this modified substrate proceeding better than that on silicon or platinum surfaces. This makes nanodiamond coatings a proper candidate for fabrication of biosensors. The chemical passivity of such coatings suggests that the immune system will not react to it, so it may become a promising substrate for future development of in vivo systems in which biocompatibility and inertness of the material impose major restrictions [6, 109].

There are proposals aimed at the development of stable and high-efficiency platforms for surface hybridization based on nanodiamond films covalently modified by oligonucleotides, which would be promising substrates for fluorescent biosensors [69].

The literature offers descriptions of various DND-modified biochips. There is, for instance, a luminescent biochip based on the enzyme luciferase adsorbed on made of DND particles [115]. Microspikes made of nanocrystalline diamond were fabricated using standard microelectromechanical (MEMS) techniques increases the efficiency of mechanical cell lysis.[c] [116]. It is pointed out that in

[a]This is the cell line derived from a pheochromocytoma of the rat adrenal medulla.
[b]This is the cell culture obtained from *Mus musculus*.
[c]Lysis is dissolution and disruption of cells and their systems, including microorganisms, initiated by various agents, for instance, enzymes, bacteriolysins, bacteriophages, and antibiotics.

this particular case surface roughness caused by microspikes on diamond nanocrystals enhances the efficiency of cell lysis by a pure mechanical effect.

10.3.2 *Nanodiamonds in vitro*

10.3.2.1 Nanodiamonds as fluorescent biomarkers

One of the important areas of nanomedicine at present is to develop methods of introducing different fluorophores[a] into the living systems. There are many organic molecules containing fluorophores being widely used for a long time, which, however, have common drawbacks: their fluorescence does not persist for long and it exhibits photoblinking. It may be also added that the wavelength of their emission is not always suitable. Indeed, the most convenient for biological studies are fluorophores that absorb radiation with wavelengths above 500 nm and emit at wavelengths in excess of 600 nm [117, 118]. That is why we are witnessing presently active interest in the development, for cellular fluorescent labeling, of the so-called quantum dots, which are stable, strong luminophors characterized by a high quantum yield and a narrow fluorescence band. They are superior to organic luminophors in a higher absorption coefficient and higher emission and photostability [119]. Quantum dots have, however, a serious drawback. Indeed, they contain such compounds as cadmium selenide, cadmium sulfide, indium arsenide, and indium phosphide, and the intrinsic toxicity of the ions of these metals makes quantum dots a potential hazard for in vivo use [120, 121]. Although many methods of their synthesis, masking of the metallic core, and surface functionalization have been explored by now [122], they remain toxic and reveal side effects, particularly in prolonged exposures [119]. Another adverse property of quantum dots is the intermittency of their fluorescence [121] or photoblinking, which interferes markedly with continuous tracking of a nanoparticle in three dimensions [123]. This stresses the need for developing nanomaterials of new

[a]A fluorophore is a fragment of a molecule that confers to the molecule fluorescent properties.

types that would be free of these limitations. This problem can certainly be solved using DNDs as such materials [120, 124].

Nanodiamonds with optical centers activated by "nitrogen vacancy" (NV) complexes are capable of luminescing in the red and green regions, which stimulates interest in their use as optical biomarkers [120, 125]. Fabrication of fluorescent nanodiamonds from natural or statically synthesized diamonds has recently been announced. This subject is covered in a review [10] and Chapter 5 of the present monograph. Nanodiamonds containing NV centers emit fluorescence at levels strong enough to allow intracellular imaging of a single particle in the cell, with no photobleaching observed in for 300 seconds, whereas for conventional organic dyes this time is about 10 seconds [10, 27]. Nanodiamonds as fluorescent probes are superior, first, in that they emit in the red region (575–750 nm) [58], thus permitting operation outside the region where inherent cellular fluorescence is observed (300–500 nm) [126]. And second, these emitters are photostable [24, 127, 128].

Fluorescence of nanodiamonds can be as strong as, or even stronger than, that of quantum dots. Just as in the case of carbon nanotubes [129], surfactants enhance markedly the stability of nanodiamond gels [128]. However, because emission is initiated by defects in the nanodiamond lattice, surface modification of the particles does in no way affect signal generation [126, 127, 130]. This is major attractive feature of nanodiamonds because modification can be used to advantage in targeting a marker to specific objects in the cell [131]. For instance, when the corresponding antibodies have to be attached to fluorescing nanodiamonds, this vehicle is delivered purposefully to intracellular structures, in particular actin filaments or mitochondria [132].

Covalent attachment of poly-L-lysine to the carboxyl-modified nanodiamonds surface, followed by noncovalent adsorption of the DNA onto the precursor, was shown not to suppress the ability of nanodiamonds to fluorescence [26]. Such modified DNDs can penetrate HeLa cells and are detected in the cytoplasm. Other researchers also point out that surface modification does not affect in any way the fluorescent property of nanodiamonds [133]. It is reported that while a graphite shell (on particles 5 nm in size) does quench the luminescence, it is restored following oxidative treatment of the particles [134].

Nanodiamond-mediated fluorophores were found instrumental in imaging capture of a nanodiamonds particle 100 nm in size by human kidney cells and tracking diffusion of 35 nm particulates in the cytoplasm of HeLa cells [26]. Use of nanodiamonds less than 50 nm in size as photoluminescent robes was also reported [135].

An emerging area of application is based essentially on using DNDs functionalized by organic molecules to produce fluorophores. To cite just one example, iron nanoparticles contained in a graphene layer on the DND surface were used as fluorescent magnetic nanoprobes, the source of the emission light being in this case the fluorescein *o*-methacrylate [136].

In another case, a DND-functionalized fluorescent biomarker was produced by covalent linking of an organic dye [137]. The DND surface is oxidized with Fenton reagent (H_2O_2 + $FeSO_4$, H_2SO_4), thus increasing strongly the concentration of hydroxyl groups on the surface (as evidenced by Fourier transform infrared spectroscopy), after which tionin, a fluorescent dye agent, is attached to the DND surface via a silyl linker. This compound features an attractive ability of penetrating through cellular membranes and reaching nuclei without incurring any damage to the cell (as shown on HeLa cells).

The concentration of the organic fluorophore may be high enough. In the case of covalent attachment of rhodamin-labeled biotin to amino-modified DND, its concentration on the surface was found to reach as high as approximately 1 molecule/nm^2 [50].

One can envisage also development of stable and highly efficient platforms for surface hybridization, which would be based on CVD nanodiamond films covalently modified by oligonucleotides, to be employed to advantage in developing fluorescent biosensors [69].

The remarkable properties of nanodiamonds revealed in spectroscopic studies of cells are not limited to their application as fluorescent markers only. Quite recently, the possibility of using nanodiamonds as specific nanosized magnetometers has been convincingly demonstrated at a conceptual level [138, 139], which holds forth considerable promise in the field of detection of electronic and nuclear spins in an object under study with nanoscale resolution and, hence, of determination of the structure of three-dimensional objects, including proteins. This possibility stems from the NV center with a spin forming in a nanodiamond. The state of the spin of a single NV center defines the optical properties (fluorescence) of the DND and, hence, can be visualized optically in the single-photon emission mode. Such a magnetometer can be readily fabricated by attaching a nanodiamond particle to a scanning nanoscale probe. A similar magnetometer can be used for precision positioning of nanodiamond particles within the cell. Indeed, the magnetic field generated by a scanning magnetic probe was shown to permit determination of the distance between spins of two nanodiamond particles residing about 100 nm from one another, with an accuracy about 20 nm, and the positions of the nanodiamonds themselves can be defined with nanometer-scale accuracy, which exceeds by far the optical resolution of a fluorescence microscope, which is about 200 nm [139]. The principles underlying optical detection of magnetic resonance in the presently emerging field of nanoscale-resolution magnetometry are discussed at depth in Chapter 5.

10.3.2.2 Other ways of nanodiamond visualization in living structures

Fluorescence is not the only phenomenon that can be applied to visualization of various cellular structures. DNDs (modified by surface carboxyl groups, size 100 nm) have been successfully used to locate a growth hormone receptor on a single tumor cell. For this purpose, molecule of the growth hormone was covalently attached to a DND particle. This modification of the hormone did not interfere with its binding with the corresponding receptor on the surface of A549 human lung epithelial cells, and the interaction itself in physiological conditions was imaged by confocal Raman spectroscopy [49].

The easily identifiable bands in Raman spectra and the fluorescence property, combined with the ability demonstrated by DNDs of internalizing into cells, offer us a possibility to use them as biomarkers for a variety of proteins within cells. It was shown, in particular, that the DND-lysozyme complex attaches to *E. coli* bacteria, which suggests that the protein retains its functions [63].

Magnetic resonance tomography (MRT) is one more method that can contribute to imaging nanodiamond particles. It is based essentially on attaching a chelating group capable of binding the Gd^{3+} atom to carboxyl-functionalized nanodiamond particles (2–8 nm in size) [140].

10.3.2.3 Nanodiamonds and cells

As mentioned in the preceding section, DNDs are capable of internalizing into cells and interacting with organelles (see A.1). This licenses their use, besides other nanoparticles (carbon nanotubes, quantum dots, gold and silicon nanoscale particulates, etc.) [141–143], for incorporation of biologically active agents into cells [136]. Viewed in this context, nanoparticles present considerable interest also because minute probes comparable in size with biomolecules and their aggregates may be appropriately employed to study the fine structure and inherent functions of cells ("to spy on cellular machinery" [144]).

Standard eukaryotic cells[a] are capable of taking up particles <1 μm in size. Significantly, these particles are internalized via endocytosis which, in its turn, may or may not involve receptors (see A.2). As shown in the particular example of gold nanoparticles, internalization, as well as the corresponding cellular response, including protein expression and cell death, depends radically on the nanoparticle size [145]. It is known also that gold nanoparticles can penetrate directly through the mammalian surface cell membranes [146, 147]. Short single-wall carbon nanotubes (SWNTs) are also assumed to uptake not via endocytosis but rather by direct diffusion through the membrane [148]. Some reports, however, suggest that short SWNTs enter cells via clathrin-mediated endocytosis [149].

Cultured nanodiamond particulates about 100 nm and 400 nm in size are also internalized spontaneously by cells [26, 27, 49]. The 48 nm DNDs are taken up by HeLa cells in less than three hours' time [128]. Nanodiamond particles modified by an arginine-containing

---

[a]Eukaryote is a superregnum of organisms whose cells contain a nucleus.

peptide (starting size about 50 nm, after modification 120–130 nm) penetrate into the Chinese hamster ovary (CHO) cells and bind to actin [48].

On the surface of many tumor cells, there are receptors for folate, but usually they do not exist on the surface of normal cells [150]. Therefore, if on the surface of nanoparticles there are residues of folic acid they can internalize into tumor cells via receptor-mediated endocytosis [151]. Normal cells can also take up, however, nanodiamonds not modified by folate residues [136], which evidences the existence of an alternate pathway for endocytosis that is not mediated by receptors. Significantly, this pathway appears to be rather a general mechanism of internalizing nanoparticles by various cells. This is the case, for instance, of the internalization of carboxylated DND by human embryo kidney cells 293T [27, 126], RAW macrophage cells [54], human epithelial lung cells A549, and normal human fibroblasts HFL-1 [152]; uptake of silicon-coated iron oxide nanoparticles by human mesenchymal stem cells (hMSCs) [153]; of gold nanoparticles by C166 (endothelial murine embryo cells) [154], etc.

It is known that conjugation, that is, linking of nanoparticles to various biological ligands, enhances in many cases both the stability of the latter and their ability to penetrate cells. This is why such conjugates were proposed to be employed to ferry drugs, genes, and vaccines to specific target cells [54, 149, 155–158]. Penetration into the cell nucleus requires usually application of a short peptide called the nuclear location signal (NLS; of the type Pro-Lys-Lys-Lys-Arg-Lys-Val) and providing of attachment to the cell nucleus membrane [159].

All of the above relate in full measure to DNDs. Even fairly large modified DND particulates about 100 nm in size, of the type of agglomerates with iron nanoparticles, whose surface is additionally functionalized by polyacrylic acid, can be internalized into HeLa cells [136]. It appears that they were imaged in the cytoplasm but not in the nucleus. Other studies suggest that carboxyl-modified DNDs can be internalized into the cytoplasm of RAW cells, while not penetrating into the nucleus [31].

Investigation of photoluminescent DNDs revealed that particles less than 50 nm in size become internalized into HeLa cells via clathrin-mediated endocytosis [25].

After the penetration, large particles or aggregates were seen to reside in vesicles, while smaller ones appeared to diffuse freely in the cytosol. Studies in which particles about 35 nm in size were used showed, however, that most of the nanodiamond particles are found to be inside endosomes (see A.2), which makes their free diffusion through the cytoplasm impossible [24, 26, 128, 160].

While HeLa cells are known to feature frequently non-receptor-mediated endocytosis [161], 100 nm DNDs conjugated with transferrin were shown to enter HeLa cells by the receptor-mediated mechanism [133].

The 100 nm DNDs are internalized into A549 and 3T3-L1 (cells embryonic fibroblasts) via macropinocytosis and clathrin-mediated endocytosis. For 10 days, cell growth appeared unaffected, with the DNDs themselves becoming distributed after division between the two daughter cells approximately evenly. After a few generations, however, single DND clusters became visible in the cells. These particles did not affect in any way gene and protein expression and the cells' ability to start the new cycle. Thus, DND particles confined to cells do not interfere with their division and differentiation [162]. At the same time, absorption of large modified DND particles (about 100 nm in size) does not pass without consequences for the HeLa cells; indeed, their viability decreases compared with that in the case of uncoated DNDs [136].

Internalization of 140 nm nanodiamonds into cells was also studied [60]. The choice of such particles was based on two considerations; first, there is the size of nonviral gene vectors used commonly in drug delivery experiments [163, 156], and second, these particles produce fluorescence of a very high intensity, which permits their detection in HeLa cells if incubated with nanoparticles in a culture with a concentration of 1 μg/mL. The 140 nm particles penetrate cells when incubated in a standard Dulbecco's modified eagle's medium (DMEM) culture with a concentration of about 100 particles/cell in the absence of fetal bovine serum (FBS). The uptake half-time was about 1.5 hours. After addition of 10% FBS into the culture, however, the number of internalized particles dropped by 90%; this is suggested to be partially caused by particle agglomeration. The above authors corroborate the earlier conclusions that internalization of nonmodified acidic DNDs occurs by the clathrin-dependent mechanism (see footnote †) having features in common with endocytosis-mediated uptake of SWNTs noncovalently bound with protein or DNA molecules [149]. If, however, one attaches to these particles covalently via polyethylene linker folic acid molecules, whose receptors are present on human tumor cells [164], these particles become internalized into the cells by the receptor-mediated mechanism and accumulate in the perinuclear region. And in this case FBS promotes penetration of particles into cells, which indicates a change in the mechanism of their entry—it becomes calveolin dependent [164, 165].

The mechanism responsible for DND internalization remains an open problem requiring additional investigation [136], although it is quite possible that there is no common mechanism that would account for internalization by different cells of DNDs differing in size. As in the study of the toxicology of nanoparticles the use the method of "in each case" is proposed [166], and for DND quite similar approach is applicable, not only in this case, but generally in the study of their biological properties, because various samples can be too differ from each other and do not always reliably standardized.

10.3.2.4 Nanodiamonds in drug delivery

It is the relative ease with which nanodiamonds become internalized into cells that accounts for the flurry of activities in the development of methods that would provide delivery of drugs into cells. Another significant factor is that a compound immobilized on DNDs (either covalently or noncovalently), be it even as complex as a protein or an enzyme, was demonstrated in several studies to retain practically completely its native activity. More frequently than not, however, one prefers noncovalent modification, which simplifies considerably release of the agent at the target.

Incidentally, a transferrin molecule conjugated with 100 nm DND particles preserves its ability toward interaction with a receptor

[133]. With antibodies attached to fluorescent DNDs, this complex can be transported purposefully to intracellular structures, for instance, actinic filaments or mitochondria, thus demonstrating that proteins retain their activity too [132]. A lysozyme molecule bound to DND does not fail in its biological functions [63]. A growth hormone molecule bound covalently to a 100 nm DND likewise retained its ability to attach to the corresponding receptor on the A549 cell surface [49]. Covalent binding of immunoglobuline, bovine serum albumin, and antibodies to DND surface produces adducts, which were also found to retain the biological activity of attached proteins [167].

A variety of molecules, not only large but relatively small as well, for instance, alpha-bungarotoxin, a specific blocker of the alpha-7-nicotin acetylcholine receptor, retain their blocking properties after noncovalent binding to DND particles [66]. This is all the more important that all steric requirements providing specific interaction of small molecules with a biological target be notably more stringent than those in the protein case. It was also shown that attachment of alpha-bungarotoxin to DND particles of about 140 nm in size increases the particle size more than twice, approximately to 300 nm.

The 2–8 nm DNDs form by ionic bonding adducts with the antitumoral drug doxorubicin. This process is reversible and can be mediated by properly varying the concentration of Cl^- ions. Functionalization with DND favors internalization of doxorubicin into murine macrophages and tumor cells, and as believed by the authors, the composition can be used for delivery of chemotherapeutic drugs into colorectal carcinoma cells [54]. The noncovalent DND–insulin complex was demonstrated to be able to release insulin predominantly at neutral pH of the medium [58].

The low solubility of active compounds that interferes with their delivery to an organ or a targeted tissue in concentrations high enough for therapeutic purposes is one of acute problems in the chemistry of drugs. This situation spawned the proposal of using DND clusters for transferring low-solubility compounds to an aqueous medium [55]. It was demonstrated that compounds such as the antitumoral drugs purvalanol A and 4-hydroxytamoxifen

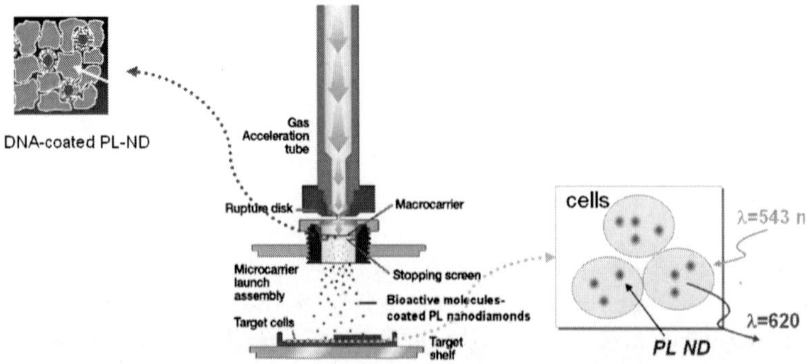

Figure 10.1 Schematic of ballistic delivery of DNA-coated photoluminescent nanodiamonds into cells (picture courtesy of V. Grichko and S. Hens, ITC, USA).

and dexametason, an anti-inflammatory drug, can become adsorbed onto DND clusters and, last but not least, preserve their therapeutic potential in these conditions [55].

Nanodiamonds can be employed also as platforms for ballistic delivery to tissues of biologically active complexes and therapeutic drugs [168]. Ballistic delivery compares favorably with other methods of transport of various DND-immobilized biomolecules, because it does not require for its realization specific agents to stimulate uptake of these complexes by a cell. The Bio-Rad PDS-1000/He instrument used in the experiments [168] generates a shock wave, which accelerates nano- or microparticles toward the target (cells or tissue) (Fig. 10.1). The shock wave parameters are chosen such that the drug platforms penetrate the cell membranes with the lowest possible damage. The ballistic methods of drug delivery are employed widely in biology (genetic transformation of plants) and medicine (vaccination) [169]. Gold or tungsten particles serve here usually as peculiar bullets. Being nontoxic and having a large sorption capacity, DNDs are particularly attractive as vehicles for ballistic delivery. The possibility of *E. coli* transformation by various fluorescent and ampicillin-resistant plasmids incorporated into cells in the form of DND complexes and of yeast transformation by DNA vectors, as well as delivery and incorporation of fluorescent DND-poly-L–fluorescein isothiocyanate

(FITC) complexes into onion cells was experimentally demonstrated [12, 168]. "Shooting" bananas with DNDs coated by ethephon- or diphenylcyclopropenone (hormones mediating acceleration or slowing down of ripening) brought the expected results, namely, the treated bananas ripened faster or remained unripe compared with reference samples, depending on the treatment they were subjected to.

Multilayer DND nanofilms capable of absorbing therapeutically active molecules were also proposed [56]. A nanofilm formed of DND and a collagen absorbs dexamethasone [170]. Such compositions can be used to advantage in delivery systems for topical application and anti-inflammatory coatings.

A similar approach proves useful in developing various complexes of therapeutics embedded into a matrix that are capable of releasing slowly the drug molecules into the surrounding medium. For instance, adsorption of doxorubicin coated on DND particles (2–8 nm in size) into a polymer film produces a depo-form system for doxorubicin, which is eluted slowly (during up to one month) out of this film [57]. More than one application of DNDs and DND films to immobilization of therapeutic molecules has been reported thus far [171–174].

Intracellular gene transport was described, in which polycation-coated nanodiamond particles were employed to incorporate the gene of green fluorescent protein (GFP) into HeLa cells [137]. The reader can be referred to another study of modified nanodiamond particles as platforms for targeted gene delivery [175].

The above illustrations exemplify the rich potential inherent in the use of DNDs for internalization of various biologically active complexes (low-molecular compounds, peptides, and nucleic acids) into cells.

10.3.3 *Nanodiamonds in vivo*

Studies of the biological and pharmacokinetic parameters of DNDs in vivo are extremely scarce. Of those that are known, few if any offer an encouraging picture. It was shown, for instance, that 50 nm DND particles introduced by intravenous injection accumulate in the liver, spleen, and lungs. Significantly, in 0.5 hours, 60% of all

DNDs can be found in the liver and 8% in lungs. These figures remain about the same on the 28th day of observation [176]. Thus, one does not practically witness any DND distribution throughout the human organism. More than that, as this should be expected, there is no DND excretion either. This study casts grave doubts on the possibility itself of designing DND-based therapeutics delivery methods applicable to in vivo systems. Incidentally, some studies suggest that uptake of large modified DND particles (about 100 nm in size) does affect, for instance, HeLa cells [136].

Among the very few observations of in vivo nanodiamond introduction, a study of the pulmonary toxicity of nanodiamonds deserves mention [177]. The particles used were 4 nm in size, and they did not produce any noticeable negative effect on pulmonary tissues. Significantly, the number of DND particles in the alveoles decreased with time, while the macrophages were burdened with nanodiamonds.

There is, however, no need altogether in introducing DND particles directly into the organism, that is, into the circulatory system. Indeed, it was demonstrated [178] that neuron stem cells deposited on a DND substrate proliferate nicely, and such prepared cells can be used for transplantation in the central nervous system. If the surface is hydrophobic, which can be reached by replacing all functional groups with hydrogen atoms, cell attachment on the surface breaks down and, accordingly, cell growth stops, but cells immobilized on a nanocrystalline diamond surface, even on a nanoscale powder, do proliferate [179]. These properties suggest strongly the use of nanocrystalline diamond-like materials as supports for implants [180–184].

DNDs enjoy application also in cosmetics both as a binder for the biological material acting as filler and for improvement of the mechanical properties of skin [185].

10.4 Biocompatibility of Nanodiamonds

All studies involving the use of DNDs in laboratory practice have a common attractive feature that there is no need of speculation about their effect on biological systems. The situation changes dramatically when one attempts to use DNDs both in vitro and, particularly, in vivo.

The potential of any material intended for use for biological purposes and in medicine is based primarily on two parameters, namely, its *toxicity* and *biocompatibility*.

Consider first, briefly, the meaning behind both these concepts.

Toxicity (from the Greek *toxicon*, or poison) is the ability of a substance to disrupt the physiological functions of the organism (or separately of some of its organs and organelles), which entails dysfunction of the organism, down to its death.

Biocompatibility is a concept more difficult to define. One of the definitions refers to it as the ability of a material to perform with an appropriate host response in a specific application [186]. This definition appears to be rather more suitable to implants. There are certainly other definitions as well. Since, however, the mechanical and tribological properties of DND-coated surfaces make them promising candidates for orthopedic implants [183, 184], we may accept the first definition of their biocompatibility as fully appropriate too.

Despite the presence of some misunderstanding and an obvious lack of certainty in formulations, it appears fairly obvious that biocompatibility is a parameter defining the ability of a material to exist inside a living organism without disrupting the surrounding cells (and tissues) or modifying noticeably their functions [187]. Said otherwise, the biocompatibility of a material means that for a *living organism*, it is not toxic, it does not damage tissues or cells, and/or it does not initiate a harmful response of the organism as a whole (e.g., of its immune reactions). Therefore, when one discusses designing DND-based implants (see above), this will be possible only if they are biocompatible.

We should like to emphasize that the term "biocompatibility" is applicable only to the organism as a whole, an essential point underlying the first definition [186]. This is why before broad-range and, most essentially, prolonged studies on whole organisms (in vivo) are performed, there will hardly be any sense in discussing seriously the biocompatibility of any material, and, in particular, of DNDs.

As obvious appears also the inherent relation between toxicity and biocompatibility; indeed, while a toxic substance (or material) just cannot be biocompatible, a nontoxic substance (or material) is not necessarily biocompatible [188, 189]. This means that one has to discriminate between biocompatibility and simply a lack of toxicity [190].

This relation can be readily illustrated with the particular case of carbon nanotubes. It was shown that disruption of the integrity of nanotubes and formation of defects in their structure increase their toxic impact [191, 192]. Other studies [193, 194] suggest that noncovalent functionalization lowers noticeably the cytotoxicity of carbon nanotubes and increases their biocompatibility. For instance, streptavidin- or biotin-conjugated nanotubes can internalize into cells [195–197]. On the basis of the above, the statement "While nanotubes are cytotoxic, they are biocompatible" sounds strange [198].

10.4.1 *Toxicity*

The only reference to diamonds being toxic we are aware of is a mention that grinded diamond is the strongest poison in existence. The latter opinion is ascribed to the wisdom of Eastern sages. It persisted, however, only until the first experimental check, when Al-Biruni (973–1048) pounded to powder several large diamonds and, upon stirring them up with water, gave the mixture to a dog to drink; the dog allegedly drank the "poison" but remained alive. This is how the ancient legend was debunked. However, "pounded" diamond, just as a number of other water-insoluble crystalline substances, should better not be taken in, because there is a danger, a purely physical one, that diamond will act as an abrasive and irritate the intestinal tissues. As Benvenuto Cellini (1500–1571) writes in his memoirs, "An idea came to their mind to add pounded diamond to food; it is known to contain no poison of any kind, but, being indescribably hard, it retains the sharpest edges one could imagine and behaves not as any other stone would do; namely, when pounded, all the sharpness of the stones degrades to dust, leaving the stones round, as it were; and only diamond retains its sharp edges; and, on entering the stomach, together with food, and being turned in all directions together with the food that performs digestion, this diamond sticks to the gristle of the stomach and of the intestines, and, as the newly coming food pushes it all the time forward, this stuck diamond will pierce them in a short time; and this will cause people die, whereas stones of any other kind or glass mixed with food will not be strong enough to stick and, thus, will leave with the food without any harmful effect inflicted."

The last example demonstrates, however, that a reason for the negative effect is not related to the toxicity: actions of a knife and a poison are rather distinct. Thus, if one would consume particles without sharp edges, there will be no harm. Thus, throughout the history of mankind, people not only have not expected diamonds to cause any health hazard problems but also, to the contrary, have believed them to be endowed with an intriguing therapeutic potential (except for criminal cases, but this is beyond the scope of the problem we are discussing here).

We also have to keep in mind that insoluble substances are practically nontoxic. An appropriate example is furnished by barium sulfate, practically insoluble in water, which is customarily used as a contrasting agent in studies of the gastrointestinal tract, despite the fact that Ba^{2+} ions are highly toxic.

The second reason for diamonds not being expected to cause any particular trouble, specifically cancerogenic hazard, is related to the fact that in diamond carbon atoms are sp^3 hybridized, which makes their reactivity fairly low. By contrast, in the vast majority of other carbon nanostructures (nanotubes, graphene, fullerenes, etc.) the carbon atom is in sp^2 hybridization, which is characteristic, in particular, of aromatic hydrocarbons. Its ability to addition reactions is relatively high, which is often the cause of manifestation of carcinogenicity. For example, the high carcinogenicity of benzo[a]pyrene is due to oxidation products of the rings, in particular 7,8-dihydroxy-9,10-epoxybenzo[a]pyrene, formed, unfortunately, from the pyrene itself by microsomal oxidation systems, which are responsible for oxidative degradation of xenobiotics in humans and animals [199]. Naturally, not all other aromatic hydrocarbons and carbon nanostructures need be necessarily cancerogenic.

Carboxylated DNDs (in concentrations of up to 1,000 μg/mL) are nontoxic for cells and are able to internalize into intracellular space

[63] (see also Ref. [200]). At the same time, as already mentioned, uptake of large modified DNDs (about 100 nm in size) does not come off without any damage to HeLa cells; indeed, cells viability decreases compared with the case of adding pure DNDs to the cell culture [136].

However, at the same time, there are many in vitro studies that did not reveal any toxic effect of DNDs [26, 27, 126, 152, 200].

10.4.2 *Metabolism and Excretion*

Problems involved in the metabolism and excretion of a substance are related closely, albeit not directly, with its toxic effect on the organism. It should be stressed that the literature is practically lacking publications dealing with the investigation of the associated problems.

In the particular case of the response of the metabolizing systems of an organism to diamond-like structures, there is relevant data on the metabolism of lower diamondoids. Many researchers point out that diamondoids are poorly oxidized by the liver microsomal oxidation system of xenobiotics [201]. In particular, the major pathway of metabolism of adamantan derivatives is via formation of hydroxylated derivatives [201, 202].

That's why diamandoids where used to study the structure of cytochrome P450, the main enzyme of metabolism of xenobiotics. Because the probes to be used for this purpose should be hydrophobic, diamondoids turned out to be an appropriate model [203–206]. The use of these compounds allowed us to determine the size of the hydrophobic cavity of this enzyme, whose diameter is equal to be 5.9 Å and depth 7.9 Å [207] (see also Ref. [208]). The sterically most appropriate substrate for cytochrome P450 is diadamantan because the adamantan itself is too small and triadamantan, conversely, too large to penetrate the cavity. If triamantan cannot fit, then higher diamandoids are not able to fit either, and consequently, DNA also cannot be metabolized by this system (unless other oxidases will be involved) [209, 210].

A critical assessment of the relevant literature suggests that the biological properties of the lower diamondoids is based primarily on the high lipophility of their molecules [211], which fundamentally distinguishes them from hydrophilic DND particles. Regrettably, studies of DND metabolism and excretion are practically lacking.

10.5 Emerging Problems and Application Potential

DND particulates may be considered as nanoscale molecules of hydrocarbons [212]; their size and characteristics permit them to be regarded as an illustration of a transition from chemistry and physics [213] to materials science [214–216]. There is, however, a significant difference between diamondoids and DNDs; indeed, whereas the former are chemical compounds with a clearly established structure, DNDs are essentially mixtures. Significantly, they are mixtures not only of isomers but of different compositions as well. In other words, while diamondoids are chemical compounds, DNDs are actually materials [217].[a] It is this difference, more specifically the inhomogeneity of nanodiamond compositions, that accounts for the difficulties encountered in their use in biology and medicine.

Therefore, one of the most hard-to-tackle problems that will emerge in its most acute form as soon as a possibility of DND application in medicine to a human organism becomes real is standardization.[b] The above examples of DND biological activity were obtained on particles of different sizes. At the same time it is well known that the biological impact of nanoparticles depends directly on their size [218]. We are not aware, however, of any systematic study of this problem. While the papers cited above seems to suggest, on the one hand, a conclusion of the absence of such a dependence, on the other hand, however, it would be wrong to draw it, because the biological models assumed in different studies are different.

Another problem emerging in in vivo studies of DNDs is their pronounced tendency to aggregation [52, 221, 222]. Although DNA is introduced into the blood as hydrogels, with particles size vary

[a]This reminds one of an old joke: Chemists use poor methods to experiment with good substances, and physicists good methods but with bad substances.

[b]Time-of-flight [219], as well as thermosorption mass spectrometry [220], may become appropriate methods for analysis and standardization of nanodiamonds.

within the range of 4–5 nm to several tens of nanometers, changing the environment, in particular, the ionic strength, may lead (and most likely will lead) to aggregation, that is, an increase in the size of the particles. Moreover, this process will be uncontrollable, and to predict the final size of the particles is quite difficult. Even if they do not reach sizes of several microns (a particle of this size could simply clog capillaries and cause thrombosis), increasing the size to several tens or hundreds of nanometers will affect their ability to penetrate cells and, most importantly, to influence cell viability.

For example, agglomerates of nanodiamonds (size of about 1 micron) can be trapped by endothelial cells [223]. But their size is comparable to the size of cells, and this is probably why some cells demonstrate changes in the structure and size. Such particles also affect cell proliferation. This conclusion is, in principle, important because it shows that even though the particles of this size can be captured by cells, it is not good. But it's not so bad if it is nontoxic nanodiamonds. Although in this case questions arise about the fate of the entire cell population, cell morphology, and their ability to proliferate, as well as their viability as a whole [223].

Despite a certain pessimism in a few preceding paragraphs, the emerging overall conclusion, based on our present-day knowledge, would imply that DND has a certain application potential in biology.

First, one of the currently most promising areas of research is application of fluorescent DNDs as an alternative to quantum dots.

Second, DNDs can be utilized to advantage in any in vitro system, in particular for delivery of various substances to cells, which appears a sound basis for development of specific diagnostics systems.

Third, the possibility of surface modification of DNDs suggests immediately their use for development of relevant sorbents, including combinations that would be highly selective to specific classes of substances or even to individual substances.

Fourth, DND-assisted targeted delivery of genes opens up a particularly appealing possibility for genetic engineering, in particular for development of new bacteria producers.

And fifth, although the problems associated with the metabolism or excretion of DNDs from the organism are still not solved, DNDs can be used in compositions intended for topical application.

Appendix

A.1 Cell

The cell is an elementary building block providing a basis for the structure and vital functions of all living organisms. It supports its own metabolism and is capable of independent existence, self-reproduction, and development. All living organisms consist of cells. A cell is made up of organelles ("minute organisms") and constant cellular structures, each accounting for a specific function, which are located in the cytoplasm, the inner medium of a living cell (Fig. A.1).

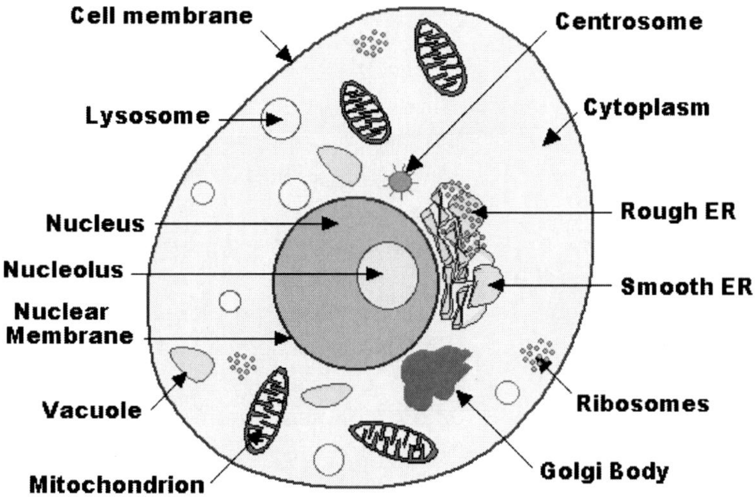

Figure A.1 Cross section of an animal cell.

A.1.1 *Organelles and Their Functions*

Organelle	Function
Cell membrane	The thin layer of protein and fat that surrounds a cell. The cell membrane is semipermeable, allowing some substances to pass into the cell and blocking others.
Centrosome (also called the "microtubule-organizing center") and centrioles	A small body located near the nucleus, it has a dense center and radiating tubules. The centrosome is where microtubules are made. During cell division (mitosis), the centrosome divides and the two parts move to opposite sides of the dividing cell. The centriole is the dense center of the centrosome and is involved in the mitotic spindle.
Cytoplasm	The jellylike material outside the cell nucleus in which the organelles are located.
Golgi body (also called Golgi apparatus or Golgi complex)	A flattened, layered, sac-like organelle that looks like a stack of pancakes and is located near the nucleus. It produces membranes that surround lysosomes. The Golgi body packages proteins and carbohydrates into membrane-bound vesicles for "export" from the cell.
Lysosome (also called cell vesicle)	Round organelles surrounded by a membrane and containing digestive enzymes. This is where digestion of cell nutrients takes place.
Mitochondrion	Spherical to rod-shaped organelles with a double membrane. The inner membrane is infolded many times, forming a series of projections (called cristae). The mitochondrion converts the energy stored in glucose into adenosine triphosphate (ATP) for the cell.
Nuclear membrane	The membrane that surrounds the nucleus.

Organelle	Function
Nucleolus	An organelle within the nucleus, it is where ribosomal ribonucleic acid (rRNA) is produced. Some cells have more than one nucleolus.
Nucleus	A spherical body containing many organelles, including the nucleolus. The nucleus controls many of the functions of the cell (by controlling protein synthesis) and contains DNA (in chromosomes). The nucleus is surrounded by the nuclear membrane.
Ribosome	Small organelles composed of RNA-rich cytoplasmic granules that are sites of protein synthesis.
Rough endoplasmic reticulum (rough ER)	A vast system of interconnected, membranous, infolded, and convoluted sacks that are located in the cell's cytoplasm (the ER is continuous with the outer nuclear membrane). Rough ER is covered with ribosomes that give it a rough appearance. Rough ER transports materials through the cell and produces proteins in sacks called cisternae (which are sent to the Golgi body or inserted into the cell membrane).
Smooth endoplasmic reticulum (smooth ER)	A vast system of interconnected, membranous, infolded, and convoluted tubes that are located in the cell's cytoplasm. The space within the ER is called the ER lumen. Smooth ER transports materials through the cell. It contains enzymes and produces and digests lipids (fats) and membrane proteins. Smooth ER buds off from rough ER, moving the newly made proteins and lipids to the Golgi body, lysosomes, and membranes.
Vacuole	A fluid-filled, membrane-surrounded cavity inside a cell. Vacuoles fill with food being digested and waste material that is on its way out of the cell.

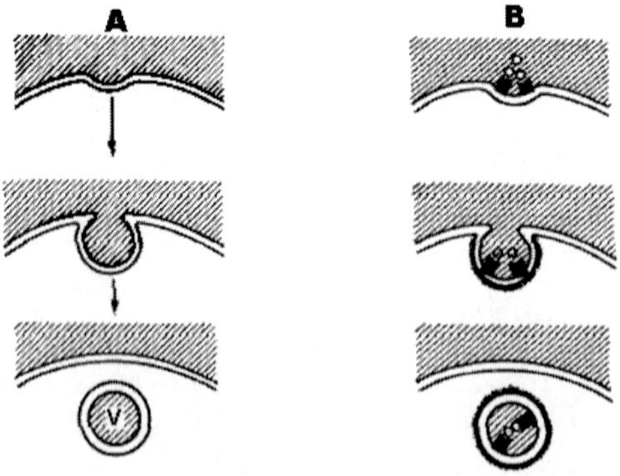

Figure A.2 Two types of endocytosis are generally recognized: (A) Nondirectional pinocytosis: Uptake by the cell of the liquid phase out of the surrounding medium, which contains soluble substances, including large molecules (proteins, polysaccharides, etc.). In pinocytosis, small sacks—vesicules (V)—detach from the membrane into the cell interior. (B) Receptor-mediated endocytosis: An active specific process in which the cell membrane bulges inward into the cell to form small bordered cavities (BCs), sometimes called caveoles. The inward-directed side of the bordered cavity bears a set of adaptive proteins, among them clathrin, adaptin, etc. Polymerization of clathrin immobilized to the cell membrane deforms it heavily, a process terminating with the formation of a hemispherical, kidney-shaped structure bulging inward. When this structure closes, vesicules form. Following the formation of a vesicule (V) and its detachment from the membrane, the clathrin shell dissociates rapidly, the released clathrin becoming free to participate again in endocytosis. This process lasts, as a rule, for about one minute. The transport vesicule merges with intracellular endosomes and delivers internal material into the cell. The particles and macromolecules attached to these specific receptors on the cell surface penetrate the cell with a velocity much higher than that of the particles entering cells by pinocytosis. This case can be illustrated, for instance, by penetration of cholesterol into the cell, a mechanism involving clathrin-containing caveoles. Receptor-mediated endocytosis is initiated to provide fast and controllable uptake of a ligand (e.g., of a nanodiamond) by the cell. *Abbreviation*: BC, bordered cavity.

Unmodified nanodiamonds can penetrate through the cell membrane, move in the cytoplasm, as endosomes and lysosomes, and reach the nuclei, being located in the perinuclear region. However, surface modification by various biologically important molecules can also lead to targeted delivery to specific cellular organelles, particularly mitochondria (see text).

A.2 Endocytosis

Endocytosis (Fig. A.2) is internalization of external material by the cell, which is mediated by formation of membrane vesicules (endosomes). Endocytosis provides hydrophilic material for the cell necessary for support of its vital functions, which otherwise cannot penetrate the lipid bilayer of the cell membrane.

References

1. Schwertfeger H., Fokin A. A., and Schreiner P. R. (2008) Diamonds are a chemist's best friend: diamondoid chemistry beyond adamantine, *Angew. Chem., Int. Ed.*, **47**, pp. 1022–1036.
2. De Clercq E. (1979) New trends in antiviral chemotherapy, *Arch. Int. Physiol. Biochim.*, **87**(2), pp. 353–395.
3. Kavirajan H. (2009) Memantine: a comprehensive review of safety and efficacy, *Expert Opin. Drug Saf.*, **8**(1), pp. 89–109.
4. Morozov I. S., Petrov V. I., and Sergeeva S. A. (2001) *Pharmacology of Adamantanes* (Volgograd) (in Russian).
5. Parkes D. (1974) Amantadine, *Adv. Drug Res.*, **8**, pp. 11–81.
6. Krueger A. (2008) Diamond nanoparticles: jewels for chemistry and physics, *Adv. Mater.*, **20**, pp. 2445–2449.
7. Krueger A. (2008) New carbon materials: biological applications of functionalized nanodiamond materials, *Chem. Eur. J.*, **14**, pp. 1382–1390.
8. Xing Y., and Dai L. (2009) Nanodiamonds for nanomedicine, *Nanomedicine*, **4**(2), pp. 207–218.
9. Baidakova M., and Vul' A. Ya. (2007) New prospects and frontiers of nanodiamond clusters, *J. Phys. D: Appl. Phys.*, **40**, pp. 6300–6311.
10. Barnard A. S. (2009) Diamond standard in diagnostics: nanodiamond biolabels make their mark, *Analyst*, **134**, pp. 1751–1764.
11. Dolmatov V. Y. (2006) *Ultrananocrystalline Diamond: Synthesis, Properties, and Applications*, eds. Shenderova, O., and Gruen, D., "Application of

detonation nanodiamond" (William Andrew, Norwich, New York, USA), pp. 513–517.

12. Holt K. B. (2007) Diamond at the nanoscale: applications of diamond nanoparticles from cellular biomarkers to quantum computing, *Phil. Trans. R. Soc. A*, **365**, pp. 2845–2861.
13. Shenderova O. A., and Hens S. A. C. (2010) *Nanodiamonds: Applications in Biology and Nanoscale Medicine*, ed. Ho, D., "Detonation nanodiamond particles processing, modification and bioapplications" (Springer), pp. 79–116.
14. Shenderova O. A., Zhirnov V. V., and Brenner D. W. (2002) Carbon nanostructures, *Crit. Rev. Solid State Mater. Sci.*, **27**(3), pp. 227–356.
15. Schrand A. M., Hens S. A. C., and Shenderova O. A. (2009) Nanodiamond particles: properties and perspectives for bioapplication, *Crit. Rev. Solid State Mater. Sci.*, **34**, pp. 18–74.
16. Ho, D. (ed.) (2010) *Nanodiamonds: Applications in Biology and Nanoscale Medicine* (Springer).
17. Jee A.-Y., and Lee M. (2009) Surface functionalization and physico-chemical characterization of diamond nanoparticles, *Curr. Appl. Phys.*, **9**, pp. e144–e147.
18. Mochalin V., Osswald S., and Gogotsi Y. (2009) Contribution of functional groups to the Raman spectrum of nanodiamond powders, *Chem. Mater.*, **21**, pp. 273–279.
19. Osswald S., Yushin G., Mochalin V., Kucheyev S. O., and Gogotsi Y. (2006) Control of sp2/sp3 carbon ratio and surface chemistry of nanodiamonds powders by selective oxidation in air, *J. Am. Chem. Soc.*, **128** (35), pp. 11635–11642.
20. Cui F. Z., and Li D. J. (2000) A review of investigations on biocompatibility of diamond-like carbon and carbon nitride films, *Surf. Coatings Technol.*, **131**(1–3), pp. 481–487.
21. Dolmatov V. Yu., and Kostrova L. N. (2000) Detonation-synthesized nanodiamonds and the develop a new generation of medicines, *J. Superhard Mater. A*, **3**, p. 79.
22. Grill A. (2003). Diamond-like carbon coatings as biocompatible materials—an overview, *Diam. Relat. Mater.*, **12**(2), pp. 166–170.
23. Sheeja D., Tay B. K., and Nung L. N. (2004) Feasibility of diamond-like carbon coatings for orthopedic applications, *Diam. Relat. Mater.*, **13**(1), pp. 184–190.

24. Chang Y.-R., Lee H.-Y., Chen K., Chang C.-C., Tsai D.-S., Fu C.-C., Lim T.-S., Tzeng Y.-K., Fang C.-Y., Han C.-C., Chang H.-C., and Fann W. (2008) Mass production and dynamic imaging of fluorescent nanodiamonds, *Nat. Nanotechnol.*, **3**, pp. 284–288.
25. Faklaris O., Joshi V., Irinopoulou T., Tauc P., Sennour M., Girard H., Gesset C., Arnault J.-C., Thorel A., Boudou J.-P., Curmi P. A., and Treussart F. (2009) Photoluminescent diamond nanoparticles for cell labeling: study of the uptake mechanism in mammalian cells, *ACS Nano*, **3**(12), pp. 3955–3962.
26. Fu C.-C., Lee H.-Y., Chen K., Lim T.-S., Wu H.-Y., Lin P.-K., Wei P.-K., Tsao P.-H., Chang H.-C., and Fann W. (2007) Characterization and application of single fluorescent nanodiamonds as cellular biomarkers, *Proc. Natl. Acad. Sci. U. S. A.*, **104**(3), pp. 727–732.
27. Yu S. J., Kang M. W., Chang H. C., Chen K. M., and Yu Y. C. (2005) Bright fluorescent nanodiamonds: no photobleaching and low cytotoxicity, *J. Am. Chem. Soc.*, **127**, pp. 17604–17605.
28. Toshihiro A., Kiyoharu N., and Mika G. (2001) Diamond surface: an interface between inorganic crystals and organic molecules, *Oyo Butsuri*, **70**, pp. 1447–1451 (in Japanese).
29. Fang X. W., Mao J. D., Levin E. M., and Schmidt-Rohr K. (2009) Nonaromatic core-shell structure of nanodiamond from solid-state NMR spectroscopy, *J. Am. Chem. Soc.*, **131**(4), pp. 1426–1435.
30. Kuznetsov V. L., Aleksandrov M. N., Zagoruiko I. V., Chuvilin A. L., Moroz E. M., Kolomiichuk V. N., Likholobov V. A., Brylyakov P. M., and Sakovitch G. V. (1991) Study of ultradispersed diamond powders obtained using explosion energy, *Carbon*, **29**(4–5), pp. 665–668.
31. Ushizawa K., Sato Y., Mitsumori T., Machinami T., Ueda T., and Ando T. (2002) Covalent immobilization of DNA on diamond and its verification by diffuse reflectance infrared spectroscopy, *Chem. Phys. Lett.*, **351**, pp. 105–108.
32. Dolmatov V. Y. (2011) *Detonation Nanodiamonds* (NPO "Professional," SPb) 535 pp. (in Russian).
33. Tapraeva F. M., Pushkin A. N., Kulakova I. I., Rudenko A. P., Elagin A. A., and Tikhomirov C. V. (1990) Functional coating of the diamond surface under various conditions of modification by gas, *Zh. Fiz. Khim.*, **64**, pp. 2445–2451.
34. Spitsyn B. V., Davidson J. L., Gradoboev M. N., Galushko T. B., Serebryakova N. V., Karpukhina T. A., Kulakova I. I., and Melnik N. N.

(2006) Inroad to modification of detonation nanodiamond, *Diam. Relat. Mater.*, **15**, pp. 296–299.

35. Tsubota T., Tanii S., Ida S., Nagata M., and Matsumoto Y. (2004) Chemical modification of diamond surface with various carboxylic acids by radical reaction in liquid phase, *Diam. Relat. Mater.*, **13**, pp. 1093–1097.
36. Tsubota T., Ohno T., Yoshida H., and Kusakabe K. (2006) Introduction of molecules containing a NO_2 group on diamond surface by using radical reaction in liquid phase, *Diam. Relat. Mater.*, **15**, pp. 668–672.
37. Ida S., Tsubota T., Tanii S., Nagata M., and Matsumoto Y. (2003) Chemical modification of the diamond surface using benzoyl peroxide and dicarboxylic acids, *Langmuir*, **19**, pp. 9693–9698.
38. Liu Y., Gu Z., Margrave J. L., and Khabashesku V. N. (2004) Functionalization of nanoscale diamond powder: fluoro-, alkyl-, amino-, and amino acid-nanodiamond derivatives, *Chem. Mater.*, **16**, pp. 3924–3930.
39. Zhong Y. L., Loh K. P., Midya A., and Chen Z.-K. (2008) Suzuki coupling of aryl organics on diamond, *Chem. Mater.*, **20** (9), pp. 3137–3144.
40. Cheng C.-L., and Chen C. (2009) Organic functionalization of ultradispersed nanodiamond: synthesis and applications, *J. Mat. Chem.*, **19**, pp. 8432–8441.
41. Krueger A., Liang Y. J., Jarre G., and Stegk J. (2006) Surface functionalisation of detonation diamond suitable for biological applications, *J. Mat. Chem.*, **16**, pp. 2322–2328.
42. Yeap W. S., and Loh K. P. (2010) *Nanodiamonds: Applications in Biology and Nanoscale Medicine*, ed. Ho, D., "Functionalization of nanodiamond for specific biorecognition" (Springer), pp. 117–125.
43. Zheng W.-W., Hsieh Y.-H., Chiu Y.-C., Cai S.-J., Cheng C.-L., and Chen C. (2009) Organic functionalization of ultradispersed nanodiamond: synthesis and applications, *J. Mat. Chem.*, **19**, pp. 8432–8441.
44. Piotrovsky L. B., and Kiselev O. I. (2006) *Fullerenes in Biology* (St. Petersburg) 335 (in Russian).
45. Chung P.-H., Perevedentseva E., Tu J.-S., Chang C. C., and Cheng C.-L. (2006) Spectroscopic study of bio-functionalized nanodiamonds, *Diam. Relat. Mater.*, **15**, pp. 622–625.
46. Kong X. L., Huang L. C. L., Hsu C. M., Chen W. H., Han C. C., and Chang H. C. (2005) High-affinity capture of proteins by diamond nanoparticles for mass spectrometric analysis, *Anal Chem.*, **77**, pp. 259–265.

47. Chang C.-K., Wu C.-C., Wang Y.-S., and Chang H.-C. (2008) Selective extraction and enrichment of multiphosphorylated peptides using polyarginine-coated diamond nanoparticles, *Anal. Chem.*, **80**, pp. 3791–3797.

48. Vial S., Mansuy C., Sagan S., Irinopoulou T., Burlina F., Boudou J. P., Chassaing G., and Lavielle S. (2008) Peptide-grafted nanodiamonds: preparation, cytotoxicity and uptake in cells, *ChemBioChem*, **9**, pp. 2113–2119.

49. Cheng C. Y., Perevedentseva E., Tu J. S., Chung P. H., Cheng C. L., Liu K. K., Chao J. I., Chen P. H., and Chang C. C. (2007) Direct and *in vitro* observation of growth hormone receptor molecules in A549 human lung epithelial cells by nanodiamond labeling, *Appl. Phys. Lett.*, **90**, 163903.

50. Hens S. C., Cunningham G., Tyler T., Moseenkov S., Kuznetsov V., and Shenderova O. (2008) Nanodiamond bioconjugate probes and their collection by electrophoresis, *Diam. Relat. Mater.*, **17**(11), pp. 1858–1866.

51. Dahoumane S. A., Nguyen M. N., Thorel A., Boudou J.-P., Chehimi M. M., and Mangeney C. (2009) Protein-functionalized hairy diamond nanoparticles, *Langmuir*, **25**(17), pp. 9633–9638.

52. Kruger A., Ozawa M., Kataoka F., Fujino T., Suzuki Y., Aleksenskii A. E., Vul' A. Ya., and Osawa E. (2005) Unusually tight aggregation in detonation nanodiamond: identification and disintegration, *Carbon*, **43**, pp. 1722–1730.

53. Takimoto T., Chano T., Shimizu S., Okabe H., Ito M., Morita M., Kimura T., Inubushi T., and Komatsu N. (2010) Preparation of fluorescent diamond nanoparticles stably dispersed under a physiological environment through multistep organic transformations, *Chem. Mater.*, **22**, pp. 3462–3471.

54. Huang H., Pierstorff E., Osawa E., and Ho D. (2007) Active nanodiamond hydrogels for chemotherapeutic delivery, *Nano Lett.*, **7**(11), pp. 3305–3314.

55. Chen M., Pierstorff E. D., Lam R., Li S. Y., Huang H., Osawa E., and Ho D. (2009) Nanodiamond-mediated delivery of water-insoluble therapeutics, *ACS Nano*, **3**(7), pp. 2016–2022.

56. Huang H., Pierstorff E., Osawa E., and Ho D. (2008) Protein-mediated assembly of nanodiamond hydrogels into a biocompatible and biofunctional multilayer nanofilm, *ACS Nano*, **2**, pp. 203–212.

57. Lam R., Chen M., Pierstorff E., Huang H., Osawa E., and Ho D. (2008) Nanodiamond-embedded microfilm devices for localized chemotherapeutic elution, *ACS Nano*, **2** (10), pp. 2095–2102.
58. Shimkunas R. A., Robinson E., Lam R., Lu S., Xu X., Zhang X., Huang H., Osawa E., and Ho D. (2009) Nanodiamond-insulin complexes as ph-dependent protein delivery vehicles, *Biomaterials*, **30**, pp. 5720–5728.
59. Yeap W. S., Tan Y. Y., and Loh K. P. (2008) Using detonation nanodiamond for the specific capture of glycoproteins, *Anal. Chem.*, **80**(12), pp. 4659–4665.
60. Zhang B., Li Y., Fang C.-Y., Chang C.-C., Chen C.-S., Chen Y.-Y., and Chang H.-C. (2009) Receptor-mediated cellular uptake of folate-conjugated fluorescent nanodiamonds: a combined ensemble and single-particle study, *Small*, **5**(23), pp. 2716–2721.
61. Hartl A., Schmich E., Garrido J. A., Hernando J., Catharino S. C. R., Walter S., Feulner P., Kromka A., Steinmuller D., and Stutzmann M. (2004). Protein-modified nanocrystalline diamond thin films for biosensor application, *Nat. Mater.*, **3**, pp. 736–742.
62. Huang, L. C., and Chang, H. C. (2004). Adsorption and immobilization of cytochrome c on nanodiamonds, *Langmuir*, **20**, pp. 5879–5884.
63. Chao J.-I., Perevedentseva E., Chung P.-H., Liu K.-K., Cheng C.-Y., Chang C.-C., and Cheng C.-L. (2007) Nanometer-sized diamond particle as a probe for biolabeling, *Biophysical J.*, **93**, pp. 2199–2208.
64. Nguyen T.-T.-B., Chang H.-C., and Wu V. W.-K. (2007) Adsorption and hydrolytic activity of lysozyme on diamond nanocrystallites, *Diam. Relat. Mater.*, **16**, pp. 872–876.
65. Kossovsky N., Gelman A., Hnatyszyn H. J., Rajguru S., Garrell R. L., Torbati S., Freitas S. S., and Chow G. M. (1995) Surface-modified diamond nanoparticles as antigen delivery vehicles, *Bioconjugate Chem.*, **6**, pp. 507–511.
66. Liu K.-K., Chen M.-F., Chen P.-Y., Lee T. J. F., Cheng C.-L., Chang C.-C., Ho Y.-P., and Chao J.-I. (2008) Alpha-bungarotoxin binding to target cell in a developing visual system by carboxylated nanodiamond, *Nanotechnology*, **19**, 205102.
67. Kruger A. (2006) Hard and soft: biofunctionalized diamond, *Angew. Chem., Int. Ed.*, **45**, pp. 6426–6427.
68. Takahashi K., Tanga M., Takai O., and Okamura H. (2003) DNA preservation using diamond chips, *Diam. Relat. Mater.*, **12**, pp. 572–576.

69. Yang W., Auciello O., Butler J. E., Cai W., Carlisle J. A., Gerbi J. E., Gruen D. M., Knickerbocker T., Lasseter T. L., Russell Jr. J. N., Smith L. M., and Hamers R. J. (2002) DNA-modified nanocrystalline diamond thin-films as stable, biologically active substrates, *Nat. Mater.*, **1**, pp. 253–257.

70. Merrifield R. B. (1963) Solid phase peptide synthesis. I. The synthesis of a tetrapeptide, *J. Am. Chem. Soc.*, **85**(14), pp. 2149–2154.

71. Ermolat'ev D. S., and Babaev E. V. (2005) Solid-phase synthesis of *N*-(pyrimidin-2-yl)amino acid amides, *ARKIVOC*, part. (iv), pp. 172–178.

72. Lazo J. S., and Wipf P. (2000) Combinatorial chemistry and contemporary pharmacology, *J. Pharmacol. Exp. Ther.*, **293**(3), pp. 705–709.

73. Drummond T. G., Hill M. G., and Barton J. K. (2003) Electrochemical DNA sensors, *Nat. Biotechnol.*, **21**, pp. 1192–1199.

74. Lee Y.-S., and Mrksich M. (2002) Protein chips: from concept to practice, *Trends Biotechnol.*, **20**, pp. s14–s18.

75. Zhu H., and Snyder M. (2003) Protein chip technology, *Curr. Opin. Chem. Biol.*, **7**, pp. 55–63.

76. Miller J. B., and Brown D. W. (1996) Photochemical modification of diamond surfaces, *Langmuir*, **12**, pp. 5809–5817.

77. Miller J. B. (1999) Amines and thiols on diamond surfaces, *Surf. Sci.*, **439**, pp. 21–33.

78. Huang T. S., Tzeng Y., Liu Y. K., Chen Y. C., Walker K. R., Guntupalli R., and Liu C. (2004) Immobilization of antibodies and bacterial binding on nanodiamond and carbon nanotubes for biosensor applications, *Diam. Relat. Mater.*, **13**, pp. 1098–1102.

79. Affoune A. M., Prasad B. L. V., Sato H., and Enoki T. (2001) Electrophoretic deposition of nanosized diamond particles, *Langmuir*, **17**(2), pp. 547–551.

80. Liu Y., Khabashesku V. N., and Halas N. J. (2005) Fluorinated nanodiamond as a wet chemistry precursor for diamond coatings covalently bonded to glass surface, *J. Am. Chem. Soc.*, **127**, pp. 3712–3713.

81. Krueger A., Stegk J., Liang Y., Lu L., and Jarre G. (2008) Biotinylated nanodiamond: simple and efficient functionalization of detonation diamond, *Langmuir*, **24** (8), pp. 4200–4204.

82. Cserhati T. (2009) Carbon-based sorbents in chromatography. New achievements., *Biomed. Chromatogr.*, **23**, pp. 111–118.

83. Nesterenko P. N., and Haddad P. R. (2010) Diamond-related materials as potential new media in separation science, *Anal. Bioanal. Chem.*, **396**, pp. 205–211.

84. Saini G., Jensen D. S., Wiest L. A., Vail M. A., Dadson A., Lee M. L., Shutthanandan V., and Linford M. R. (2010) Core-shell diamond as a support for solid-phase extraction and high-performance liquid chromatography, *Anal. Chem.*, **82**, pp. 4448–4456.

85. Kutsay O., Loginova O., Gontar A., Perevertallo V., Zanevskyy O., Katrusha A., Ivakhnenko S., Gorokhov V., Stank S., Tkach V., and Novikov N. (2008) Surface properties of amorphous carbon films, *Diam. Relat. Mater.*, **17**, pp. 1689–1691.

86. Bondar', V. S., Puzyr', A. P., Purtov, K. V., Mogil'naya, O. A., Vidryakova, G. A., Tul'kova, N. A., Rodicheva, A. K., Medvedeva, S. I., Degermendzhi, A. G., and Gitel'zon, I. I. (2008) Detonation nanodiamond: creation of new materials and technologies for the isolation of proteins, *Russ. Nanotechnol.*, **3**(5–6), pp. 42–45 (in Russian).

87. Chen W. H., Lee S. C., Sabu S., Fang H. C., Chung S. C., Han C. C., and Chang H. C. (2006) Solid-phase extraction and elution on diamond (SPEED): a fast and general platform for proteome analysis with mass spectrometry, *Anal. Chem.*, **78**, pp. 4228–4234.

88. Saini G., Yang L., Lee M. L., Dadson A., Vail M. A., and Linford M. R. (2008) Amino-modified diamond as a durable stationary phase for solid-phase extraction, *Anal. Chem.*, **80**, pp. 6253–6259.

89. Spitsyn B. V., Denisov S. A., Skorik N. A., Chopurova A. G., Parkaeva S. A., Belyakova L. D., and Larionov O. G. (2010) The physical–chemical study of detonation nanodiamond application in adsorption and chromatography, *Diam. Relat. Mater.*, **19**(2-3), pp. 123–127.

90. Nesterenko P. N., and Fedyanina O. N. (2010) Properties of microdispersed sintered nanodiamonds as a stationary phase for normal-phase high performance liquid chromatography, *J. Chromat. A*, **1217**(4), pp. 498–505.

91. Kong X., Huang L. C. L., Liau S.-C. V., Han C.-C., and Chang H.-C. (2005) Polylysine-coated diamond nanocrystals for MALDI-TOF mass analysis of DNA oligonucleotides, *Anal. Chem.*, **77**, pp. 4273–4277.

92. Nesterenko P. N., Fedyanina O. N., and Volgin Y. V. (2007) Microdispersed sintered nanodiamonds as a new stationary phase for high performance liquid chromatography, *Analyst*, **132**, pp. 403–405.

93. Nesterenko P. N., Fedyanina O. N., Volgin Y. V., and Jones P. (2007) Ion chromatographic investigation of the ion-exchange properties of microdisperse sintered nanodiamonds, *J. Chromatogr. A*, **1155**(1), pp. 2–7.

94. Sakurai H., Ebihara N., Osawa E., Takahashi M., Fujinami M., and Oguma K. (2006) Adsorption characteristics of a nanodiamond for oxoacid anions and their application to the selective preconcentration of tungstate in water samples, *Anal. Sci.*, **22**, pp. 357–362.
95. Patel B. A., Rutt K. J., Padalko V. I., and Mikhalovsky S. V. (2002) Use of industrial diamonds in HPLC, *J. Superhard Mater.*, **6**, pp. 51–54.
96. Belyakova L. D., Kudinova A. N., Larionova A. O., Larionov O. G., and Spitsyn B. V. (2008) The surface properties of nanodiamonds by gas chromatography, *Russ. J. Phys. Chem.*, **82**, pp. 420–423.
97. Bondar' V. S., Pozdnyakova I. O., Puzyr' A. P. (2004) Applications of nanodiamonds for separation and purification of proteins, *Phys. Solid State*, **46**(4), pp. 758–760.
98. Puzyr A. P., Purtov K. V., Shenderova O. A., Luo M., Brenner D. W., and Bondar V. S. (2007) The adsorption of aflatoxin B1 by detonationsynthesis nanodiamonds, *Dokl. Biochem. Biophys.*, **417**, pp. 299–301.
99. Purtov K. V., Puzyr' A. P., and Bondar' V. S. (2008) Nanodiamond sorbents: new carriers for column chromatography of proteins, *Dokl. Biochem. Biophys.*, **419**(4), pp. 72–74.
100. Sabu S., Yang F. C., Wang Y. S., Chen W. H., Chou M. I., Chang H. C., and Han C. C. (2007) Peptide analysis: solid phase extraction-elution on diamond combined with atmospheric pressure matrix-assisted laser desorption/ionization–Fourier transform ion cyclotron resonance mass spectrometry, *Anal. Biochem.*, **367**, pp. 190–200.
101. Wu V. W. K. (2006) Adsorption reaction constants between nanosilica/nanodiamond and lysozyme molecule at pH=11.0, *Chem. Lett.*, **35**, pp. 1380–1381.
102. Valov R. I., Larionova I. S., Hanina M. G., Rodin A. P., and Hanina M. A. (2010) Sorption of natural biologically active substances on nanodiamonds, *Pharmaciya*, **6**, pp. 28–31 (in Russian).
103. Quershi A., Kallempudi Saravan S., Gurbuz Y., Howell M., Kang W. P., and Davidson J. L. (2009) A new nanocrystalline diamond-based biosensor for the detection of cardiovascular risk markers, *Procedia Chem.*, **1**(1), pp. 1079–1082.
104. Qureshi A., Gurbuz Y., Howell M., Kang W. P., and Davidson J. L. (2010) Nanocrystalline diamond film for biosensor applications, *Diam. Relat. Mater.*, **19**, pp. 457–461.

105. Raina S., Kang W. P., and Davidson J. L. (2009) Nitrogen incorporated nanodiamond film with "ridge" surface morphology for detection of bio-analyte, *Diam. Relat. Mater.*, **18**(2-3), pp. 574–577.

106. Poghossian A., Abouzar M. H., Razavi A., Bäcker M., Bijnens N., Williams O. A., Haenen K., Moritz W., Wagner P., and Schöning M. J. (2009) Nanocrystalline-diamond thin-films with high pH and penicillin sensitivity prepared on a capacitive Si-SiO_2 structure, *Electrochim. Acta*, **54**(25), pp. 6010–6014.

107. Chevallier E., Scorsonea E., and Bergonzo P. (2010) New sensitive coating based on modified diamond nanoparticles for chemical SAW sensors, *Sens. Actuators, B*, **154**, pp. 238–244.

108. Fortin E., Chane-Tune J., Delabouglise D., Bouvier P., Livache T., Mailley P., Marcus B., Mermoux M., Petit J.-P., Szunerits S., and Vieil E. (2005) Interfacing boron doped diamond and biology: an insight on its use for bioanalytical applications, *Electroanalysis*, **17**(5–6), pp. 517–526.

109. Bajaj P., Akin D., Gupta A., Sherman D., Shi B., Auciello O., and Bashir R. (2007) Ultrananocrystalline diamond film as an optimal cell interface for biomedical applications, *Biomed. Microdevices*, **9**, p. 787.

110. Rezek B., Shin D. C., Nakamura T., and Nebel C. E. (2006). Geometric properties of covalently bonded DNA on single-crystalline diamond, *J. Am. Chem. Soc.*, **128**, pp. 3884–3885.

111. Steenackers M., Lud S. Q., Niedermeier M., Bruno P., Gruen D. M., Feulner P., Stutzmann M., Garrido J. A., and Jordan R. (2007) Structured polymer grafts on diamond, *J. Am. Chem. Soc.*, **9**(50), pp. 15655–15661.

112. Chatterjee A., and Foord J. S. (2009) Biological applications of diamond electrodes: electrochemical studies of riboflavin, *Diam. Relat. Mater.*, **18**(5-8), pp. 899–903.

113. Gao Z., Carabelli V., Carbone E., Colombo E., Demaria F., Dipalo M., Gosso S., Manfredotti Ch., Pasquarelli A., Rossi S., Xu Y., Vittone E., and Kohn E. (2010) Transparent diamond microelectrodes for biochemical application. *Diam. Relat. Mater.*, **19**, pp. 1021–1026.

114. Kong X.-L. (2008) Nanodiamonds used as a platform for studying noncovalent interaction by MALDI-MS, *Chin. J. Chem.*, **26**, pp. 1811–1815.

115. Puzyr A. P., Pozdnyakova I. O., and Bondar V. S. (2004) Design of a luminescent biochip with nanodiamonds and bacterial luciferase, *Phys. Solid State*, **46**(4), pp. 761–763.

116. Khanna P., Ramachandran N., Yang J., Wang J., Kumar A., Jaroszeski M., and Bhansali S. (2009) Nanocrystalline diamond microspikes increase

the efficiency of ultrasonic cell lysis in a microfluidic lab-on-a-chip, *Diam. Relat. Mater.*, **18**, pp. 606–610.

117. Lim Y. T., Kim S., Nakayama A., Stott N. E., Bawendi M. G., and Frangioni J. V. (2003) Selection of quantum dot wavelengths for biomedical assays and imaging, *Mol. Imaging*, **2**, pp. 50–64.
118. Mansfield J. R., Gossage K. W., Hoyt C. C., and Levenson R. M. (2005) Autofluorescence removal, multiplexing, and automated analysis methods for in-vivo fluorescence imaging, *J. Biomed. Opt.*, **10**, p. 041207.
119. Derfus A. M., Chan W. C. W., Bhatia S. N. (2004) Probing the cytotoxicity of semiconductor quantum dots, *Nano Lett.*, **4**, pp. 11–18.
120. Chang H.-C. (2010) *Nanodiamonds: Applications in Biology and Nanoscale medicine*, ed. Ho D., "Development and use of fluorescent nanodiamonds as cellular markers" (Springer), pp. 127–150.
121. Yao J., Larson D. R., Vishwasrao H. D., Zipfel W. R., and Webb W. W. (2005) Blinking and nonradiant dark fraction of water-soluble quantum dots in aqueous solution, *Proc. Natl. Acad. Sci. U. S. A.*, **102**, pp. 14284–14289.
122. Medintz I. L., Uyeda H. T., Goldman E. R., and Mattoussi H. (2005) Quantum dot bioconjugates for imaging, labeling and sensing, *Nat. Mater.*, **4**, pp. 435–446.
123. Cui B. X., Wu C. B., Chen L., Ramirez A., Bearer E. L., Li W. P., Mobley W. C., and Chu S. (2007) One at a time, live tracking of NGF axonal transport using quantum dots, *Proc. Natl. Acad. Sci. U. S. A.*, **104**, pp. 13666–13671.
124. Chao J. I., Perevedentseva E., Chang C. C., Cheng C. Y., Liu K. K., Chung P. H., Tu J. S., Chu C. D., Cai S. J., and Cheng C. L. (2010) *Nanodiamonds: Applications in Biology and Nanoscale Medicine*, ed. Ho D., "Protein-nanodiamond complexes for cellular surgery. Nanodiamond and its bioapplications using the spectroscopic properties as probe for biolabeling" (Springer), pp. 189–224.
125. Smith B. R., Niebert M., Plakhotnik T., and Zvyagin A. V. (2007) Transfection and imaging of diamond nanocrystals as scattering optical labels, *J. Lumin.*, **127**, pp. 260–263.
126. Gruber A., Drabenstedt A., Tietz C., Fleury L., Wrachtrup J., and Von Borczyskowsky C. (1997) Scanning confocal optical microscopy and magnetic resonance on single defect centers, *Science*, **276**, pp. 2012–2014.

127. Faklaris O., Garrot D., Joshi V., Boudou J.-P., Sauvage T., Curmi P. A., and Treussart F. (2009) Comparison of the photoluminescence properties of semiconductor quantum dots and non- blinking diamond nanoparticles and observation of the diffusion of diamond nanoparticles in living cells, *J. Eur. Opt. Soc. Rap.*, **4**, 09035.

128. Neugart F., Zappe A., Jelezko F., Tietz C., Boudou J. P., Krueger A., and Wrachtrup J. (2007) Dynamics of diamond nanoparticles in solution and cells, *Nano Lett.*, **7**(12), pp. 3588–3591.

129. O'Connell M. J., Bachilo S. M., Huffman C. B., Moore V. C., Strano M. S., Haroz E. H., Rialon K. L., Boul P. J., Noon W. H., Kittrell C., Ma J. P., Hauge R. H., Weisman R. B., and Smalley R. E. (2002) Band gap fluorescence from individual single-walled carbon nanotubes, *Science*, **297**(5581), pp. 593–596.

130. Wee T.-L., Mau Y.-W., Fang C.-Y., Hsu H.-L., Han C.-C., and Chang H.-C. (2009) Preparation and characterization of green fluorescent nanodiamonds for biological applications, *Diam. Relat. Mater.*, **18**, pp. 567–573.

131. Vaijayanthimala V., and Chang H. C. (2009) Functionalized fluorescent nanodiamonds for biomedical applications, *Nanomedicine (London)*, **4**(1), pp. 47–55.

132. Mkandawire M., Pohl A., Gubarevich T., Lapina V., Appelhans D., Rödel G., Pompe W., Schreiber J., and Opitz J. (2009) Selective targeting of green fluorescent nanodiamond conjugates to mitochondria in HeLa cells, *J. Biophotonics*, **2**(10), pp. 596–606.

133. Weng M.-F., Chiang S.-Y., Wang N.-S., and Niu H. (2009) Fluorescent nanodiamonds for specifically targeted bioimaging: application to the interaction of transferrin with transferrin receptor, *Diam. Relat. Mater.*, **18**(2–3), pp. 587–591.

134. Smith B. R., Gruber D., and Plakhotnik T. (2010) The effects of surface oxidation on luminescence of nano diamonds, *Diam. Relat. Mater.*, **19**(4), pp. 314–318.

135. Cuche A., Sonnefraud Y., Faklaris O., Garrot D., Boudou J.-P., Sauvage T., Roch J.-F., Treussart F., and Huant S. (2009) Diamond nanoparticles as photoluminescent nanoprobes for biology and near-field optics, *J. Lumin.*, **129** (12), pp. 1475–1477.

136. Chang I. P., Hwang K. C., and Chiang C.-S. (2008) Preparation of fluorescent magnetic nanodiamonds and cellular imaging, *J. Am. Chem. Soc.*, **130**(46), pp. 15476–15481.

137. Martin R., Alvaro M., Herance J. R., and Garcia H. (2010). Fenton-treated functionalized diamond nanoparticles as gene delivery system, *ACS Nano*, **4** (1), pp. 65–74.

138. Balasubramanian G., Chan I. Y., Kolesov R., Al-Hmoud M., Tisler J., Shin C., Kim C., Wojcik A., Hemmer P. R., Krueger A., Hanke T., Leitenstorfer A., Bratschitsch R., Jelezko F., and Wrachtrup J. (2008) Nanoscale imaging magnetometry with diamond spins under ambient conditions, *Nature*, **455**, pp. 648–651.

139. Maze J. R., Stanwix P. L., Hodges J. S., Hong S., Taylor J. M., Cappellaro P., Jiang L., Gurudev Dutt, M. V., Togan E., Zibrov A. S., Yacoby A., Walsworth R. L., and Lukin M. D. (2008) Nanoscale magnetic sensing with an individual electronic spin in diamond, *Nature*, **455**, pp. 644–647.

140. Manus L. M., Mastarone D. J., Waters E. A., Zhang X.-Q., Schultz-Sikma E. A., MacRenaris K. W., Ho D., and Meade T. J. (2010) Gd(III)-nanodiamond conjugates for MRI contrast enhancement, *Nano Lett.*, **10** (2), pp. 484–489.

141. Lee J.-H., Huh Y.-M., Jun Y.-W., Seol J.-W., Jang J.-T., Song H.-T., Kim S., Cho E.-J., Yoon H.-G., Suh J.-S., and Cheon J. (2007) Artificially engineered magnetic nanoparticles for ultra-sensitive molecular imaging, *Nat. Med.*, **13**, pp. 95–99.

142. Lewinski N., Colvin V., and Drezek R. (2008) Cytotoxicity of nanoparticles, *Small*, **4**, pp. 26–49.

143. Yang W., Thordarson P., Gooding J. J., Ringer S. P., and Braet F. (2007) Carbon nanotubes for biological and biomedical applications, *Nanotechnology*, **18**, p. 412001.

144. Alivisatos P. (2004) The use of nanocrystals in biological detection, *Nat. Biotechnol.*, **22**, pp. 47–52.

145. Jiang W., Kim B. S., Rutka J. T., and Chan W. C. W. (2008) Nanoparticle-mediated cellular response is size-dependent, *Nat. Nanotechnol.*, **3**, pp. 145–150.

146. Verma A., Uzun O., Hu Y., Han H.-S., Watson N., Chen S., Irvine D. J., and Stellacci F. (2008) Surface-structure-regulated cell-membrane penetration by monolayer-protected nanoparticles, *Nat. Mater.*, **7**, pp. 588–595.

147. Xia T., Rome L., and Nel A. (2008) Nanobiology: particles slip cell security, *Nat. Mater.*, **7**, pp. 519–520.

148. Pantarotto D., Briand J., Prato M., and Bianco A. (2004) Translocation of bioactive peptides across cell membranes by carbon nanotubes, *Chem. Commun.*, pp. 16–17.

149. Kam N. W. S., Liu Z., and Dai H. J. (2006) Carbon nanotubes as intracellular transporters for proteins and DNA: an investigation of the uptake mechanism and pathway, *Angew. Chem., Int. Ed.*, **45**, pp. 577–581.

150. Gabizon A., Horowitz A. T., and Goren D. (1999) Targeting folate receptor with folate linked to extremities of poly(ethylene glycol)-grafted liposomes: *in vitro* studies, *Bioconjugate Chem.*, **10**, pp. 289–298.

151. Zhang Q. I., Xiang G., Zhang Y., Yang K., Fan W., Lin J., Zeng F., and Wu J. (2006) Increase of doxorubicin sensitivity for folate receptor positive cells when given as the prodrug N-(phenylacetyl) doxorubicin in combination with folate-conjugated PGA, *J. Pharm. Sci.*, **95**, pp. 2266–2275.

152. Liu K. K., Cheng C. L., Chang C. C., and Chao J. I. (2007) Biocompatible and detectable carboxylated nanodiamond on human cell, *Nanotechnology*, **18**, 325102.

153. Lu, C.-W., Hung, Y., Hsiao, J.-K., Yao, M., Chung, T.-H., Lin, Y.-S., Wu, S.-H., Hsu, S.-C., Liu, H.-M., Mou, C.-Y., Yang, C.-S., Huang, D.-M., and Chen, Y.-C. (2007). Bifunctional magnetic silica nanoparticles for highly efficient human stem cell labeling, *Nano Lett.*, **7**, pp. 149–154.

154. Rosi N. L., Giljohann D. A., Thaxton C. S., Lytton-Jean A. K. R., Han M. S., and Mirkin C. A. (2006) Oligonucleotide-modified gold nanoparticles for intracellular gene regulation, *Science*, **312**, pp. 1027–1030.

155. Roy I., Ohulchanskyy T. Y., Bharali D. J., Pudavar H. E., Mistretta R. A., Kaur N., and Prasad P. N. (2005) Optical tracking of organically modified silica nanoparticles as DNA carriers: a nonviral, nanomedicine approach for gene delivery, *Proc. Natl. Acad. Sci. U. S. A.*, **102**, pp. 279–284.

156. Sahoo S. K., and Labhasetwar V. (2003) Nanotech approaches to drug delivery and imaging, *Drug Discov. Today*, **8**, pp. 1112–1120.

157. Sokolova V., and Epple M. (2008) Inorganic nanoparticles as carriers of nucleic acids into cells, *Angew. Chem., Int. Ed.*, **47**, pp. 1382–1395.

158. Torney F., Trewyn B. G., Lin V. S.-Y., and Wang K. (2007) Mesoporous silica nanoparticles deliver DNA and chemicals into plants, *Nat. Nanotechnol.*, **2**, pp. 295–300.

159. Kalderon D., Roberts B. L., Richardson W. D., and Smith A. E. (1984) A short amino acid sequence able to specify nuclear location, *Cell*, **39**, pp. 499–509.

160. Faklaris O., Garrot D., Joshi V., Druon F., Boudou J.-P., Sauvage T., Georges P., Curmi P. A., and Treussart F. (2008) Detection of single photoluminescent diamond nanoparticles in cells and study of the internalization pathway, *Small*, **4**, pp. 2236–2239.

161. Olson J. K., and Grose C. (1997) Endocytosis and recycling of Varicella-Zoster virus Fc receptor glycoprotein gE: internalization mediated by a YXXL motif in the cytoplasmic tail, *J. Virol.*, **71**, pp. 4042–4054.

162. Liu K. K., Wang C. C., Cheng C. L., and Chao J. I. (2009) Endocytic carboxylated nanodiamond for the labeling and tracking of cell division and differentiation in cancer and stem cells, *Biomaterials*, **30**(26), pp. 4249–4259.

163. Bausinger R., von Gersdorff K., Braeckmans K., Ogris M., Wagner E., Brauchle C., and Zumbusch A. (2006). The transport of nanosized gene carriers unraveled by live-cell imaging, *Angew. Chem., Int. Ed.*, **45**, pp. 1568–1572.

164. Sudimack J., and Lee R. J. (2000) Targeted drug delivery via the folate receptor, *Adv. Drug Delivery Rev.*, **41**, pp. 147–162.

165. Pelkmans L., and Helenius A. (2002) Endocytosis via caveolae, *Traffic*, **3**, pp. 311–320.

166. European Commission Health & Consumer Protection. SCENIHR opinion on "the appropriateness of existing methodologies to assess the potential risks associated with engineered and adventitious products of nanotechnologies," September 2005.

167. Purtov K. V., Petunin A. I., Burov A. E., Puzyr A. P., and Bondar V. S. (2010) Nanodiamonds as carriers for address delivery of biologically active substances, *Nanoscale Res Lett.*, **5**(3), pp. 631–636.

168. Grichko O., Grishko V., and Shenderova O. (2006) Nanodiamond bullets and their biological targets, *NanoBiotechnology*, **2**, pp. 1551–1555.

169. Taylor N. J., and Fauquet C. M. (2002) Microparticle bombardment as a tool in plant science and agricultural biotechnology, *DNA Cell Biol.*, **21**, pp. 963–977.

170. Huang H., Chen M., Bruno P., Lam R., Robinson E., Gruen D., and Ho D. (2009) Ultrananocrystalline diamond thin films functionalized with therapeutically active collagen networks, *J. Phys. Chem. B*, **113**(10), pp. 2966–2971.

171. Huang H., Pierstorff E., Liu K., Osawa E., and Ho D. (2010) *Nanodiamonds: Applications in Biology and Nanoscale Medicine*, ed. Ho D., "Nanodiamond-mediated delivery of therapeutics via particle and thin film architectures" (Springer), pp. 151–174.

172. Lam R., Chen M., Huang H., Osawa E., and Ho D. (2010) *Nanodiamonds: Applications in Biology and Nanoscale Medicine*, ed. Ho, D., "Polymeric encapsulation of nanodiamond—chemotherapeutic complexes for localized cancer treatment" (Springer), pp. 175–188.

173. Liu W. K., Adnan A., Kopacz A. M., Hallikainen M., Ho D., Lam R., Lee J., Belytschko T., Schatz G., Tzeng Y., Kim Y.-J., Baik S., Kim M. K., Kim T., Lee, J., Hwang E.-S., Im S., Osawa E., Barnard A., Chang H.-C., Chang C.-C., and Onate E. (2010) *Nanodiamonds: Applications in Biology and Nanoscale Medicine*, ed. Ho, D., "Design of nanodiamond based drug delivery patch for cancer therapeutics and imaging applications" (Springer), pp. 249–284.

174. Loh O., Lam R., Chen M., Ho D., and Espinosa H. (2010) *Nanodiamonds: Applications in Biology and Nanoscale Medicine*, ed. Ho D., "Microfluidic platforms for nanoparticle delivery and nanomanufacturing in biology and medicine" (Springer), pp. 225–234.

175. Zhang X.-Q., Chen M., Lam R., Xu X., Osawa E., and Ho D. (2009) Polymer-functionalized nanodiamond platforms as vehicles for gene delivery, *ACS Nano*, **3**(9), pp. 2609–2616.

176. Yuan Y., Chen Y., Liu J.-H., Wang H., and Liu Y. (2009) Biodistribution and fate of nanodiamonds *in vivo*, *Diam. Relat. Mater.*, **18**(1), pp. 95–100.

177. Yuan Y., Wang X., Jia G., Liu J.-H., Wang T., Gu Y., Yang S.-T., Zhen S., Wang H., and Liu Y. (2010) Pulmonary toxicity and translocation of nanodiamonds in mice, *Diam. Relat. Mater.*, **19**, pp. 291–299.

178. Chen Y.-C., Lee D.-C., Hsiao C.-Y., Chung Y.-F., Chen H.-C., Thomas J. P., Pong W.-F., Tai N.-H., Lin I.-N., and Chiu I.-M. (2009) The effect of ultra-nanocrystalline diamond films on the proliferation and differentiation of neural stem cells, *Biomaterials*, **30**(20), pp. 3428–3435.

179. Lechleitner T., Klauser F., Seppi T., Lechner J., Jennings P., Perco P., Mayer B., Steinmüller-Nethl D., Preiner J., Hinterdorfer P., Hermann M., Bertel E., Pfaller K., and Pfaller W. (2008) The surface properties of nanocrystalline diamond and nanoparticulate diamond powder and their suitability as cell growth support surfaces, *Biomaterials*, **29**(32), pp. 4275–4284.

180. Grausova L, Bacakova L, Kromka A, Potocky S, Vanecek M, Nesladek M., and Lisa V. (2009) Nanodiamond as promising material for bone tissue engineering, *J. Nanosci. Nanotechnol.*, **9**(6), pp. 3524–3534.

181. Kalbacova M., Rezek B., Baresova V., Wolf-Brandstetter C., and Kromka A. (2009) Nanoscale topography of nanocrystalline diamonds

promotes differentiation of osteoblasts, *Acta Biomater.*, **5**(8), pp. 3076–3085.

182. Thalhammer A., Edgington R. J., Cingolani L. A., Schoepfer R., and Jackman R. B. (2010) The use of nanodiamond monolayer coatings to promote the formation of functional neuronal networks, *Biomaterials*, **31**(8), pp. 2097–2104.
183. Yang L., Sheldon B. W., and Webster T. J. (2009) The impact of diamond nanocrystallinity on osteoblast functions, *Biomaterials*, **30**(20), pp. 3458–3465.
184. Yang L., Sheldon B. W., and Webster T. J. (2009) Orthopedic nano diamond coatings: control of surface properties and their impact on osteoblast adhesion and proliferation, *J. Biomed. Mater. Res.*, **91A**, pp. 548–556.
185. Sung C.-M., Sung M., and Sung E. (2007) Healthcare and cosmetic compositions containing nanodiamond, US Patent 7294340.
186. Williams D. F. (1999) *The Williams Dictionary of Biomaterials* (Liverpool University Press).
187. Lee K. Y., and Mooney D. J. (2001) Hydrogels for tissue engineering, *Chem. Rev.*, **101**(7), pp. 1869–1880.
188. Colvin V. L. (2003) The potential environmental impact of engineered nanomaterials, *Nat. Biotechnol.*, **21**, pp. 1166–1170.
189. European Committee for Standardization (1996) *EN 540: Clinical Investigation of Medical Devices for Human Subjects*, Brussels.
190. Ikada Y. (1993) Interfacial biocompatibility, polymers of biological and biomedical significance, *ACS Symp. Ser.*, **540**, pp. 35–48.
191. Kang S., Mauter M. S., and Elimelech M. (2008) Physicochemical determinants of multiwalled carbon nanotube bacterial cytotoxicity, *Environ. Sci. Technol.*, **42**, pp. 7528–7534.
192. Muller J., Huaux F., Fonseca A., Nagy J. B., Moreau N., Delos M., Raymundo-Pinero E., Beguin F., Kirsch-Volders M., Fenoglio I., Fubini B., and Lison D. (2008) Structural defects play a major role in the acute lung toxicity of multiwall carbon nanotubes: toxicological aspects, *Chem. Res. Toxicol.*, **21**, pp. 1698–1705.
193. Prato M., Kostarelos K., and Bianco A. (2008) Functionalized carbon nanotubes in drug design and discovery, *Acc. Chem. Res.*, **41**, pp. 60–68.
194. Sayes C. M., Liang F., Hudson J. L., Mendez J., Guo W., Beach J. M., Moore V. C., Doyle C. D., West J. L., Billups W. E., Ausman K. D., and Colvin V. L.

(2006) Functionalization density dependence of single-walled carbon nanotubes cytotoxicity *in vitro*, *Toxicol. Lett.*, **16**, pp. 135–142.

195. Kam N. W. S., and Dai H. (2005) Carbon nanotubes as intracellular protein transporters: generality and biological functionality, *J. Am. Chem. Soc.*, **127**, pp. 6021–6026.

196. Kam N. W. S., Jessop T. C., Wender P. A., and Dai H. (2004) Nanotube molecular transporters: internalization of carbon nanotube-protein conjugates into mammalian cells, *J. Am. Chem. Soc.*, **126**, pp. 6850–6851.

197. Shim M., Kam N. W. S., Chen R. J., Li Y., and Dai H. (2002) Functionalization of carbon nanotubes for biocompatibility and biomolecular recognition, *Nano Lett.*, **2**, pp. 285–288.

198. Smart S. K., Cassady A. I., Lu G. Q., and Martin D. J. (2006) The biocompatibility of carbon nanotubes, *Carbon*, **44**, pp. 1034–1047.

199. Denissenko M. F., Pao A., Tang M., and Pfeifer G. P. (1996) Preferential formation of benzo[*a*]pyrene adducts at lung cancer mutational hotspots in P53, *Science*, **274**(5286), pp. 430–432.

200. Schrand A. M., Huang H., Carlson C., Schlager J. J., Osawa E., Hussain M. S., and Dai L. (2007) Are diamond nanoparticles cytotoxic? *J. Phys. Chem. B*, **111**, pp. 2–7.

201. Loh A. C., Szuna A. J., Williams T. H., Sasso G. J., and Leinweber F. J. (1991) The metabolism of [14C]rimantadine hydrochloride in rats and dogs, *Drug Metab. Dispos.*, **19**(2), pp. 381–387.

202. Hoffman H. E., Gaylord J. C., Blasecki J. W., Shalaby .L. M., and Whitney C. C. Jr. (1988) Pharmacokinetics and metabolism of rimantadine hydrochloride in mice and dogs, *Antimicrob. Agents Chemother.*, **32**(11), pp. 1699–1704.

203. Hodek P., Karabec M., Sulc M., Sopko B., Smrcek S., Martínek V., Hudecek J., and Stiborová M. (2007) Mapping of cytochrome P450 2B4 substrate binding sites by photolabile probe 3-azidiamantane: identification of putative substrate access regions, *Arch. Biochem. Biophys.*, **468**(1), pp. 82–91.

204. Hodek P., and Smrcek S. (1999) Evaluation of 3-azidiamantane as photoaffinity probe of cytochrome P450, *Gen. Physiol. Biophys.*, **18**(2), pp. 181–198.

205. Lewis D. F., Lake B. G., Ito Y., and Anzenbacher P. (2006) Quantitative structure-activity relationships (QSARs) within cytochromes P450 2B (CYP2B) subfamily enzymes: the importance of lipophilicity for binding and metabolism, *Drug Metabol. Drug Interact.*, **21** (3-4), pp. 213–231.

206. Sulc M., Hudecek J., Stiborova M., and Hodek P. (2008) Structural analysis of binding of a diamantoid substrate to cytochrome P450 2B4: possible role of Arg 133 in modulation of function and activity of this enzyme, *Neuro Endocrinol. Lett.*, **29**(5), pp. 722–727.

207. Hodek P., Bortek-Dohalská L., Sopko B., Sulc M, Smrcek S., Hudecek J., Jank J., and Stiborová M. (2005) Structural requirements for inhibitors of cytochromes P450 2B: assessment of the enzyme interaction with diamondoids, *J. Enzyme Inhib. Med. Chem.*, **20**(1), pp. 25–33.

208. de Groot M. J., Vermeulen N. P., Kramer J. D., van Acker, F. A., and Donné-Op den Kelder G. M. (1996). A three-dimensional protein model for human cytochrome P450 2D6 based on the crystal structures of P450 101, P450 102, and P450 108, *Chem. Res. Toxicol.*, **9**(7), pp. 1079–1091.

209. Hodek P., Burkhard J., and Jank J. (1995) Probing the cytochrome P-450 2B1 active site with diamantoid compounds, *Gen. Physiol. Biophys.*, **14**(3), pp. 225–239.

210. Hodek P., Janscák P., Anzenbacher P., Burkhard J., Jank J., and Vodicka L. (1988) Metabolism of diamantane by rat liver microsomal cytochromes P-450, *Xenobiotica*, **18**(10), pp. 1109–1118.

211. Schwab R. S., Poskanzer D. C., England A. C., and Young R. R. (1972) Amantadine in the treatment of Parkinson's disease. Review of more than two years' experience, *J. Amer. Med. Assoc.*, **222**, pp. 792–795.

212. Danilenko V. V. (2004) On the history of the discovery of nanodiamond synthesis, *Phys. Solid State*, **46**, pp. 595–599.

213. Klein C. A. (2005) Flexural strength of diamond: comparison of nanocrystalline thin films, massive chemical vapor deposits, and natural single crystals, *J. Appl. Phys.*, **97**, pp. 016105–016106.

214. Carlisle J. A. (2004) Diamond films: precious biosensors, *Nat. Mater.*, **3**, pp. 668–669.

215. Sumant A. V., Grierson D. S., Gerbi J. E., Birrell J., Lanke U. D., Auciello O., Carlisle J. A., and Carpick R. W. (2005) Toward the ultimate tribological interface: surface chemistry and nanotribology of ultrananocrystalline diamond, *Adv. Mater.*, **17**, pp. 1039–1045.

216. Zimmermann T., Kubovic M., Denisenko A., Janischowsky K., Williams O. A., Gruen D. M., and Kohn E. (2005) Ultra-nano-crystalline/single crystal diamond heterostructure diode, *Diam. Relat. Mater.*, **14**, pp. 416–420.

217. Turner S., Lebedev O. I., Shenderova O., Vlasov I. I., Verbeeck J., and Van Tendeloo G. (2009) Determination of size, morphology, and nitrogen impurity location in treated detonation nanodiamond by transmission electron microscopy, *Adv. Funct. Mater.*, **19**, pp. 2116–2124.

218. Oberdorster G., Maynard A., Donaldson K., Castranova V., Fitzpatrick J., Ausman K., Carter J., Karn B., Kreyling W., Lai D., Olin S., Monteiro-Riviere N., Warheit D., and Yang H. (2005) Principles for characterizing the potential human health effects from exposure to nanomaterials: elements of a screening strategy, *Particle Fibre Toxicol.*, **2**, p. 8.

219. Houska J., Panyala N. R., Pena-Mendez E. M., and Havel J. (2009) Mass spectrometry of nanodiamonds, *Rapid Commun. Mass Spectrom.*, **23**, pp. 1125–1131.

220. Koscheev A. P. (2008) Termal desorption mass spectrometry in the light of the problem of certification and harmonization of the surface properties of detonation nanodiamonds. *Ros. Khim. Zh. (J. Mendeleev Russ. Chem. Soc.)*, **LII**(5), pp. 88–96 (in Russian).

221. Barnard A. S. (2008) Self-assembly in nanodiamond agglutinates, *J. Mater. Chem.*, **18**, pp. 4038–4041.

222. Barnard A. S., and Sternberg M. (2007) Crystallinity and surface electrostatics of diamond nanocrystals, *J. Mater. Chem.*, **17**, pp. 4811–4819.

223. Walkowiak B., Okroj W., Jerczynska H., and Pawlowska Z. (2009) Micron-sized diamond particles are internalized by endothelial cells, *Diam. Relat. Mat.*, **18**(4), pp. 651–656.

Index